AF411342

Histopathology of Aquatic Animals

Histopathology of Aquatic Animals

Editors

Panagiotis Berillis
Božidar Rašković

MDPI • Basel • Beijing • Wuhan • Barcelona • Belgrade • Manchester • Tokyo • Cluj • Tianjin

Editors
Panagiotis Berillis   Božidar Rašković
University of Thessaly   University of Belgrade
Greece   Serbia

Editorial Office
MDPI
St. Alban-Anlage 66
4052 Basel, Switzerland

This is a reprint of articles from the Special Issue published online in the open access journal *Applied Sciences* (ISSN 2076-3417) (available at: https://www.mdpi.com/journal/applsci/special_issues/histopathology_of_aquatic_animals).

For citation purposes, cite each article independently as indicated on the article page online and as indicated below:

LastName, A.A.; LastName, B.B.; LastName, C.C. Article Title. *Journal Name* **Year**, *Volume Number*, Page Range.

ISBN 978-3-0365-3833-4 (Hbk)
ISBN 978-3-0365-3834-1 (PDF)

Contents

About the Editors

Panagiotis Berillis has a PhD in Biophysics and is an Associate Professor at the Department of Ichthyology & Aquatic Environment, University of Thessaly. He has published more than 90 articles and conference papers and is the Chief Editor of 2 international scientific journals, as well as a member of the Editorial Board of 3 international scientific journals and reviewer for 29 international scientific journals. He has been scientific leader of 3 research projects and team member in 10 research projects (National and EU). He is a member of the Hellenic Technology Platform for Aquaculture (HETEPA) and of the Hellenic Society for Biological Sciences. His research interests are focused on aquaculture, aquaponics, histology/histopathology of aquatic organism, and collagen and tissue mechanical properties.

Božidar Rašković, Associate Professor, Since 2008, I have worked at the Faculty of Agriculture at the University of Belgrade (Belgrade, Serbia). However, my background is in biology, and I completed my PhD in histo(patho)logy of fish in aquaculture. Currently, my research interests are heading in two directions: toxicopathology of fish and the effects of various feeds/feed ingredients on the histology of the digestive tract in nutritional assays with fish (mainly common carp). I am very much involved in the study of metals in aquatic environments and using fish as a target organism in environmental risk assessment.

Editorial

Special Issue on the Histopathology of Aquatic Animals

Božidar Rašković [1,2,*] and Panagiotis Berillis [3]

[1] Institute of Animal Science, Faculty of Agriculture, University of Belgrade, 11080 Belgrade, Serbia

[2] Department of Microscopy, Institute of Biomedical Sciences Abel Salazar (ICBAS), University of Porto, 4050-313 Porto, Portugal

[3] Department of Ichthyology and Aquatic Environment, School of Agricultural Sciences, University of Thessaly, 384 46 Volos, Greece; pveril@uth.gr

* Correspondence: raskovic@agrif.bg.ac.rs; Tel.: +381-11-441-32-89

1. Introduction

Histopathology is the study of changes in any tissue associated with a disease or disorder. The word came from ancient Greek words: ἱστός (histos) = tissue, πάθος (pathos) = disease, and λογία (logia) = word. It involves the examination of tissues and cells under a microscope in order to identify the alterations to the normal structure of tissues (and perhaps their etiology). Information obtained from the study of histopathological/histomorphological lesions in aquatic animals can be a useful addition when determining the general state of health of aquatic animals, especially if chronic stressors and/or pathogens are present. Compared to mammals, postmortem autolysis progresses very rapidly in most aquatic organisms. This fact makes histopathological examinations quite complex and demanding, not only in a histotechnical sense. A prerequisite for a successful study is the baseline knowledge of physiological processes and histological architecture of the studied species.

The aim of this Special Issue was to contribute to the current state of knowledge on the histopathology of aquatic animals and to provide a collection of novel articles for biologists/veterinarians/animal scientists/other experts. The topic is extremely broad, and therefore, the published manuscripts are also diverse in terms of research areas, employed methodology and investigated species. Eight original studies and two review articles are part of this Special Issue and their short overview by two Guest Editors is given below.

2. Fish Nutrition

In fish nutrition trials, histopathology is frequently used to describe subtle alterations in tissues and cells of the digestive system upon the addition of novel ingredients or the replacement of fish meal in fish feed [1,2]. If the growth parameters of fish are not satisfactory or are lagging compared to the control group of fish or available literature data, the histological approach could provide researchers with possible answers for this phenomenon. Often, new ingredients in fish feeds will not cause significant histopathological alterations but will have an effect on the appearance of cells involved in the absorption or transformation of molecules that originate from the feed, such as enterocytes in the intestine or hepatocytes in the liver, respectively [3]. Their surface area or volume can be determined using various histological techniques, and a decreased size of cells (or their nuclei) is a marker for metabolic and physiological differences of organs in fish that are fed alternative feeds [4]. However, some antinutritional factors in plant-based fish feed have inflammatory effects on the distal intestine of fish, causing enteritis. This happens in both carnivorous and omnivorous fish [5,6], although the intensity and extent of enteritis is more pronounced in carnivorous fish species.

Rombenso et al. [7] studied the origin of enteritis in Atlantic salmon (*Salmo salar*) fed soybean meal derived from specific soybean genotypes, which had reduced content of lipoxygenases and oligosaccharides, as well as an altered glycinin profile. They concluded that neither of these substances is a driver of distal intestine inflammation, as they did not

Citation: Rašković, B.; Berillis, P. Special Issue on the Histopathology of Aquatic Animals. *Appl. Sci.* **2022**, *12*, 971. https://doi.org/10.3390/app12030971

Received: 11 January 2022

Accepted: 13 January 2022

Published: 18 January 2022

Publisher's Note: MDPI stays neutral with regard to jurisdictional claims in published maps and institutional affiliations.

influence the extent of enteritis in Atlantic salmon. Psofakis et al. [8] conducted a feeding study of gilthead seabream (*Sparus aurata*) for 56 days with feeds in which fishmeal was replaced by poultry by-product meal and hydrolyzed feather meal in various concentrations. The nutritional trial showed that higher levels (>50%) of replacement caused severe liver pathologies, such as cirrhosis and necrosis of pancreatic acini. On the other hand, there was no intestine inflammation, but the histological structure of the intestine remained without pathological features, which is a nice demonstration of why two (or possibly more) organs should be included in histological studies. Both studies published in this Special Issue used a semi-quantitative scoring system as a means to evaluate histopathological changes in the intestines and livers of experimental fish.

3. Exposure Studies

Histopathology is one method of choice in exposure assays, as it could give mechanistic insight into toxicological pathology [9]. Several guidance documents are published by the Organisation for Economic Co-operation and Development (OECD) and the US Environmental Protection Agency (EPA) for conducting standardized exposure assays on aquatic animals. The guidelines cover a number of experimental species, different life stages and exposure scenarios, thus giving the possibility of inter-laboratory comparison. Depending on the choice of stressor in exposure assays, researchers will target different organs and assess the change of its micro-architecture.

One group of authors published two manuscripts in this Special Issue from a single experiment in which an amphibian, *Xenopus laevis*, was exposed to atrazine, a frequently used herbicide (at least in the United States, since the European Union banned it) in a dose-dependent approach. The effects of exposure are evaluated on the histology of the heart and cerebellum [10] and liver and kidney [11]. Immunohistochemical markers were used on the same types of tissue, and the authors opted for caspase-3 in the liver and kidney and Inositol 1,4,5-trisphosphate in the heart and cerebellum. Both studies showed significant adverse effects of the herbicide on the tissues, with a range of different histopathological alterations most frequently observed in the liver and kidney of this frog. Histomorphometrical methods were further employed in order to show changes on a range of different cells and structures in the mentioned tissues. Gómez de Anda et al. [12] exposed rainbow trout (*Oncorhynchus mykiss*) fingerlings to a bath containing several species of *Aeromonas* spp. They found several histopathological alterations of nervous tissues (nasal epithelium, telencephalon, and oblongata medulla), which were distinct to the fish infected with *A. lusitana* and *A. salmonicida* and with the higher severity grades. Fish infected with other bacterial species had middle-to-less severity and distribution of histopathological alterations.

4. Environmental Studies

The state of freshwater or marine ecosystems worldwide is frequently monitored in order to assess the general health of the aquatic animals inhabiting it. Usually, a battery of biomarkers is employed during monitoring, with histopathology being one of them, in order to obtain as much information as possible from the sampled fish [13]. A similar approach is used for studies published in this Special Issue. Three published manuscripts used histopathological methodology to investigate tissues of interest on various cyprinid fish from ecosystems of the Balkan Peninsula. The case study of lipoma found in the liver of the black barbel (*Barbus balcanicus*) in the Bregalnica River (North Macedonia) was described by Rebok et al. [14]. This agglomerate of adipocytes was assessed microscopically after being serially sectioned, while the authors also reconstructed a 3D image after applying software. Tsakoumis et al. [15] studied the histopathology of gills in the endemic fish species *Alburnus vistonicus* over one year in Vistonis Lake (Greece). The ecosystem of this lake is, according to the authors, "peculiar", in the sense that it is a mixture of freshwater (originated from three rivers) and seawater (from the Aegean Sea). Therefore, salinity levels fluctuate due to the seasonal activity of rivers, and the authors monitored the mentioned fish species using a battery of biomarkers, including histopathology, in order to establish a correlation between salinity and alterations of fish gills. A similar use of histopathology as a biomarker in the study of fish in a natural ecosystem is conducted in the Tamiš river (Serbia) [16]. This river is burdened with the presence of metals that originate from anthropogenic activity. The authors used a semi-quantitative scoring system to assess the gills of the common bream (*Abramis brama*) in order to study the effects of water quality during two seasons (autumn and spring aspect).

5. Review Manuscripts

The review papers are focused on two hot topics in environmental pollution: pesticides and micro/nanoplastics. In both articles, histopathology is used as a marker of effects, and it is one of the several biomarkers that were reviewed. Tresnakova et al. [17] presented an overview of the effects of glyphosate and its major metabolite AMPA in different formulations to various species of fish, amphibians and aquatic invertebrates. Endpoints included: hematological parameters, blood biochemistry, gene expression, oxidative stress biomarkers and antioxidant enzymes. Glyphosate is the most frequently used herbicide worldwide, but recently, it sparked controversy in the EU [18]. Guerrera et al. [19] wrote a comprehensive and detailed review paper of the effects of micro- and nanoplastics on aquatic organisms, accounting for ten chapters in total. Special focus is given to the blood–brain barrier in fish, neurotoxicity, adverse effects in fish gills, intestine, liver and gills, while one chapter is dedicated to toxicity of micro- and nanoplastics to zebrafish (*Danio rerio*).

6. Conclusions

Histopathological techniques can provide a more analytical view of disease and its effect on tissues and cells. Histopathology helps ichthyopathologists to reach a diagnosis by examining a small piece of tissue from various organs. This Special Issue confirms statements that the histopathological method does not have limitations in terms of laboratory setup, choice of ecosystem, climate or aquatic species investigated. This method is employed on very different species in terms of phylogeny, age, sex and physiology, with a range of different tissues investigated and various research scopes, in both laboratories and the field. The different evaluation techniques of sections were chosen depending on the goals of the particular study. We hope that this collection of manuscripts will be useful to readers and scientists working in the area of histology/histopathology of aquatic organisms and that it will further increase interest in this scientific area.

Author Contributions: Conceptualization, B.R. and P.B.; methodology, B.R. and P.B.; investigation, B.R.; writing—original draft preparation, B.R.; writing—review and editing, P.B. All authors have read and agreed to the published version of the manuscript.

Funding: This research received no external funding.

Institutional Review Board Statement: Not applicable.

Conflicts of Interest: The authors declare no conflict of interest.

References

1. Rašković, B.; Stanković, M.; Marković, Z.; Poleksić, V. Histological methods in the assessment of different feed effects on liver and intestine of fish. *J. Agric. Sci.* **2011**, *56*, 87–100.
2. Randazzo, B.; Zarantoniello, M.; Gioacchini, G.; Cardinaletti, G.; Belloni, A.; Giorgini, E.; Faccenda, F.; Cerri, R.; Tibaldi, E.; Olivotto, I. Physiological response of rainbow trout (*Oncorhynchus mykiss*) to graded levels of *Hermetia illucens* or poultry by-product meals as single or combined substitute ingredients to dietary plant proteins. *Aquaculture* **2021**, *538*, 736550. [CrossRef]
3. Ostaszewska, T.; Dabrowski, K.; Palacios, M.E.; Olejniczak, M.; Wieczorek, M. Growth and morphological changes in the digestive tract of rainbow trout (*Oncorhynchus mykiss*) and pacu (*Piaractus mesopotamicus*) due to casein replacement with soybean proteins. *Aquaculture* **2005**, *245*, 273–286. [CrossRef]
4. Rašković, B.; Cruzeiro, C.; Poleksić, V.; Rocha, E. Estimating volumes from common carp hepatocytes using design-based stereology and examining correlations with profile areas: Revisiting a nutritional assay and unveiling guidelines to microscopists. *Microsc. Res. Tech.* **2019**, *82*, 861–871. [CrossRef] [PubMed]

5. Baeverfjord, G.; Krogdahl, Å. Development and regression of soybean meal induced enteritis in Atlantic salmon, *Salmo salar* L., distal intestine: A comparison with the intestines of fasted fish. *J. Fish Dis.* **1996**, *19*, 375–387. [CrossRef]
6. Urán, P.A.; Gonçalves, A.A.; Taverne-Thiele, J.J.; Schrama, J.W.; Verreth, J.A.J.; Rombout, J.H.W.M. Soybean meal induces intestinal inflammation in common carp (*Cyprinus carpio* L.). *Fish Shellfish Immunol.* **2008**, *25*, 751–760. [CrossRef] [PubMed]
7. Rombenso, A.N.; Blyth, D.; James, A.T.; Nikolaou, T.; Simon, C.J. Lipoxygenase Enzymes, Oligosaccharides (Raffinose and Stachyose) and 11sA4 and A5 Globulins of Glycinin Present in Soybean Meal Are Not Drivers of Enteritis in Juvenile Atlantic Salmon (*Salmo salar*). *Appl. Sci.* **2021**, *11*, 9327. [CrossRef]
8. Psofakis, P.; Meziti, A.; Berillis, P.; Mente, E.; Kormas, K.A.; Karapanagiotidis, I.T. Effects of Dietary Fishmeal Replacement by Poultry By-Product Meal and Hydrolyzed Feather Meal on Liver and Intestinal Histomorphology and on Intestinal Microbiota of Gilthead Seabream (*Sparus aurata*). *Appl. Sci.* **2021**, *11*, 8806. [CrossRef]
9. Wester, P.W.; Van der Ven, L.T.M.; Vethaak, A.D.; Grinwis, G.C.M.; Vos, J.G. Aquatic toxicology: Opportunities for enhancement through histopathology. *Environ. Toxicol. Pharmacol.* **2002**, *11*, 289–295. [CrossRef]
10. Asouzu Johnson, J.; Nkomozepi, P.; Opute, P.; Mbajiorgu, E.F. Cardiac and Cerebellar Histomorphology and Inositol 1,4,5-Trisphosphate (IP3R) Perturbations in Adult *Xenopus laevis* Following Atrazine Exposure. *Appl. Sci.* **2021**, *11*, 10006. [CrossRef]
11. Sena, L.; Asouzu Johnson, J.; Nkomozepi, P.; Mbajiorgu, E.F. Atrazine-Induced Hepato-Renal Toxicity in Adult Male *Xenopus laevis* Frogs. *Appl. Sci.* **2021**, *11*, 11776. [CrossRef]
12. Gómez de Anda, F.R.; Vega-Sánchez, V.; Reyes-Rodríguez, N.E.; Martínez-Juárez, V.M.; Ángeles-Hernández, J.C.; Acosta-Rodríguez, I.; Campos-Montiel, R.G.; Zepeda-Velázquez, A.P. The Nasal Epithelium as a Route of Infection and Clinical Signs Changes, in Rainbow Trout (*Oncorhynchus mykiss*) Fingerlings Infected with *Aeromonas* spp. *Appl. Sci.* **2021**, *11*, 9159. [CrossRef]
13. Van der Oost, R.; Beyer, J.; Vermeulen, N.P. Fish bioaccumulation and biomarkers in environmental risk assessment: A review. *Environ. Toxicol. Pharmacol.* **2003**, *13*, 57–149. [CrossRef]
14. Rebok, K.; Jordanova, M.; Azevedo, J.; Rocha, E. First Report and 3D Reconstruction of a Presumptive Microscopic Liver Lipoma in a Black Barbel (*Barbus balcanicus*) from the River Bregalnica in the Republic of North Macedonia. *Appl. Sci.* **2021**, *11*, 8392. [CrossRef]
15. Tsakoumis, E.; Tsoulia, T.; Feidantsis, K.; Mouchlianitis, F.A.; Berillis, P.; Bobori, D.; Antonopoulou, E. Cellular Stress Responses of the Endemic Freshwater Fish Species *Alburnus vistonicus* Freyhof & Kottelat, 2007 in a Constantly Changing Environment. *Appl. Sci.* **2021**, *11*, 11021.
16. Marinović, Z.; Miljanović, B.; Urbányi, B.; Lujić, J. Gill Histopathology as a Biomarker for Discriminating Seasonal Variations in Water Quality. *Appl. Sci.* **2021**, *11*, 9504. [CrossRef]
17. Tresnakova, N.; Stara, A.; Velisek, J. Effects of Glyphosate and Its Metabolite AMPA on Aquatic Organisms. *Appl. Sci.* **2021**, *11*, 9004. [CrossRef]
18. Kudsk, P.; Mathiassen, S. Pesticide regulation in the European Union and the glyphosate controversy. *Weed Sci.* **2020**, *68*, 214–222. [CrossRef]
19. Guerrera, M.C.; Aragona, M.; Porcino, C.; Fazio, F.; Laurà, R.; Levanti, M.; Montalbano, G.; Germanà, G.; Abbate, F.; Germanà, A. Micro and Nano Plastics Distribution in Fish as Model Organisms: Histopathology, Blood Response and Bioaccumulation in Different Organs. *Appl. Sci.* **2021**, *11*, 5768. [CrossRef]

Review

Micro and Nano Plastics Distribution in Fish as Model Organisms: Histopathology, Blood Response and Bioaccumulation in Different Organs

Maria Cristina Guerrera [1], Marialuisa Aragona [1,†], Caterina Porcino [1,†], Francesco Fazio [2,*], Rosaria Laurà [1], Maria Levanti [1], Giuseppe Montalbano [1], Germana Germanà [1], Francesco Abbate [1] and Antonino Germanà [1]

[1] Zebrafish Neuromorphology Lab, Department of Veterinary Sciences, University of Messina, 98168 Messina, Italy; mariacristina.guerrera@unime.it (M.C.G.); mlaragona@unime.it (M.A.); caterina.porcino@unime.it (C.P.); laurar@unime.it (R.L.); mblevanti@unime.it (M.L.); gmontalbano@unime.it (G.M.); pgermana@unime.it (G.G.); abbatef@unime.it (F.A.); agermana@unime.it (A.G.)

[2] Department of Veterinary Sciences, University of Messina, 98168 Messina, Italy

* Correspondence: ffazio@unime.it

† These authors equally contributed to this work.

Featured Application: Current evidence indicates that micro- and nano-plastics can be absorbed by aquatic organisms as well as mammals. The topic focuses on the important role of the biodistribution of micro- and nano-plastics (MP/NPs) and the role of blood biomarkers in the teleosts used as model organisms.

Citation: Guerrera, M.C.; Aragona, M.; Porcino, C.; Fazio, F.; Laurà, R.; Levanti, M.; Montalbano, G.; Germanà, G.; Abbate, F.; Germanà, A. Micro and Nano Plastics Distribution in Fish as Model Organisms: Histopathology, Blood Response and Bioaccumulation in Different Organs. *Appl. Sci.* **2021**, *11*, 5768. https://doi.org/10.3390/app11135768

Academic Editor: Panagiotis Berillis

Received: 24 May 2021
Accepted: 16 June 2021
Published: 22 June 2021

Publisher's Note: MDPI stays neutral with regard to jurisdictional claims in published maps and institutional affiliations.

Abstract: Micro- and nano-plastic (MP/NP) pollution represents a threat not only to marine organisms and ecosystems, but also a danger for humans. The effects of these small particles resulting from the fragmentation of waste of various types have been well documented in mammals, although the consequences of acute and chronic exposure are not fully known yet. In this review, we summarize the recent results related to effects of MPs/NPs in different species of fish, both saltwater and freshwater, including zebrafish, used as model organisms for the evaluation of human health risk posed by MNPs. The expectation is that discoveries made in the model will provide insight regarding the risks of plastic particle toxicity to human health, with a focus on the effect of long-term exposure at different levels of biological complexity in various tissues and organs, including the brain. The current scientific evidence shows that plastic particle toxicity depends not only on factors such as particle size, concentration, exposure time, shape, and polymer type, but also on co-factors, which make the issue extremely complex. We describe and discuss the possible entry pathways of these particles into the fish body, as well as their uptake mechanisms and bioaccumulation in different organs and the role of blood response (hematochemical and hematological parameters) as biomarkers of micro- and nano-plastic water pollution.

Keywords: micro-nano plastics; fish; organism model; histopathology; blood biomarkers

1. Introduction

Micro-nano plastic pollution is a global problem that has generated great interest among scientists and attracted public attention because of the potential health risks associated with exposure to MP/NPs through air [1], water [2,3], and food [4,5]. In the past years, we have seen an increase in the number of scientific works regarding the topic with the declared purpose of understanding the mechanisms underlying pathological manifestations in various organs, through the study of toxicity on cells and tissues to extrapolate risks to mammals [6,7]. This is a very complex topic due to the difficulty involved in analyzing the effects of various types of plastic. In fact, plastic particles discharged into the environment are often contaminated with different chemical pollutants [8,9] and other contaminants [10–13], including pathogens [14]. In this regard, in fact, some authors argue that microplastics can act as a vector to convey pathogens from the environment to organisms [15,16]. Therefore, adverse effects of microplastics and nano plastics may result from a combination of intrinsic toxicity of plastics and ability to absorb, concentrate, and release environmental pollutants into living organisms [17]. In addition, it has become well documented that the different particles sizes exert effects through different physiological pathways, in relation to quantity and quality in different species at different developmental stage and in different tissues [18]. Micro plastics are small synthetic solid particles mostly composed of blends of polymers and functional additives with different chemical compositions, shapes, colors, sizes, and densities. The European Food Safety Authority (EFSA) [19] defined "micro plastics" all plastic particles having a diameter < 5 mm. MPs including nano plastics (NPs), which are particles with dimensions below 0.1 μm (1–100 nm) [20]. They are classified as primary and secondary based on their origin. Primary micro plastics are any plastic particles that are directly released into the environment. These include microfibers, fragments, microbeads, and plastic pellets. Secondary micro plastics, on the contrary, originate from the degradation of larger-sized plastic products once exposed to the marine environment with regular or irregular shape [21]. They are derived from abrasion, mechanical wear as a wave action, photo oxidation and biological degradation of small pieces of plastic [22,23]. MP/NPs, once released into the environment do not degrade and therefore accumulate in aquatic organisms entering the food chain until they are ingested by humans through food. MPs/NPs acute toxicity are correlated with plastic size [24], concentration [25], and cellular/tissue uptake and accumulation [26]. In general, small size, high dose, and the presence of toxic additives or pollutants in the micro/nano plastics appear to induce cellular toxicity [27]. In fact, small sizes of the microplastics enhance their translocation across the biological membranes with various effects on aquatic organisms via endocytosis-like mechanisms resulting in internalization into cells and tissues with potential adverse consequences on health [20]. In contrast, particles larger than 100 μm do not have any effect [28,29]. In terrestrial vertebrates, the main entry route of microplastics into the body, is the digestive pathway through food and water with pathological changes consisting of gut barrier dysfunction, intestinal inflammation, and gut micro biota dysbiosis [30–33]. In fishes, the absorption routes are various, beyond the digestive way, absorption occurs through the gills and the skin. Therefore, scientists started to document the effects of microplastics on fish [34] and the signs of toxicity observed in these animals are, chain can be illustrated [35]. More specifically, the purpose of this review is not only to investigate the uptake routes and accumulation of MPs in fish, including zebrafish, but also to reveal the toxic effects of MPs in different organs (gut, liver, gills, and kidney), including the brain, with implications on growth and food consumption. In addition, we focused our attention on the central nervous system to study the neurotoxicity effects due to the crossing of the blood-brain barrier as well as on the blood response to MP/NPs. Microplastics cause anemia and alterations in hemato-biochemical parameters. Therefore, representing important blood biomarkers of water pollution by micro- and nano-plastics. Blood tests represent useful experimental tools to monitor the response of fish to the aquatic load of micro and nanoplastics. We have reviewed 219 publications until 2021. Below, we report the data collected for each organ, focusing on structural and molecular effects caused by the exposure to micro and nano plastics.

2. Micro- and Nano-Plastics in Aquatic Ecosystems Can Be Taken-Up by Fish and Reside in Their Brain

Since microplastics (MPs) and nanoplastics (NPs) ingestion by aquatic organisms is reported by laboratory and field studies with a regard to commercial species too [36–38], an unavoidable concern is whether MPs and NPs, one ingested by fish, can be absorbed by the gastrointestinal (GI) tract and can pass into other body districts. However, this field of knowledge just started to be investigated recently. A proof of the MPs and NPs movement from intestine to other body compartments is represented by the evidence that, once these plastics are ingested, they can be found in different organs, outside the GI tract [39]. Once the gastrointestinal barrier is bypassed, also the brain may be at risk of exposure when micro and nanoplastics are able to pass the blood brain barrier (BBB). According to Ding's study on *Oreochromis niloticus* [40], MPs reach the fish brain along with blood circulation. This agrees with Sökmen [41] and his team, hypothesizing that since nanoplastics can pass into red blood cells, as shown by Geiser's in vitro study [42], they can reach the brain through the blood. Indeed, nanoplastic particles have been identified in the brain of fish after waterborne [43] or food-mediated exposure [35], indicating that they are capable of crossing a highly selective permeability barrier as the blood brain one [44]. Mattsson et al. [35] demonstrated that, once ingested, 53 and 180 nm sized polystyrene plastics can arrive into the brain of Crucian carp (*Carassius carassius*). Ding et al. [45] demonstrated that 0.3, 5, and even 70−90 μm polystyrene MPs could be accumulated in the brain, indicating their tissue translocations within red tilapia (*O. niloticus*). In their reviews, Barboza [46] and Campanale [47] agree that about 10 μm or smaller microplastics might penetrate organs and cross the blood-brain barrier reaching the brain (Table 1). Furthermore, micro and nanoplastics could reach the brain before the BBB is established and this could result in an uptake of NPs by the brain at early stages of fish life. Indeed, Sökmen et al. [41] demonstrated that 20 nm diameter polystyrene nanoplastics (PS-NPs) injected to yolk sac can reach the brain and bioaccumulate there. Even when zebrafish (*Danio rerio*) embryos are just exposed to NPs, these ones can accumulate in the yolk sac and migrate to brain [48]. However, Sökmen [41] and his team also hypothesized that the presence of PNPs in the brain may be due to the alteration of the hemodynamic physiology induced by these plastics. These and other pathways traveled by MPs and NPs to the brain add up: NPs can reach the head of young zebrafish that have just undergone an in-water exposure [49] and MPs can cross the gills, reach the fish head, and accumulate there [50]. Therefore, it is necessary to investigate the effects micro and nanoplastics have on the brain.

Table 1. Micro and nanoplastics crossing the blood brain barrier in fish.

Specimens	Micro and Nano-Plastics Type	Micro and Nanoplastics Size	References
Data not referred to a particular species	Data not referred to a particular micro and nano-plastics type	<10 μm	[46,47]
Red tilapia (*Oreochromis niloticus*)	Polystyrene	0.1 μm	[40]
Red tilapia (*Oreochromis niloticus*)	Polystyrene	0.3; 5 μm 70−90 μm	[45]
See-through medaka (*Oryzias latipes*)	Polystyrene	39.4 nm	[43]
Crucian carp (*Carassius carassius*)	Polystyrene	53 and 180 nm	[35]

3. Neurotoxicity of Micro and Nano-Plastics in Fish

Nanoplastics are small hydrophobic particles that can reach the brain by crossing the blood-brain barrier [35,43,51]. Nanoplastic particles have been identified in the brain of fish after waterborne or food-mediated exposure, indicating that they are capable of crossing the blood-brain barrier [44]. Then, in aquatic models, the brain is essential for evaluating the toxic effect of nanoparticles (NPs) [52,53]. In fish at different stages of development, NPs accumulate in the tissues, resulting in multiple negative effects including the nervous system [54]. Ding et al. [45] suggest that the effects of NPs and MPs on the nervous system depend not only on the size and time of exposure but also on indirect mechanisms such as oxidative stress. In this regard, Barboza et al. [51] observed higher levels of LPO and greater AChE activity in the brains of *Dicentrachus labrax*, *Trachurus*, and *Scomber colias* exposed to plastics than in unexposed specimens of the same species, probably because higher concentrations of LPO induce the break of acetylcholine-containing vesicles resulting in increased neurotransmitter released in synaptic clefts and increased production of acetylcholine as a compensation mechanism [51]. Wan et al. [55] showed that neurotoxicity following polystyrene microplastics (PS-MPs) exposure in zebrafish larvae could be associated with some metabolic changes. Instead, Ding et al. [45] show, in adult specimens of tilapia, a downregulation of the amino acids phenylalanine and tyrosine, associated with neurological functions and consequently the polystyrene nanoplastics (PS-NPs) PS-MPs exposure. According to Ding et al. [45], amino acids associated with neurological functions could interact with nanoplastics and microplastics resulting in alteration of the metabolic pathways of these amino acids and alterations in the formation of various neurotransmitters. Furthermore, it has been observed in tilapia that PS between one and 100 micrometers in size can induce more severe stress than 0.3 micrometers particles and this stress seems to be mainly due to oxidative stress and damage caused by LPO, more severe in MPs than NPs exposure. This could be explained by the fact that medium-sized MPs can move freely between cells and induce mechanical injury and alterations in biochemical pathways such as oxidative damage and inflammatory response [45]. Furthermore, NPPs can favor the accumulation of reactive oxygen species (ROS) and cause negative effects at the cellular level [43,56,57]. In adult specimens of medaka fish (*Oryzias latipes*) polystyrene nanoplastics in both organs and blood have been found. Furthermore, after exposure for seven days at 10 mg/L, polystyrene nanoplastics crossed the blood brain barrier in this species and get to the brain [43]. The central nervous system is particularly vulnerable to oxidative stress, in fact it determines developmental and motility alterations and neurotoxicity [58,59]. The results of Chen et al. [60,61], who evaluated specific biomarkers of oxidative stress, the activity of catalase (CAT), glutathione, peroxidase activity (GPx), and the reduced form of glutathione (GSH), confirm the role of oxidative stress in neurotoxicity [60,61]. Behavioral anomalies in fish exposed to pollution by micro and nano plastics have been observed [62]. It has been observed that zebrafish exposed to plastic particles show uncontrolled movements probably due to an improper activation of neurons, due to an altered expression of important neurotransmitters such as acetylcholine, glutamate, and γ-aminobutyric acid [63–67]. Such alterations can be considered indicators of neurotoxicity [68,69]. In fish, exposure to microplastics triggers the activation of cellular oxidative stress processes with consequent peroxidation of cell membranes [46,70]. The damage from lipid peroxidation in the brain can cause the break of the vesicles membranes containing neurotransmitters with consequent increase of the neurotransmitter concentration in the synaptic junctions [71,72]. The neurotoxic effects of microplastics were confirmed by measuring the increase in activity of the enzyme acetylcholinesterase [46,73]. These results are severe because the activity of cholinesterase (ChE) enzymes is essential for cholinergic neurotransmission in neuromuscular junctions and in cholinergic brain synapses [51,74]. Acetylcholinesterase (AChE) activity regulates brain function and is considered an important biomarker for neurotoxicity, also in zebrafish [75,76]. A decrease in AchE activity [61,77], as well as the inhibition of neurotransmitters dopamine, melatonin, aminobutyric acid, serotonin, vasopressin, kisspeptin, and oxytocin [77], in zebrafish exposed to NPs was found [78]. Inhibition of acetylcholinesterase (AChE) enzymatic activities also in Medaka, *O. latipes*, *Pomatoschistus microps* exposed to micro and nano plastics has been reported [10,54,79]. It has been also observed that, in zebrafish larvae, the inhibition of AChE activity interferes with the functioning of the nervous system, causing growth retardation, paralysis, and death [75,76]. Since EE2 (17α-ethinylestradiol) modulates the development of the neuroendocrine system, it was considered a positive control in the evaluation of neurotoxicity [60,80,81]. In groups of larvae exposed to MPP or coexposed with MPP and EE2, the activity of AChE decreased [60]. Moreover, polyethylene microplastics (1–5 μm) inhibited AChE activity in *P. microps* [60,73]. The small particles can inhibit AChE activity causing neurotoxicity with adverse effects in the cholinergic system. Therefore, the suppression of locomotor capacity in zebrafish exposed to nanoparticles and nanoplastics can be explained by the inhibition of acetylcholinesterase activity [60]. Downregulation of genes associated with neuronal functioning (glutamate receptors, potassium channels, synapsin), genes implicated in neuron differentiation and axon genesis (*neurod4*, neuronal differentiation 4, Gene ID: 266958; *lrnn2*, leucine rich repeat neuronal 2, Gene ID: 558051) support the neurotoxic effects of plastics in zebrafish [82] and plastics seem to influence the visual system by upregulating the gene expression of *zfrho* (rhodopsin, Gene ID: 30295) [60,78]. In zebrafish, *gfap* (glial fibrillary acidic protein, Gene ID: 30646) and *α1-tubulin* (alpha-1 tubulin, Gene ID: 842782) genes are expressed in the central nervous system during the early stages of development and are considered important biomarkers of neurotoxicity [83]. *α1-tubulin* is essential in the formation of microtubules, while *gfap* intervenes in the development of astrocytes [84,85]. In zebrafish, the *gfap* gene and the corresponding protein are highly conserved and exhibit functions similar to those of mammals [86]. Moreover, the up regulation of *gfap* is an indicator of neurotoxicity, also observed in mammals [86,87]. The use of *gfap*'s upregulation in neurotoxicity screenings is now accepted, and its use by the US Environmental Protection Agency (EPA) has been recommended [88]. Instead, the up regulation of *gfap* and *α1-tubulin* in the central nervous system promotes an increase in the expression of zfrho and zfblue and this alteration may be a side effect of neurotoxicity [87]. Zebrafish larvae exposed to nanoplastics showed more severe hypoactivity phenomena when co-exposed to NPPs E EE2 due to oxidative damage [60]. In treatments with MPP and MPP + EE2, the up-regulation of sight-related genes such as opsin which is expelled by photoreceptor cells of the retina [89] has been observed [60]. Upregulation of the opsin gene cannot indicate alterations in movement [90].

In *Clarias gariepinus* the effects of PVC microparticles, considering a battery of biomarkers of oxidative stress (superoxide dismutase, catalase, glutathione peroxidase), neurotoxicity (acetylcholinesterase) and lipid peroxidation, have been evaluated [53]. It has been shown that in the presence of PVC the activity of GPx, Sod, Cat, and AchE is inhibited while the levels of LPO have increased, all progressively with time [53].

Specimens of red tilapia were exposed to polystyrene particles of three different sizes (0.3, 5 and 70–90 μm) to evaluate their accumulation, oxidative stress, p450 enzyme activity, metabolic changes, and neurotoxicity. It was shown that only the 5 μm polystyrene microplastics significantly inhibited the activity of acetyl cholinesterase in the brain. In this species, in treatments with PS-N and MPs (0.3, 5 and 70–90 μm) a decrease in the activity of the enzyme acetylcholinesterase in the brain has been observed [45], confirming previous studies in the same species [40] and in *P. microps* [91], as well as in zebrafish larva [60]. It has been shown that plastic materials are capable of conveying pollutants to the body's tissues [60,92]. Although some authors give little importance to this vector function [93–96], it is known that the toxicity of plastics and their interaction with others pollutants can affect the bioavailability and toxicity of these pollutants for the biota [92]. Plastics have shown the ability to absorb various pollutants (IPA, PCB, HCH) [97]. Furthermore, all additives added to plastics must be taken into consideration: plasticizers, flame retardants and antimicrobials whose toxicity can be altered by co-exposure with the transported toxicants [98,99]. In epoxy resins and polycarbonate plastics, the most used additive is bisphenol A (BPA). This additive can be released from plastic materials into the aquatic environment and can reach a concentration of 4 μg/L [98,100,101]. While for years it has been considered a cause of obesity and disorders of the immune and endocrine systems [102,103], Saili et al. [104] highlighted the negative effects of BPA on the brain. The accumulation of BPA and NPPs in various zebrafish organs including the brain has been demonstrated [43,105]. Chen et al. [106] evaluated the different expression of genes and proteins as biomarkers of neurotoxicity during the development of the zebrafish central nervous system. Chen et al. [61] demonstrated that the co-exposure of NPPs and BPA not only leads to increased accumulation of BPA in zebrafish organs, but is responsible for neurotoxic effects on the central nervous and dopaminergic systems. Indeed, during co-exposure to NPPs and BPA, not only the upregulation of myelin and tubulin basic protein genes in the central nervous system and increased dopamine, but also upregulation of astrocyte-derived neurotrophic factor (MANF) expression in the midbrain and significant inhibition of acetyl cholinesterase enzyme activity [61] were evident. The NPs have shown the ability to absorb organic pollutants (IPA, PCB, DDT, PBDE, PFOA) and heavy metals (Ni, Cu, Zn, Pb) and act as a vector for them, thus increasing their bioavailability [6,16,107,108]. They may also contain additives such as bisphenol A (BPA) or phthalates [109,110], whose negative effects on the biota are known [39,108,111]. In the work of Chen et al. [61], the effects of exposure of NNPs and co-exposure of NNPs and EE2 were evaluated. It has been observed that exposure to PS-NPs as well as the co-exposure of NNPs with EE2 inhibits the activity of the acetylcholinesterase enzyme (AChE) with consequent reduction of locomotor activity, and induces an increase in expression of the gfap and alpha tubulin genes related to the central nervous system [60]. In the work of Lee et al. [112], zebrafish embryos were exposed only to fluorescent NNPs and combined with Au ions, and an increase in deformity and mortality rate was observed as well as the activation of the inflammatory response (increase in the expression of IL6 and IL 1β). Furthermore, ROS levels have increased resulting in mitochondrial damage [6,112]. Equally important are the negative effects of micro and nanoplastics on fish development. In particular, the negative effects of micro and nanoplastics during embryonic development and the protective function of the chorion in the early stages of development have been evaluated [113,114]. The protective effect of chorion against pollutants, in both de-chorionate embryos [60,115–117] and elsewhere [118], has been observed [119]. Furthermore, absorption, bioaccumulation and toxicity in zebrafish embryos and larvae have been described using 44 nm model nanoplastics [120].

4. Occurrence of Micro and Nano-Plastics in Fish Gills and Consequent Effects

The occurrence of microplastics in fish gills is reported in different studies on different fish species. Studies reporting NPs and MPs detection in gills refer to animals naturally exposed to these plastics [121–124] and exposed to microplastics in laboratory condition [43,69,125,126]. Like fish collected in fish farm and mariculture areas, those exposed to MPs in laboratory conditions show a higher or, at most, a comparable accumulation of MPs in gills than in gut [127,128]. Therefore the MPs accumulation in gills could be influenced by microplastics shape [126], and the entity of accumulation in gills increases with increasing time of exposure [40] and is related to MPs size [129]. It is intriguing to notice that, once taken up, microplastics still exist in gill fishes after depuration for seven days [130]. However, microplastics elimination time seems to be a specie-specific parameter as shown on larvae by Zhang et al. [131]. The effects of microplastics on gills (Table 2) were tested with in vitro and in vivo studies.

Table 2. Effects of micro and nano plastics on gills.

Specimens	Micro- and Nano-Plastics Sizes and Type	Concentration	Exposition Time	Effects	References
Japanese medaka (*Oryzias latipes*)	10–20 µm sized PES and 50–60 µm sized PP fiber	10,000 Microplastic fiber/L	21 days	- increased mucus, - denuding of epithelium on arches, - fusion of primary lamellae. - aneurysms in secondary lamellae - epithelial lifting, swellings of inner opercular membrane that altered morphology of rostral most gill lamellae	[132]
Goldfish (*Carassius auratus*)	from 0.7 mm to 5.0 mm fiber, ranging between 2.5–3.0 mm and 4.9/5.0 mm, respectively fragments and pellets	0.03 g of fiber/ 15 commercial food pellets; 0.96%, 1.36%, 1.94% and 3.81% (g (food + MPs)/g ww fish).	Six weeks	- fragmentation of gills filaments	[126]

Table 2. *Cont.*

Specimens	Micro- and Nano-Plastics Sizes and Type	Concentration	Exposition Time	Effects	References
Goldfish (*Carassius auratus*)	0.100–1000 μm sized PVC	0.1 or 0.5 mg/L	4 days	- undamaged gills - limited or no effect to antioxidant activity - unaffected ion-regulation function	[133]
Goldfish larvae (*Carassius auratus*)	70 nm and 5 μm sized PS	10, 100 and 1000 μg/L	1, 3 and 7 days	- disordered cell arrangement	[134]
European seabass (*Dicentrarchus labrax*)	1–5 μm polymer microspheres (an average of 2 μm diameter)	0.26 and 0.69 mg/L	96 h	- increased superoxide dismutase (SOD) activity - induction of CAT, GST and SOD, increasing of LPO levels	[46]
Guppy (*Poecilia reticulata*)	32−40 μm sized PS	100 and 1000 μg/L	28 days	weakened Na+/K+-ATPase activity in gills	[135]
Marine medaka, (*Oryzias melastigma*)	10 μm sized PS	20 and 200 mg/L	60 day	- production of ROS - inhibition of SOD, CAT and GST activities - abscission in gill lamellas, loose arrangement of gill filaments, shortening and thickening of gill lamellas	[127]
rainbow trout (*Oncorhynchus mykiss*) gill epithelial cells	220 nm sized polystyrene	0.3 and 3 mM	48 h	- cell viability decrease	[136]
rainbow trout (*Oncorhynchus mykiss*) fingerlings and wildtype Zebrafish (*Danio rerio*)	1 μm, 2 μm, 20 μm, 40 μm and 90 μm sized PS	2×10^5 particles/L	2 h	- up or down regulation of some genes related to immune response - 0.2 μm MPs induced ifnγ gene up-regulation in trout - 1 μm MPs enhanced up-regulation of three immune genes in zebrafish	[31]
Zebrafish (*Danio rerio*)	~70 μm (mean) sized PA, PE, PP, PVC	0.001–10.0 mg/L	10 days	no histological damage	[137]
Zebrafish (*Danio rerio*)			20 days	- alterations of gill epithelium, - adhesion and partial fusion of secondary lamellae and mucous hypersecretion, a higher density of neutrophils	[138]
Common carp larvae (*Cyprinus carpio*)	90 μm; 50 μm and 25 μm sized HD-PE and PS	100 and 1000 μg/L			
Black rockfish (*Sebastes schlegelii*)	0.5 μm and 15 μm polystyrene	190 μg/L	14 days	hypoxia respiratory in gill	[139]
European seabass (*Dicentrarchus labrax*)	3 μm; 3–1.2 μm and 1.2–0.45 μm sized EMPs	0.33 mg/g of feed (corresponding to 5 mg of EMPs/150 g of feed)	3 and 5 days	increase in GST and SOD activities	[129]

Table abbreviations: polystyrene (PS), polyester (PES), polypropylene (PP), polyamides (PA), polyethylene (PE), hidh density polyethylene (HD-PE), polypropylene (PP), polyvinyl chloride (PVC), environmental microplastics (EMPs).

Bussolaro et al. [136] found that polystyrenes NPs decreased the viability of rainbow trout gill epithelial cells in a concentration-dependent manner. On the other hand, Hu et al. [132] conducted an in vivo study by exposing adult Japanese medaka (*O. latipes*) to polyester (PES) and polypropylene (PP) MPs fibers. By means of scanning electron microscopy (SEM), observations of gills increased mucus, denuding of epithelium on arches, and fusion of primary lamellae were observed. Histologic sections revealed: aneurysms in secondary lamellae, epithelial lifting, swellings of the inner opercular membrane that altered morphology of rostral gill lamellae. SEM and histochemical analyses showed increased mucous cells. Similar results were obtained by Jabeen et al. [126]. They observed that microplastics fibers, retained in gills, cause the filaments fragmentation. According to the authors, this is probably due to the direct contact, and it is correlated to plastic particle size and shape. On zebrafish gills, MPs effects are similar to those induced on Japanese medaka, but Limonta and his team [138] revealed also a higher density of neutrophils in gill tissue of MPs-exposed fish. Microplastics effects on gills also depend on their exposure concentration. Barboza et al. [46] found that lipid oxidative damage in gills occurs after exposure to the higher MPs concentration they tested, but not to the lower. An oxidative stress status induced by microplastics was also detected on larvae gills after dietary exposures [140]. According to Zitouni et al. [129], oxidative alterations were highly correlated with MPs size range. Moreover, the authors emphasized that the toxicity of smaller MPs was closely related to other different factors, including the target tissue, exposure duration, and MPs chemical properties. Oxidative stress induced by MPs leads to consequent gill structural damage and histological changes [127]: abscission in gill lamellae, loose arrangement of gill filaments, shortening and thickening of gill lamellae. As noticed for oxidative stress, also gene expression could be affected by MPs and NPs exposure in different ways based on the plastic sizes. Lu et al. [31] found the smallest MPs particles, used in their study, induced ifnγ gene up-regulation in *Oncorhynchus mykiss*, S100a gene up-regulation in zebrafish gills and up-regulation of three immune genes in zebrafish gills (ifnγ, il1β, and igm). So, the scientists speculated that immune gene expression starts in gill epithelia and associated cells after small microplastics are taken up by them. For Huang et al. [135], another effect of PSMPs exposure was a weakened Na+/K+-ATPase activity in gills that could affect the osmotic balance of these organs. Besides the above-mentioned damages on the gills, a changed oxygen consumption that is an indicator of respiration stress, can occur. Therefore, bigger microplastics have a greater impact than smaller ones, as Yin et al. [139] demonstrated. On the other hand, Yang et al. [134] argue that nanoplastics are more dangerous for gills than microplastics. According to these authors, NPs effects on histological damage have an entity that is proportional to the exposure concentration. Interestingly, the authors found that, by detecting gill at both microscopic and ultra-micro levels, no enrichment of MPs or NPs was found within their tissues. Despite this, the gill tissues of larvae showed damage phenomenon. Differently from what reported above, some authors found no damage or alteration on gills after MPs and NPs exposure [133,137]. The inconsistencies in previous studies may be due to the different experimental concentrations employed (Table 2). In addition to MPs and NPs effects on gills, there are other issues. One of the concerns about MPs and NPs is represented by the evidence that, once absorbed by the gill, they can be internalized and spread in fish body [43,141]. Another growing concern is represented by the effects of the co-exposure to microplastics with other contaminants. This is due to the evidence that MPs and NPs, not only can increase the bioavailability and uptake of contaminants but can also enhance their toxic effect on gills [46,61,142–144]. For example, Karami and his team [145], demonstrated that low-density polyethylene (LDPE) fragments can cause toxicity and modulate the adverse impacts of phenanthrene (Phe), by influencing among other parameters, the degree of tissue changes (DTC) in the gills. In the gill tissues, histological alterations were observed, e.g., some basal cell hyperplasia, sloughing, necrosis in the connective tissue. By increasing the concentration of MPs, the following phenomena were observed: epithelial lifting, hyperplasia, necrosis of the connective tissue, extensive cell sloughing, desquamation and consequently blood vessel exposure and, in some instances, total loss of the secondary lamellae. This corroborates the concentration-dependent impact of MPs that other scientists [46,136] encountered too. Besides the oxidative stress induction, microplastics associated with other substances can enhance the expression of immune-related genes in gills as shown by Xu et al. [144] with phenanthrene on zebrafish.

5. Toxicity of Microplastics and Nanoplastics in Digestive Tract: Intestinal Retention Time, Uptake and Elimination

The food-borne route is the major contamination by MP/NPs pathway, with the gut being the target organ [146]. Plastic fragments, in the form of microfibers or microbeads, may accumulate in digestive tract of fish via food and water [147,148]. The bioavailability of MP/NPs and consequently the effects at different levels depend mainly on their size, shape, concentration, and the presence of chemical pollutants [18,149]. The contaminants, species, developmental stage are considered as co-factors for plastic particle toxicity (PPT) [18]. As evidenced by numerous studies carried out under experimental conditions, the ingestion of microplastics can generate two different types of impacts on marine organisms: physical injuries to the organs where accumulation occurs and chemical damage by transfer and accumulation of pollutants with an enhancement of toxicant uptake or an increase in their bioavailability [17]. Conflicting results have been reported in literature regarding the effects of MPs and NPs accumulation in the gut. It has been demonstrated that the physical presence alone of MPs in the gut may be toxic due to their intrinsic ability to induce intestinal blockages or tissue abrasions, which may cause injury of the gut lining, morbidity, and mortality. Experiments carried out on *Dicentrarchus labrax* fed with microplastics for 90 days showed physical damage in the intestinal tract, both in specimens treated with contaminated plastics and in those treated with virgin microplastics [46]. Grigorakis et al. [146] reported that MPs might scratch gut tissues. While Jovanovic et al. [150] reported that MPs did not cause imminent harm in gilt-head seabream after 45 days of exposure, other studies reported abrasions, ulcers, false satiation, blockages in the digestive tract [151], dysbiosis in the gut [30,31], inflammation, bowel wall thinning, villi and epithelial damage, and oxidative stress in the gut tissues [152–154]. Histopathological changes in intestine were observed in goldfish (*Carassius auratus*) larvae by Yang et al. [134] using PS MP/NPs of size < 5 μm with negative repercussions on larval growth and elevated oxidative stress markers. A reduction in weight and body length was also observed in larvae of *Cyprinus carpio* treated with PVC MPs [140]. MPs were found in gastrointestinal tract of wild fishes also, with higher lipid peroxidation levels in the different organs, including the gut [46]. Oxidative stress and structural damages by PS nanoparticles (10 μm at 2–200 μg/L) accumulation in intestine were observed in adult *Oryzias melastigma* [127] as well as in *O. latipes* treated with PS MPs, 10 μm. In this species, histopathological alterations with swollen enterocytes were observed [122]. In the literature there are also different opinions about the crossing of the intestinal wall by micro-nano plastics. Some authors claim that MPs in fish remain in the gastrointestinal tract and expelled with the feces [146,155]. Other authors found instead MPNs in fish fillets assuming rather that smallest plastic particles (<0.1 μm) can transfer from the intestinal tract to the circulatory system and thus spread to the muscle as to all other organs and tissues [145]. In this regard, severe toxicity usually arise from smaller plastic particles with potential impacts on living organisms due to crossing biological barriers such as cell walls or intestinal barrier [145]. MPs intake during 90 days to *Sparus aurata* induced an increase in oxidative damage and a pro-inflammatory response in gut of *S. aurata*, and were able to recover after the exposure to MPs was removed [156]. Ohkubo et al. [157] have estimated the uptake and gut retention of microplastics in juvenile stage of *Fundulus heteroclitus* and red seabream, *Pagrus major* [157]. Various factors affect MPs bioavailability and level of risks at cellular and molecular level on aquatic organisms. Biomarkers for antioxidant response (superoxide dismutase, catalase, glutathione peroxidase, reductase, and glutathione S-transferase), neurotoxic impairment (acetylcholinesterase), lysosomal activity alteration, and genotoxicity have been analyzed by Suman et al. [158]. Virgin microplastics are not causing imminent harm to fish after dietary exposure in *S. aurata* [150] and *O. mykiss* [28], and did not present evident tissue damage in *Barbodes gonionotus* exposed to PVC fragments (0.2, 0.5 and 1.0 mg/L) for 96 h [159]. Little is known about their uptake, translocation, and accumulation within fish organs [23]. MP/NPs showed that oxidative stress and its responding pathways, including inflammatory responses, could play the role of key events [132]. Several different types of effects have been observed in both freshwater [40,82,125,160] and saltwater species [127] in the larval and adult stage. Polystyrene nanoparticles (PS NPs) were found accumulated in larvae or adult gut of Medaka (*O. melastigma*) with increased mortality and decrease in average lengths and weights [161]. Because of their microscopic size, micro- and nanoparticles could find in phagocytosis or endocytosis the preferred route of absorption. The occurrence of microplastics (<5 mm in diameter) and mesoplastics (5–20 mm in diameter) in the gastrointestinal tract of some commercially pelagic, demersal and reef fishes collected from Southeast coast of the Bay of Bengal was observed by Karuppasamy et al. [162]. According to the custom of the Indian population to consume dry fish, the presence of microplastics in the fish gut is, in this case, a potential serious human health concern, as they are directly consumed. The transfer of microplastics of various types and sizes, through the intestine, to the lymphatic system has also been demonstrated. Some researchers have observed that the mechanism transporting MNPs through physiological barriers might be influenced by para-physiological conditions of the animal. In this regard, an increase in MPs transport in the colon of patients with inflammatory bowel disease related to an increase in intestinal permeability was observed [17], whereas a reduction of NPs transport in diabetes condition was seen [163]. NPs (100–1000 nm) stimulated the secretion of cytokines such as IL-6 and IL-8 from monocytes and they act as stressors to the innate immune system of fish [164]. Larger particles at about 100 μm or above were proven to have no significant effect [28]. NPs can negatively affect fish at different stages of development, accumulating in tissues, decreasing locomotor and foraging activities. However, mortality, malformation or effects on hatching related to NPs have not been reported. High concentration of MPs/NPs are usually highly cytotoxic. ROS production and pro-inflammatory responses are the most frequently encountered. MPs/NPs toxicity is evaluated based on the response of gut lipid peroxidation biomarkers, as an indication of the response to oxidative stress, inflammation, epithelial barrier integrity and changes in gut microbiota [60,61]. Therefore, the key role of exposure routes in the uptake, localization, and subsequent distribution of nanoparticles was highlighted by Zhang et al. [131]. The ingestion and effects of MPs in fish are largely dependent on their shape and size [126].

6. Toxicity of Microplastics and Nanoplastics in Digestive Tract of Zebrafish: An Emerging Model to Study the Bioaccumulation and Toxicity of Environmental Contaminants

Zebrafish represents an ideal animal model to study microplastic and nanoplastic toxicity. Due to the transparency of the eggs, embryos and post-hatch larvae, the localization of fluorescent-labelled MPs/NPs particles may be followed to determine the distribution, uptake, trafficking, degradation, and translocation using confocal microscopy [78,165]. On Medline from 2015 to 2021, using key words "microplastics" and "zebrafish", we have n.78 available results, concerning the effects of micro/nanoplastics in different organs and tissues. Typing the keyword "gut" to get only 10 results. In general, the negative effect of microplastics on zebrafish was related to the concentration and particle size. Intestinal toxicity of different length of microplastic fibers (50 ± 26 μm and 200 ± 90 μm) has been studied in both larvae and adults of zebrafish proving that longer microfibers caused very serious intestinal damage compared to shorter microfibers causing a significant reduction in food intake [166]. Studies of metabolomics were also carried out to unravel the underlying mechanisms of microplastics toxicity through the study of chemical processes involving all the endogenous and exogenous metabolites of enteric cells that cause oxidative stress and inflammation [166]. The impact of polystyrene (PS) NPs was studied at the cellular, molecular, and systemic levels by Sendra and collegues [165]. It has been demonstrated by the same authors that the absorption of PS NPs takes place by endocytosis and phagocytosis respectively for 50-nm and 1-micron nanoparticles. Regardless of the pathway of internalization, the particles will be degraded by lysosomes resulting in alkalinization and increased of reactive oxygen species (ROS) that increase their susceptibility to pathogens [165,167]. Oxidative damage and inflammation induced by co-exposure to MPs and cadmium have been studied in zebrafish by Lu et al. [11,31]

demonstrating that MPs enhanced the toxicity of heavy metals in fish tissues, including the gut. Using zebrafish, the influence of different polymer types of MPs was evaluated to understand the toxicology of microplastics (MPs) in combination with other organic pollutants (biocidal compounds or polycyclic aromatic hydrocarbons) with accumulation and consequent increasing serious toxic effects even further leading to oxidative stress and lipid peroxidation also in the liver as well as enhanced neurotoxicity in the brain and metabolic disorders [144,168]. Uptake and tissue accumulation of polystyrene microplastics (PS-MPs) studies in zebrafish have been conducted. Both 5 μm and 20 μm diameter MPs were found in the intestine after seven days of exposure, associated with inflammation, lipid accumulation and alterations of metabolic profiles in fish liver [125]. Inflammation and oxidative stress were observed in the zebrafish gut exposed to polystyrene MPs 5-μm beads in size, at different concentration (50 μg/L and 500 μg/L), for 21 days, associated with alterations in the gut microbiome and tissue metabolic profiles [154]. The shape-depended effects cannot be underestimated. The recent scientific studies have shown that microplastic fibers resulted in more severe intestinal toxicity than microplastic fragments and beads did. Their accumulation caused mucosal damage and increased permeability, inflammation, and metabolism disruption as well as gut microbiota dysbiosis and specific bacteria alterations interfere with the immune system deteriorating host health [154,169]. PS particles < 5 μm in size, polyamides, polyethylene, polypropylene, and polyvinyl chloride cause intestinal damage of adult zebrafish gut [137] although the most serious effects were found in the larval stage [111,125]. In this regard polyethylene (PE) particles 25 nm and 50 nm in sizes were found in cells of the intestinal epithelium in embryos of 0, 24, and 72 hpf [141]. Changes in gut microbiota with disorders of the metabolome and microbiome, as well as changes in the expression of genes correlated with epithelium integrity, were observed in both larval [55] and adult zebrafish [154,165,170]. Upregulation of intestinal Cytochrome P450 gene (cyp1a) was shown by Mak et al. [69] in adult zebrafish. In zebrafish, Limonta et al. [138] reported transcriptomic and histological alterations on the mucosal epithelium of zebrafish although it recognizes plastic particles as inedible materials but ingests them when mixed with food or for accidental ingestion due to small particles (1–5 μm) [23]. Fluorescent MPs were observed in the villi, in the apical surface of the enterocytes, in the lamina propria and very frequently inside the goblet cells. Microfragments and microfibres did not cross the intestinal barrier. Based on the histological analysis, significant changes in the MPs-treated gut tissues were observed: epithelial and villi damage has been described in zebrafish [154,171,172] as bowel wall thinning, cracking of villi, epithelial damage with splitting of enterocytes and increased volume of mucus [30,137]. In contrast, De Sales-Ribeiro et al. [23] argue that the ingestion of microplastics does not induce any histopathological changes in zebrafish.

7. Translocation of Micro and Nanoplastics to Liver

MP/NPs cross the intestinal barrier, travels through the bloodstream to the organs such as liver and kidney [125,173]. In the liver, oxidative stress determines enzymatic and metabolic changes [125]. Histologically, in the liver, MPs particles were detected in the cytoplasm of the hepatocytes ranged between 1416 and 1634 μm, with moderate lipid-like vacuolar degeneration translated into an enlargement of the hepatocyte cytoplasm due to the presence of one or several vacuoles that displaced the nuclei to the periphery. Congestion, sinusoidal dilatation, glycogen-depletion [107] and cellular necrosis were also found [107,126]. Signs of inflammation and lipid accumulation or lipid profile changes were found in African catfish, *C. gariepinus* by Karami et al. [145]. MPs has been observed into the livers of *Engraulis encrasicolus*, *Sardina pilchardus* and *Clupea harengus* after degradation of the hepatic tissue and in the livers cryosections observed in polarized light microscopy [174]. The livers contained relatively large MPs that ranged from 124 μm to 438 μm. High levels of bisphenol A and analogous compounds in numerous fish species of commercial interest were found. The potential relationship between MPs and bisphenol contamination of fish was investigated [175]. Higher concentrations of bisphenols were found in fish with microplastics while they were absent in liver samples of fish without microplastics. It has been established that a large part of the quantities of microplastics ingested are translocated to fish liver [176]. Different opinion about the size of the microplastics that would be able to move into the intestine, from gut to circulatory system before further translocation to liver bare present in literature. Some authors refer that 20 μm microplastics cannot translocate to fish liver in contrast to microplastics less than 5 μm [125]. In zebrafish, 5 μm diameter MPs accumulated in fish liver were found while 20 μm diameter MPs were not found in the liver but only in the gills and intestines. Histopathological analysis showed inflammation and lipid accumulation in zebrafish liver also. Alterations of metabolic profiles in fish liver and alteration of lipid and energy metabolism were also found. However, the exact route through which MPs reach the liver is still unknown.

8. Translocation of Micro and Nanoplastics to Kidney

As in other organs, micro and nanoplastics were found in fish kidney too [177]. They were found in kidney of 11 commercial fish species collected from the marine fish market [178], but also in fish exposed to them in laboratory [43]. On the other hand, Choi and his team [179] did not detect microplastics in kidney fish after exposure in their laboratory. This inconsistency may be due to the different plastic sizes employed by Kashiwada [43] and Choi's teams [179]: respectively, 39.4 nm and 150–180 μm (Table 3).

Table 3. Micro and nanoplastics effect on kidney fish.

Specimens	Micro- and Nano-Plastics Sizes and Type	Concentration	Exposition Time	Effects	References
sheepshead minnow (*Cyprinodon variegatus*) larvae	150–180 μm PE microspheres	50 and 250 mg/L	4 days	- No effects	[179]
Japanese medaka (*Oryzias latipes*)	10 μm PS	500, 1000 and 2000 μg/g	10 weeks	- glomerulopathy - nephrogenesis - glomerulomegaly	[180]
Japanese medaka (*Oryzias latipes*)	10–20 μm sized PES and 50–60 μm sized PP fiber	10,000 Microplastic fiber/L	21 days	- No alteration	[181]
Gilthead seabream (*Sparus aurata*)	From 40 to 150 μm PVC-MPs	100 mg/kg and 500 mg/kg	15 and 30 days	- alteration of creatine kinase, albumin and globulin in serum	[182]
Zebrafish (*Danio rerio*)	~70 μm (mean) sized PA, PE, PP, PVC	0.001–10.0 mg/L	10 days	- No histological damage	[137]
Zebrafish Larvae (*Danio rerio*)	Fluorescent PSNPs (25 nm; 1.05 g cm^{-3}) internally dyed with FirefliTM Fluorescent Green (468/508 nm)	0.2, 2, and 20 mg/L	120 hpf	- transcriptional alterations of pck1	[115]

Table abbreviations. PVC-MPs: polyvinylchloride microparticles, polystyrene (PS), polyester (PES), polypropylene (PP), polyamides (PA), polyethylene (PE), polypropylene (PP), polyvinyl chloride (PVC), polystyrene nanoplastic (PSNPs).

The literature reports that after MPs and NPs exposure, a damage in kidney can occur and this can also involve the immunogenic part of this organ: the head kidney. For instance,

Zhu et al. [183] found, through the histological analysis, alterations in kidney after MPs exposure, like glomerulopathy and nephrogenesis. Glomerulopathy was showed as expansion and congestion of glomerular capillaries, with severe cases with increased glomerular size (i.e., glomerulomegaly) and expansion of Bowman's space while nephrogenesis was evident as tubular and/or glomerular generation or regeneration. The authors found that glomerulopathy and nephrogenesis, in exposed fish, increases in severity with exposure level. Kidney health status can be also evaluated by assessing creatinine and uric acid that can be considered as an index of the glomerular filtration rate and as a biomarker for kidney dysfunction. So, Hamed et al. [184], after 15 days exposure to MPs, detected a significant increase of fish kidney function (creatinine and uric acid). MPs-dependent kidney damage was also demonstrated by Espinosa et al. [182], testing creatine kinase, albumin, and globulin in fish. These parameters were significantly increased in serum from fish fed with PVC-MPs at 100 mg/kg for 15 days and 30 days. Alterations in plasma globulin and albumin were also obtained on juvenile *C. gariepinus* by Karami et al. [145], but with different kind and size of MPs, time of exposure, and administered doses (Table 3). In accordance with the results of Brun et al. [115], another effect of NPs could be the disruption of processes related to energy metabolism. The authors demonstrated that NPs affect pck1 transcription in kidney by inducing alterations in glucose homeostasis. According to Zhu et al. [183], head kidney is one of the most altered sites when fish kidney is exposed to MPs and NPs and head-kidney leucocytes are affected too. Brandts et al. [185] found that NPs altered the expression of target genes related to the immune function, cellular stress, and stress-related hormone secretion in *D. labrax* head kidney. Among these genes *tnfα* (tumour necrosis factor alpha-like, Gene ID: 100136034), *tgfb* (transforming growth factor beta 1, Gene ID: 7040), *hsp70* (heat shock protein 8, Gene ID: 115594641), *mc2r* (melanocortin 2 receptor, Gene ID: 115567821), and *gr* (glutathione reductase, Gene ID: 115568704) were interested. With regard to head–kidney leucocytes, Espinosa et al. [186] found that their phagocytosis and respiratory burst respectively decrease and increase on gilthead seabream (*S. aurata*) and European sea bass (*D. labrax*) after polyvinylchloride (PVC) and polyethylene (PE) microplastics exposure. Moreover, according to the authors, these microplastics induced *nrf2* (nuclear factor erythroid 2-related factor 2-like, Gene ID: 109102552) up-regulation in seabream HKLs. In addition, Espinosa et al. [182] tested the expression of stress genes in head kidney. The gene expression of *prdx5* (peroxiredoxin 5, Gene ID: 115593974) decreased already after 15 days of exposure while the expression of *prdx1* (peroxiredoxin 1, Gene ID: 115573364) and *prdx3* (peroxiredoxin 3, Gene ID: 115596707) increased after 30 days of PVC-MPs. As for other organs (i.g. gills), micro and nanoplastics can exacerbate other compounds toxicity in kidney. For example, Brandts et al. [185] found that NPs can amplify humic acid effects. The combination of this compound and nanoplastics induces an alteration of target genes involved in stress-related functions such as il10 (interleukin 10, Gene ID: 100136835), tgfb, mc2r, and gr gene expression in head kidney of *D. labrax*. Microplastics can affect paraquat effects too [187] and influence blood biochemical parameters in common carp (*C. carpio*). For example, exposure to microplastics was followed by an increase in creatine phosphokinase (CPK) activity in blood that may signify among, other things, renal failure [188]. As reported for gills, Lei et al. [137] found no histological damage to kidneys of zebrafish after microplastics exposure. In the same way, Hu et al. [132] wrote that hematoxylin eoxin staining showed no alterations in internal organs (among the others, kidney) of their samples of exposed individuals to MPs. However, all the mentioned studies reported above employed different species, MPs type and sizes, concentrations, and exposure times (see Table 3). This could explain the different effects obtained by the different authors.

9. Impact of Micro and Nano Plastics Bioaccumulation on Blood Response in Fish

In recent years, the investigation of blood parameters in fish have become of considerable importance and aroused enormous interest not only in aquaculture, but also in the evaluation of fish as organisms sensitive to anthropogenic pollutant load. In fish, the hematological and hematochemical parameters represent important laboratory indexes useful in the diagnosis of many diseases and therefore for the evaluation of the state of health [189,190], physiological state of fish, food conditions, and quality of the water in which they live. The values of blood parameters provide useful information in various body processes and are of great importance in evaluating the harmful effects of anthropogenic pollution on aquatic environments. Fish live in intimate contact with the aquatic environment that significantly influences their blood homeostasis. In the past few years, there has been a significant increase of the experimental studies conducted in the hope of achieving a better understanding of the impact of micro- and nano-plastics (MP/NPs) on diverse organisms including fish. The fish homeostasis is altered by the exposure to polluting substances and materials of various kinds (including micro-nano plastics), causing a series of protective mechanisms such as changes in blood parameters [191,192]. Therefore, the hematological and hematochemical parameters are useful bioindicators in evaluating the effect of different pollutants in fish [192]. In fish, the complete blood count or CBC (complete blood count), is a laboratory test of the blood (fast, inexpensive and practical) which contains all the information necessary for the evaluation of hematopoiesis [193]. The parameters evaluated are the following: WBC (white blood cells), RBC (red blood cells), Hb (hemoglobin), Hct (hematocrit), MCV (mean corpuscolar volume), MCH (mean corpuscolar hemoglobin), MCHC (mean corpuscolar hemoglobin concentration), and TC (thrombocytes). Currently, an improvement in the techniques of haematological investigations in fish has been achieved. New methods of blood analysis are now used routinely and applied in in the study of fish species. Two of these new techniques are represented by the hematological determination by means of an automated electronic system and by flow cytometry. The hematological determination in fish by automated method uses an electronic blood cell counter with an impendence analysis system. This apparatus was previously used for mammals and subsequently modified in the software to carry out the hematological analysis also in fish. For this purpose, a suitable fish lysing solution is used in the instrument. This method has been validated for various fish species [189,194–197]. Flow cytometry is one of the most modern analysis techniques for the detection and quantification of fish blood cells. It allows to identify, count, and characterize individual cells in mixed populations with rapidity and high precision [190]. These new diagnostic methods have brought great benefits to the veterinary medicine and to the study of fish haematology. Previous studies on various fish species, have shown a marked variation in haematological and hematochemical parameters due to exposure to various pollutants [13,184,190,198,199]. Among the various polluting substances and materials, MPs represent the most plentiful pollutants on Earth [200] due to their ubiquitous presence within the aquatic environment [201]. In the last years, the studies conducted to monitor the impact of MP/NPs on aquatic organisms have widely increased. A large variety of effects in various fish species were reported [128,202]. Due to their small size, microplastics and nanoplastics can be ingested (directly or through the food chain) by fish. Once inside the body, the plastic particles mainly accumulate in the digestive tract and transferred to the circulatory system or to adjacent tissues [36], and can persist in the organism for a long time [203–205]. The harmful effects caused by microplastics are due to additives and chemical substances used during their production or utilization of pollutants absorbed from the environment [51,206,207]. MP/NPs induce physical and chemical toxic effects in various organisms including fish. These effects have an impact on the blood response of the exposed organism. In this regard, a study conducted by Hamed et al. [184] on Nile Tilapia (*O. niloticus*) showed variations in hematological and hematochemical parameters of this species after exposure to MPs. Regarding to the hematological profile, the obtained results revealed a significant reduction in red blood cell (RBC), hemoglobin concentration (Hb), hematocrit (Ht), mean corpuscular hemoglobin concentration (MCHC), thrombocytes (PT), white blood cell (WBC), and percent of monocytes after exposure to MPs. Percent of lymphocytes, eosinophils and neutrophils showed fluctuation after exposure to MPs. On the contrary, mean corpuscular volume (MCV) and mean corpuscular hemoglobin

(MCH) exhibited significant increase. These changes in hematological parameters represent a defense mechanism as a response to MPs toxicity [208]. Similar results were previously found by Mukherjee and Shiha [209] in a study on Indian freshwater major carp exposed to cadmium. Moreover, the immune system may be influenced by toxic chemicals in the micro and nano plastics [107]. The immune system of fish is a useful indicator of negative response to environmental stressors [210]. Effects of MP/NPs on chemokines, cytokines and phagocytes, and increased levels of globulins and immunoglobulins [145] were demonstrated in a previous study by Jacob et al. [13]. Changes in hematochemical parameters due to exposure to micro- and nano-plastics were detected in different species of fish [73,182,184,187,211]. A significant increment of creatinine, uric acid, AST, ALT, ALP, glucose, cholesterol, total protein, albumin, globulin, and A/G ratio was found in Nile Tilapia *O. niloticus* [184] and common goby *P. microps* [73,211] after exposure to MPs. Increased levels of ALT, AST, ALP, creatinine, glucose, also occur in common carp (*C. carpio*) exposed to sublethal concentrations of micro-plastic [187]. Furthermore, it was shown that nano-plastic intake in fish can alter the proportion of triglycerides/cholesterol of blood [212]. Contrary to what was observed in Nile tilapia and common goby the exposure to MPs causes a decrease of the cholesterol, high-density lipoprotein and triglyceride levels in African catfish and common carp [145,187]. The increased albumin levels observed in fish after MPs exposure indicate a liver or kidney demage [213] caused by the harmful effects of micro-plastics [182]. All the variations of blood parameters that occur in fish exposed to MP/NPs, represent an effective physiological response of the organism to toxic effects of plastic particles. These blood changes are important biomarkers in toxicological research, environmental monitoring, and prediction of fish health conditions. In the future, it will be important to refine the diagnostic techniques of fish hematology in order to define the potential blood biomarkers related to micro and nano plastics bioaccumulation.

10. Conclusions

This review provides crucial multidimensional characterization of NPs impacts on human and animal health, suggesting the need for deeper investigations following longer exposure times. The current scientific evidence shows the main factors determining the toxicity of plastic particles, including particle size, concentration, exposure time, particle condition, and shape.

Plastic particles smaller than 10 μm cause more toxic effects than larger plastic particles with negative effects, such as decreased survival [214], decreased activity of a neurotransmission biomarker, AChE [10,73,91,215], decreased energy storage of glycogen [79,107,216], aberration of liver energy metabolism [145], effects on heart and lipid tissues [125,217], effects on heart rate [48,118], increased feeding time [35,62,212], inflammation [125], oxidative damage [61], necrosis [61], effects on body length [60], intestinal bacterial composition [30], and texture of brain and muscle including impact on the water balance in the brain [35,62,212]. Recently, MPs/NPs effects on fish started to be investigated through transcriptomic approaches [138,166,218,219]. Nevertheless, further investigations are needed to understand MPs and NPs effects on omics fish.

Author Contributions: Writing—original draft preparation, M.C.G., M.A., and C.P.; writing—review and editing, F.F., M.L., G.M., G.G., R.L., and F.A.; visualization, M.C.G. and F.F.; supervision, A.G. All authors have read and agreed to the published version of the manuscript.

Funding: This research received no external funding.

Institutional Review Board Statement: Not applicable.

Informed Consent Statement: Not applicable.

Data Availability Statement: Not report data.

Conflicts of Interest: The authors declare no conflict of interest.

References

1. Prata, J.C.; da Costa, J.P.; Girão, A.V.; Lopes, I.; Duarte, A.C.; Rocha-Santos, T. Identifying a quick and efficient method of removing organic matter without damaging microplastic samples. *Sci. Total Environ.* **2019**, *686*, 131–139. [CrossRef] [PubMed]
2. Pivokonsky, M.; Cermakova, L.; Novotna, K.; Peer, P.; Cajthaml, T.; Janda, V. Occurrence of microplastics in raw and treated drinking water. *Sci. Total Environ.* **2018**, *643*, 1644–1651. [CrossRef]
3. Koelmans, A.A.; Mohamed Nor, N.H.; Hermsen, E.; Kooi, M.; Mintenig, S.M.; De France, J. Microplastics in freshwaters and drinking water: Critical review and assessment of data quality. *Water Res.* **2019**, *155*, 410–422. [CrossRef] [PubMed]
4. Hale, R.C.; Seeley, M.E.; La Guardia, M.J.; Mai, L.; Zeng, E.Y. A Global Perspective on Microplastics. *J. Geophys. Res. Oceans* **2020**, *125*, e2018JC014719. [CrossRef]
5. Horton, A.A.; Barnes, D.K.A. Microplastic pollution in a rapidly changing world: Implications for remote and vulnerable marine ecosystems. *Sci. Total Environ.* **2020**, *738*, 140349. [CrossRef]
6. Barría, C.; Brandts, I.; Tort, L.; Oliveira, M.; Teles, M. Effect of nanoplastics on fish health and performance: A review. *Mar. Pollut. Bull.* **2020**, *151*, 110791. [CrossRef]
7. Rubio, L.; Marcos, R.; Hernández, A. Potential adverse health effects of ingested micro-and nanoplastics on humans. Lessons learned from in vivo and in vitro mammalian models. *J. Toxicol. Environ. Health Part B* **2020**, *23*, 51–68. [CrossRef]
8. Hartmann, N.B.; Rist, S.; Bodin, J.; Jensen, L.H.; Schmidt, S.N.; Mayer, P.; Meibom, A.; Baun, A. Microplastics as vectors for environmental contaminants: Exploring sorption, desorption, and transfer to biota. *Integr. Environ. Assess. Manag. (IEAM)* **2017**, *13*, 488–493. [CrossRef] [PubMed]
9. Liu, M.; Lu, S.; Song, Y.; Lei, L.; Hu, J.; Lv, W.; Zhou, W.; Cao, C.; Shi, H.; Yang, X.; et al. Microplastic and mesoplastic pollution in farmland soils in suburbs of Shanghai, China. *Environ. Pollut.* **2018**, *242*, 855–862. [CrossRef] [PubMed]
10. Ferreira, P.; Fonte, E.; Soares, M.E.; Carvalho, F.; Guilhermino, L. Effects of multi-stressors on juveniles of the marine fish Pomatoschistus microps: Gold nanoparticles, microplastics and temperature. *Aquat. Toxicol.* **2016**, *170*, 89–103. [CrossRef]
11. Lu, L.; Luo, T.; Zhao, Y.; Cai, C.; Fu, Z.; Jin, Y. Interaction between microplastics and microorganism as well as gut microbiota: A consideration on environmental animal and human health. *Sci. Total Environ.* **2019**, *667*, 94–100. [CrossRef]
12. Miranda, T.; Vieira, L.R.; Guilhermino, L. Neurotoxicity, Behavior, and Lethal Effects of Cadmium, Microplastics, and Their Mixtures on Pomatoschistus microps Juveniles from Two Wild Populations Exposed under Laboratory Conditions—Implications to Environmental and Human Risk Assessment. *Int. J. Environ. Res. Public Health* **2019**, *16*, 2857. [CrossRef]
13. Jacob, H.; Besson, M.; Swarzenski, P.W.; Lecchini, D.; Metian, M. Effects of Virgin Micro- and Nanoplastics on Fish: Trends, Meta-Analysis, and Perspectives. *Environ. Sci. Technol.* **2020**, *54*, 4733–4745. [CrossRef]
14. Wu, X.; Pan, J.; Li, M.; Li, Y.; Bartlam, M.; Wang, Y. Selective enrichment of bacterial pathogens by microplastic biofilm. *Water Res.* **2019**, *165*, 114979. [CrossRef]
15. Alimba, C.G.; Faggio, C. Microplastics in the marine environment: Current trends in environmental pollution and mechanisms of toxicological profile. *Environ. Toxicol. Pharmacol.* **2019**, *68*, 61–74. [CrossRef] [PubMed]
16. Ferreira, I.; Venâncio, C.; Lopes, I.; Oliveira, M. Nanoplastics and marine organisms: What has been studied? *Environ. Toxicol. Pharmacol.* **2019**, *67*, 1–7. [CrossRef] [PubMed]
17. Yong, C.Q.Y.; Valiyaveettil, S.; Tang, B.L. Toxicity of Microplastics and Nanoplastics in Mammalian Systems. *Int. J. Environ. Res. Public Health* **2020**, *17*, 1509. [CrossRef]
18. Kögel, T.; Refosco, A.; Maage, A. Surveillance of Seafood for Microplastics. In *Handbook of Microplastics in the Environment*; Springer: Cham, Switzerland, 2020. [CrossRef]
19. EFSA. *European Food Safety Authority-Annual Accounts*; European Food Safety Authority: Parma, Italy, 2016.
20. Gigault, J.; Ter Halle, A.; Baudrimont, M.; Pascal, P.-Y.; Gauffre, F.; Phi, T.-L.; El Hadri, H.; Grassl, B.; Reynaud, S. Current opinion: What is a nanoplastic? *Environ. Pollut.* **2018**, *235*, 1030–1034. [CrossRef]
21. Boucher, J.; Friot, D. *Primary Microplastics in the Oceans: A Global Evaluation of Sources*; IUCN: Gland, Switzerland, 2017.
22. Lambert, S.; Wagner, M. Characterisation of nanoplastics during the degradation of polystyrene. *Chemosphere* **2016**, *145*, 265–268. [CrossRef]
23. De Sales-Ribeiro, C.; Brito-Casillas, Y.; Fernandez, A.; Caballero, M.J. An end to the controversy over the microscopic detection and effects of pristine microplastics in fish organs. *Sci. Rep.* **2020**, *10*, 12434. [CrossRef]
24. Walczak, A.P.; Hendriksen, P.J.; Woutersen, R.A.; van der Zande, M.; Undas, A.K.; Helsdingen, R.; van den Berg, H.H.; Rietjens, I.M.; Bouwmeester, H. Bioavailability and biodistribution of differently charged polystyrene nanoparticles upon oral exposure in rats. *J. Nanopart. Res.* **2015**, *17*, 1–13. [CrossRef]
25. Prokić, M.D.; Radovanović, T.B.; Gavrić, J.P.; Faggio, C. Ecotoxicological effects of microplastics: Examination of biomarkers, current state and future perspectives. *TrAC Trends Anal. Chem.* **2019**, *111*, 37–46. [CrossRef]
26. Fiorentino, G.; Ripa, M.; Protano, G.; Hornsby, C.; Ulgiati, S. Life Cycle Assessment of Mixed Municipal Solid Waste: Multi-input versus multi-output perspective. *Waste Manag.* **2015**, *46*, 599–611. [CrossRef]
27. Burgos-Aceves, M.A.; Abo-Al-Ela, H.G.; Faggio, C. Physiological and metabolic approach of plastic additive effects: Immune cells responses. *J. Hazard. Mater.* **2021**, *404*, 124114. [CrossRef]
28. Ašmonaitė, G.; Larsson, K.; Undeland, I.; Sturve, J.; Carney Almroth, B. Size Matters: Ingestion of Relatively Large Microplastics Contaminated with Environmental Pollutants Posed Little Risk for Fish Health and Fillet Quality. *Environ. Sci. Technol.* **2018**, *52*, 14381–14391. [CrossRef]

29. Jacob, H.; Gilson, A.; Lanctôt, C.; Besson, M.; Metian, M.; Lecchini, D. No Effect of Polystyrene Microplastics on Foraging Activity and Survival in a Post-larvae Coral-Reef Fish, *Acanthurus triostegus*. *Bull. Environ. Contam. Toxicol.* **2019**, *102*, 457–461. [CrossRef]

30. Jin, Y.; Xia, J.; Pan, Z.; Yang, J.; Wang, W.; Fu, Z. Polystyrene microplastics induce microbiota dysbiosis and inflammation in the gut of adult zebrafish. *Environ. Pollut.* **2018**, *235*, 322–329. [CrossRef]

31. Lu, C.; Kania, P.W.; Buchmann, K. Particle effects on fish gills: An immunogenetic approach for rainbow trout and zebrafish. *Aquaculture* **2018**, *484*, 98–104. [CrossRef]

32. Li, Y.; Li, M.; Li, Z.; Yang, L.; Liu, X. Effects of particle size and solution chemistry on Triclosan sorption on polystyrene microplastic. *Chemosphere* **2019**, *231*, 308–314. [CrossRef]

33. Luo, H.; Xiang, Y.; He, D.; Li, Y.; Zhao, Y.; Wang, S.; Pan, X. Leaching behavior of fluorescent additives from microplastics and the toxicity of leachate to *Chlorella vulgaris*. *Sci. Total Environ.* **2019**, *678*, 1–9. [CrossRef]

34. Guzzetti, E.; Sureda, A.; Tejada, S.; Faggio, C. Microplastic in marine organism: Environmental and toxicological effects. *Environ. Toxicol. Pharmacol.* **2018**, *64*, 164–171. [CrossRef] [PubMed]

35. Mattsson, K.; Johnson, E.V.; Malmendal, A.; Linse, S.; Hansson, L.-A.; Cedervall, T. Brain damage and behavioural disorders in fish induced by plastic nanoparticles delivered through the food chain. *Sci. Rep.* **2017**, *7*, 1–7. [CrossRef]

36. Lusher, A.; Hollman, P.; Mendoza-Hill, J. *Microplastics in Fisheries and Aquaculture: Status of Knowledge on Their Occurrence and Implications for Aquatic Organisms and Food Safety*; FAO: Rome, Italy, 2017.

37. Savoca, S.; Capillo, G.; Mancuso, M.; Faggio, C.; Panarello, G.; Crupi, R.; Bonsignore, M.; D'Urso, L.; Compagnini, G.; Neri, F.; et al. Detection of artificial cellulose microfibers in *Boops boops* from the northern coasts of Sicily (Central Mediterranean). *Sci. Total Environ.* **2019**, *691*, 455–465. [CrossRef]

38. Savoca, S.; Bottari, T.; Fazio, E.; Bonsignore, M.; Mancuso, M.; Luna, G.M.; Romeo, T.; D'Urso, L.; Capillo, G.; Panarello, G.; et al. Plastics occurrence in juveniles of *Engraulis encrasicolus* and *Sardina pilchardus* in the Southern Tyrrhenian Sea. *Sci. Total Environ.* **2020**, *718*, 137457. [CrossRef] [PubMed]

39. Jovanović, B. Ingestion of microplastics by fish and its potential consequences from a physical perspective. *Integr. Environ. Assess. Manag.* **2017**, *13*, 510–515. [CrossRef] [PubMed]

40. Ding, J.; Zhang, S.; Razanajatovo, R.M.; Zou, H.; Zhu, W. Accumulation, tissue distribution, and biochemical effects of polystyrene microplastics in the freshwater fish red tilapia (*Oreochromis niloticus*). *Environ. Pollut.* **2018**, *238*, 1–9. [CrossRef] [PubMed]

41. Sökmen, T.Ö.; Sulukan, E.; Türkoğlu, M.; Baran, A.; Özkaraca, M.; Ceyhun, S.B. Polystyrene nanoplastics (20 nm) are able to bioaccumulate and cause oxidative DNA damages in the brain tissue of zebrafish embryo (*Danio rerio*). *Neurotoxicology* **2020**, *77*, 51–59. [CrossRef]

42. Geiser, M.; Rothen-Rutishauser, B.; Kapp, N.; Schürch, S.; Kreyling, W.; Schulz, H.; Semmler, M.; Hof, V.I.; Heyder, J.; Gehr, P. Ultrafine particles cross cellular membranes by nonphagocytic mechanisms in lungs and in cultured cells. *Environ. Health Perspect.* **2005**, *113*, 1555–1560. [CrossRef]

43. Kashiwada, S. Distribution of nanoparticles in the see-through medaka (*Oryzias latipes*). *Environ. Health Perspect.* **2006**, *114*, 1697–1702. [CrossRef]

44. Brandts, I.; Barría, C.; Martins, M.; Franco-Martínez, L.; Barreto, A.; Tvarijonaviciute, A.; Tort, L.; Oliveira, M.; Teles, M. Waterborne exposure of gilthead seabream (*Sparus aurata*) to polymethylmethacrylate nanoplastics causes effects at cellular and molecular levels. *J. Hazard. Mater.* **2021**, *403*, 123590. [CrossRef]

45. Ding, J.; Huang, Y.; Liu, S.; Zhang, S.; Zou, H.; Wang, Z.; Zhu, W.; Geng, J. Toxicological effects of nano- and micro-polystyrene plastics on red tilapia: Are larger plastic particles more harmless? *J. Hazard. Mater.* **2020**, *396*, 122693. [CrossRef]

46. Barboza, L.G.A.; Vieira, L.R.; Branco, V.; Figueiredo, N.; Carvalho, F.; Carvalho, C.; Guilhermino, L. Microplastics cause neurotoxicity, oxidative damage and energy-related changes and interact with the bioaccumulation of mercury in the European seabass, *Dicentrarchus labrax* (Linnaeus, 1758). *Aquat. Toxicol.* **2018**, *195*, 49–57. [CrossRef] [PubMed]

47. Campanale, C.; Massarelli, C.; Savino, I.; Locaputo, V.; Uricchio, V.F. A detailed review study on potential effects of microplastics and additives of concern on human health. *Int. J. Environ. Res. Public Health* **2020**, *17*, 1212. [CrossRef]

48. Pitt, J.A.; Kozal, J.S.; Jayasundara, N.; Massarsky, A.; Trevisan, R.; Geitner, N.; Wiesner, M.; Levin, E.D.; Di Giulio, R.T. Uptake, tissue distribution, and toxicity of polystyrene nanoparticles in developing zebrafish (*Danio rerio*). *Aquat. Toxicol.* **2018**, *194*, 185–194. [CrossRef]

49. Skjolding, L.M.; Ašmonaitė, G.; Jølck, R.I.; Andresen, T.L.; Selck, H.; Baun, A.; Sturve, J. An assessment of the importance of exposure routes to the uptake and internal localisation of fluorescent nanoparticles in zebrafish (*Danio rerio*), using light sheet microscopy. *Nanotoxicology* **2017**, *11*, 351–359. [CrossRef] [PubMed]

50. Su, L.; Nan, B.; Hassell, K.L.; Craig, N.J.; Pettigrove, V. Microplastics biomonitoring in Australian urban wetlands using a common noxious fish (*Gambusia holbrooki*). *Chemosphere* **2019**, *228*, 65–74. [CrossRef] [PubMed]

51. Barboza, L.G.A.; Lopes, C.; Oliveira, P.; Bessa, F.; Otero, V.; Henriques, B.; Raimundo, J.; Caetano, M.; Vale, C.; Guilhermino, L. Microplastics in wild fish from North East Atlantic Ocean and its potential for causing neurotoxic effects, lipid oxidative damage, and human health risks associated with ingestion exposure. *Sci. Total Environ.* **2020**, *717*, 134625. [CrossRef]

52. Pala, A.; Serdar, O. Seasonal variation of acetylcholinesterase activity as a biomarker in brain tissue of *Capoeta umbla* in Pülümür Stream. *J. Limnol. Freshw. Fish. Res.* **2018**, *4*, 98–102. [CrossRef]

53. Iheanacho, S.C.; Odo, G.E. Neurotoxicity, oxidative stress biomarkers and haematological responses in African catfish (*Clarias gariepinus*) exposed to polyvinyl chloride microparticles. *Comp. Biochem. Physiol. Part C Toxicol. Pharmacol.* **2020**, *232*, 108741. [CrossRef] [PubMed]
54. Peng, L.; Fu, D.; Qi, H.; Lan, C.Q.; Yu, H.; Ge, C. Micro-and nano-plastics in marine environment: Source, distribution and threats—A review. *Sci. Total Environ.* **2020**, *698*, 134254. [CrossRef]
55. Wan, Z.; Wang, C.; Zhou, J.; Shen, M.; Wang, X.; Fu, Z.; Jin, Y. Effects of polystyrene microplastics on the composition of the microbiome and metabolism in larval zebrafish. *Chemosphere* **2019**, *217*, 646–658. [CrossRef] [PubMed]
56. Bhattacharya, P.; Lin, S.; Turner, J.P.; Ke, P.C. Physical adsorption of charged plastic nanoparticles affects algal photosynthesis. *J. Phys. Chem. C* **2010**, *114*, 16556–16561. [CrossRef]
57. Browne, M.A.; Niven, S.J.; Galloway, T.S.; Rowland, S.J.; Thompson, R.C. Microplastic moves pollutants and additives to worms, reducing functions linked to health and biodiversity. *Curr. Biol.* **2013**, *23*, 2388–2392. [CrossRef] [PubMed]
58. Salminen, L.E.; Paul, R.H. Oxidative stress and genetic markers of suboptimal antioxidant defense in the aging brain: A theoretical review. *Rev. Neurosci.* **2014**, *25*, 805–819. [CrossRef] [PubMed]
59. Zhu, B.; Wang, Q.; Shi, X.; Guo, Y.; Xu, T.; Zhou, B. Effect of combined exposure to lead and decabromodiphenyl ether on neurodevelopment of zebrafish larvae. *Chemosphere* **2016**, *144*, 1646–1654. [CrossRef] [PubMed]
60. Chen, Q.; Gundlach, M.; Yang, S.; Jiang, J.; Velki, M.; Yin, D.; Hollert, H. Quantitative investigation of the mechanisms of microplastics and nanoplastics toward zebrafish larvae locomotor activity. *Sci. Total Environ.* **2017**, *584–585*, 1022–1031. [CrossRef] [PubMed]
61. Chen, Q.; Yin, D.; Jia, Y.; Schiwy, S.; Legradi, J.; Yang, S.; Hollert, H. Enhanced uptake of BPA in the presence of nanoplastics can lead to neurotoxic effects in adult zebrafish. *Sci. Total Environ.* **2017**, *609*, 1312–1321. [CrossRef] [PubMed]
62. Mattsson, K.; Ekvall, M.T.; Hansson, L.-A.; Linse, S.; Malmendal, A.; Cedervall, T. Altered behavior, physiology, and metabolism in fish exposed to polystyrene nanoparticles. *Environ. Sci. Technol.* **2015**, *49*, 553–561. [CrossRef] [PubMed]
63. Ozawa, S.; Kamiya, H.; Tsuzuki, K. Glutamate receptors in the mammalian central nervous system. *Prog. Neurobiol.* **1998**, *54*, 581–618. [CrossRef]
64. Atzori, M.; Kanold, P.; Pineda, J.C.; Flores-Hernandez, J. Dopamine-Acetylcholine Interactions in the Modulation of Glutamate Release. *Ann. N. Y. Acad. Sci.* **2003**, *1003*, 346–348. [CrossRef] [PubMed]
65. Silva, A.J. Molecular and cellular cognitive studies of the role of synaptic plasticity in memory. *J. Neurobiol.* **2003**, *54*, 224–237. [CrossRef] [PubMed]
66. Werner, F.-M.; Coveñas, R. Classical neurotransmitters and neuropeptides involved in generalized epilepsy: A focus on antiepileptic drugs. *Curr. Med. Chem.* **2011**, *18*, 4933–4948. [CrossRef]
67. Shaikh, M.; Sancheti, J.; Sathaye, S. Effect of Eclipta alba on acute seizure models: A GABAA-mediated effect. *Indian J. Pharm. Sci.* **2013**, *75*, 380. [CrossRef]
68. Berg, A.T.; Berkovic, S.F.; Brodie, M.J.; Buchhalter, J.; Cross, J.H.; van Emde Boas, W.; Engel, J.; French, J.; Glauser, T.A.; Mathern, G.W. Revised terminology and concepts for organization of seizures and epilepsies: Report of the ILAE Commission on Classification and Terminology, 2005–2009. *Epilepsia* **2010**. [CrossRef]
69. Mak, C.W.; Yeung, K.C.-F.; Chan, K.M. Acute toxic effects of polyethylene microplastic on adult zebrafish. *Ecotoxicol. Environ. Saf.* **2019**, *182*, 109442. [CrossRef] [PubMed]
70. Alomar, C.; Sureda, A.; Capó, X.; Guijarro, B.; Tejada, S.; Deudero, S. Microplastic ingestion by *Mullus surmuletus* Linnaeus, 1758 fish and its potential for causing oxidative stress. *Environ. Res.* **2017**, *159*, 135–142. [CrossRef] [PubMed]
71. Hilfiker, S.; Pieribone, V.A.; Czernik, A.J.; Kao, H.-T.; Augustine, G.J.; Greengard, P. Synapsins as regulators of neurotransmitter release. *Philos. Trans. R. Soc. Lond. Ser. B Biol. Sci.* **1999**, *354*, 269–279. [CrossRef] [PubMed]
72. Bradbury, S.P.; Carlson, R.W.; Henry, T.R.; Padilla, S.; Cowden, J. Toxic responses of the fish nervous system. In *The Toxicology of Fishes*; CRC Press-Taylor: Boca Raton, FL, USA, 2008; pp. 417–455. [CrossRef]
73. Oliveira, M.; Ribeiro, A.; Hylland, K.; Guilhermino, L. Single and combined effects of microplastics and pyrene on juveniles (0+ group) of the common goby *Pomatoschistus microps* (Teleostei, Gobiidae). *Ecol. Indic.* **2013**, *34*, 641–647. [CrossRef]
74. Massoulié, J.; Pezzementi, L.; Bon, S.; Krejci, E.; Vallette, F.-M. Molecular and cellular biology of cholinesterases. *Prog. Neurobiol.* **1993**, *41*, 31–91. [CrossRef]
75. Worek, F.; Reiter, G.; Eyer, P.; Szinicz, L. Reactivation kinetics of acetylcholinesterase from different species inhibited by highly toxic organophosphates. *Arch. Toxicol.* **2002**, *76*, 523–529. [CrossRef] [PubMed]
76. Pereira, V.M.; Bortolotto, J.W.; Kist, L.W.; de Azevedo, M.B.; Fritsch, R.S.; da Luz Oliveira, R.; Pereira, T.C.B.; Bonan, C.D.; Vianna, M.R.; Bogo, M.R. Endosulfan exposure inhibits brain AChE activity and impairs swimming performance in adult zebrafish (*Danio rerio*). *Neurotoxicology* **2012**, *33*, 469–475. [CrossRef] [PubMed]
77. Sarasamma, S.; Audira, G.; Siregar, P.; Malhotra, N.; Lai, Y.-H.; Liang, S.-T.; Chen, J.-R.; Chen, K.H.-C.; Hsiao, C.-D. Nanoplastics cause neurobehavioral impairments, reproductive and oxidative damages, and biomarker responses in zebrafish: Throwing up alarms of wide spread health risk of exposure. *Int. J. Mol. Sci.* **2020**, *21*, 1410. [CrossRef] [PubMed]
78. Bhagat, J.; Zang, L.; Nishimura, N.; Shimada, Y. Zebrafish: An emerging model to study microplastic and nanoplastic toxicity. *Sci. Total Environ.* **2020**, 138707. [CrossRef] [PubMed]

79. Rochman, C.M.; Kurobe, T.; Flores, I.; Teh, S.J. Early warning signs of endocrine disruption in adult fish from the ingestion of polyethylene with and without sorbed chemical pollutants from the marine environment. *Sci. Total Environ.* **2014**, *493*, 656–661. [CrossRef]
80. Reyhanian, N.; Volkova, K.; Hallgren, S.; Bollner, T.; Olsson, P.-E.; Olsén, H.; Hällström, I.P. 17α-Ethinyl estradiol affects anxiety and shoaling behavior in adult male zebra fish (*Danio rerio*). *Aquat. Toxicol.* **2011**, *105*, 41–48. [CrossRef] [PubMed]
81. Maranho, L.A.; Baena-Nogueras, R.M.; Lara-Martín, P.A.; DelValls, T.A.; Martín-Díaz, M.L. Bioavailability, oxidative stress, neurotoxicity and genotoxicity of pharmaceuticals bound to marine sediments. The use of the polychaete *Hediste diversicolor* as bioindicator species. *Environ. Res.* **2014**, *134*, 353–365. [CrossRef]
82. LeMoine, C.M.; Kelleher, B.M.; Lagarde, R.; Northam, C.; Elebute, O.O.; Cassone, B.J. Transcriptional effects of polyethylene microplastics ingestion in developing zebrafish (*Danio rerio*). *Environ. Pollut.* **2018**, *243*, 591–600. [CrossRef] [PubMed]
83. Fan, C.-Y.; Cowden, J.; Simmons, S.O.; Padilla, S.; Ramabhadran, R. Gene expression changes in developing zebrafish as potential markers for rapid developmental neurotoxicity screening. *Neurotoxicol. Teratol.* **2010**, *32*, 91–98. [CrossRef] [PubMed]
84. Fausett, B.V.; Goldman, D. A Role for α1 Tubulin-Expressing Müller Glia in Regeneration of the Injured Zebrafish Retina. *J. Neurosci.* **2006**, *26*, 6303–6313. [CrossRef]
85. Veldman, M.B.; Bemben, M.A.; Goldman, D. Tuba1a gene expression is regulated by KLF6/7 and is necessary for CNS development and regeneration in zebrafish. *Mol. Cell. Neurosci.* **2010**, *43*, 370–383. [CrossRef]
86. Nielsen, A.L.; Jørgensen, A.L. Structural and functional characterization of the zebrafish gene for glial fibrillary acidic protein, GFAP. *Gene* **2003**, *310*, 123–132. [CrossRef]
87. Chen, L.; Yu, K.; Huang, C.; Yu, L.; Zhu, B.; Lam, P.K.S.; Lam, J.C.W.; Zhou, B. Prenatal Transfer of Polybrominated Diphenyl Ethers (PBDEs) Results in Developmental Neurotoxicity in Zebrafish Larvae. *Environ. Sci. Technol.* **2012**, *46*, 9727–9734. [CrossRef] [PubMed]
88. McGrath, P.; Li, C.-Q. Zebrafish: A predictive model for assessing drug-induced toxicity. *Drug Discov. Today* **2008**, *13*, 394–401. [CrossRef] [PubMed]
89. Morrow, J.M.; Lazic, S.; Chang, B.S. A novel rhodopsin-like gene expressed in zebrafish retina. *Vis. Neurosci.* **2011**, *28*, 325–335. [CrossRef] [PubMed]
90. Chen, L.; Huang, Y.; Huang, C.; Hu, B.; Hu, C.; Zhou, B. Acute exposure to DE-71 causes alterations in visual behavior in zebrafish larvae. *Environ. Toxicol. Chem.* **2013**, *32*, 1370–1375. [CrossRef] [PubMed]
91. Luís, L.G.; Ferreira, P.; Fonte, E.; Oliveira, M.; Guilhermino, L. Does the presence of microplastics influence the acute toxicity of chromium(VI) to early juveniles of the common goby (*Pomatoschistus microps*)? A study with juveniles from two wild estuarine populations. *Aquat. Toxicol.* **2015**, *164*, 163–174. [CrossRef]
92. Ma, Y.; Huang, A.; Cao, S.; Sun, F.; Wang, L.; Guo, H.; Ji, R. Effects of nanoplastics and microplastics on toxicity, bioaccumulation, and environmental fate of phenanthrene in fresh water. *Environ. Pollut.* **2016**, *219*, 166–173. [CrossRef]
93. Koelmans, A.A. Modeling the Role of Microplastics in Bioaccumulation of Organic Chemicals to Marine Aquatic Organisms. A Critical Review. In *Marine Anthropogenic Litter*; Bergmann, M., Gutow, L., Klages, M., Eds.; Springer International Publishing: Cham, Switzerland, 2015; pp. 309–324. [CrossRef]
94. Herzke, D.; Anker-Nilssen, T.; Nøst, T.H.; Götsch, A.; Christensen-Dalsgaard, S.; Langset, M.; Fangel, K.; Koelmans, A.A. Negligible Impact of Ingested Microplastics on Tissue Concentrations of Persistent Organic Pollutants in Northern Fulmars off Coastal Norway. *Environ. Sci. Technol.* **2016**, *50*, 1924–1933. [CrossRef]
95. Koelmans, A.A.; Bakir, A.; Burton, G.A.; Janssen, C.R. Microplastic as a Vector for Chemicals in the Aquatic Environment: Critical Review and Model-Supported Reinterpretation of Empirical Studies. *Environ. Sci. Technol.* **2016**, *50*, 3315–3326. [CrossRef]
96. Beckingham, B.; Ghosh, U. Differential bioavailability of polychlorinated biphenyls associated with environmental particles: Microplastic in comparison to wood, coal and biochar. *Environ. Pollut.* **2017**, *220*, 150–158. [CrossRef]
97. Lee, H.; Shim, W.J.; Kwon, J.-H. Sorption capacity of plastic debris for hydrophobic organic chemicals. *Sci. Total Environ.* **2014**, *470–471*, 1545–1552. [CrossRef] [PubMed]
98. Makris, K.C.; Andra, S.S.; Jia, A.; Herrick, L.; Christophi, C.A.; Snyder, S.A.; Hauser, R. Association between Water Consumption from Polycarbonate Containers and Bisphenol A Intake during Harsh Environmental Conditions in Summer. *Environ. Sci. Technol.* **2013**, *47*, 3333–3343. [CrossRef] [PubMed]
99. Koelmans, A.A.; Besseling, E.; Foekema, E.M. Leaching of plastic additives to marine organisms. *Environ. Pollut.* **2014**, *187*, 49–54. [CrossRef]
100. Nam, S.-H.; Seo, Y.-M.; Kim, M.-G. Bisphenol A migration from polycarbonate baby bottle with repeated use. *Chemosphere* **2010**, *79*, 949–952. [CrossRef] [PubMed]
101. Huang, Y.Q.; Wong, C.K.C.; Zheng, J.S.; Bouwman, H.; Barra, R.; Wahlström, B.; Neretin, L.; Wong, M.H. Bisphenol A (BPA) in China: A review of sources, environmental levels, and potential human health impacts. *Environ. Int.* **2012**, *42*, 91–99. [CrossRef]
102. Vom Saal, F.S.; Nagel, S.C.; Coe, B.L.; Angle, B.M.; Taylor, J.A. The estrogenic endocrine disrupting chemical bisphenol A (BPA) and obesity. *Mol. Cell. Endocrinol.* **2012**, *354*, 74–84. [CrossRef] [PubMed]
103. Michałowicz, J. Bisphenol A—Sources, toxicity and biotransformation. *Environ. Toxicol. Pharmacol.* **2014**, *37*, 738–758. [CrossRef] [PubMed]

104. Saili, K.S.; Corvi, M.M.; Weber, D.N.; Patel, A.U.; Das, S.R.; Przybyla, J.; Anderson, K.A.; Tanguay, R.L. Neurodevelopmental low-dose bisphenol A exposure leads to early life-stage hyperactivity and learning deficits in adult zebrafish. *Toxicology* **2012**, *291*, 83–92. [CrossRef]
105. Lindholst, C.; Wynne, P.M.; Marriott, P.; Pedersen, S.N.; Bjerregaard, P. Metabolism of bisphenol A in zebrafish (*Danio rerio*) and rainbow trout (*Oncorhynchus mykiss*) in relation to estrogenic response. *Comp. Biochem. Physiol. Part C Toxicol. Pharmacol.* **2003**, *135*, 169–177. [CrossRef]
106. Chen, L.; Huang, C.; Hu, C.; Yu, K.; Yang, L.; Zhou, B. Acute exposure to DE-71: Effects on locomotor behavior and developmental neurotoxicity in zebrafish larvae. *Environ. Toxicol. Chem.* **2012**, *31*, 2338–2344. [CrossRef]
107. Holmes, L.A.; Turner, A.; Thompson, R.C. Adsorption of trace metals to plastic resin pellets in the marine environment. *Environ. Pollut.* **2012**, *160*, 42–48. [CrossRef] [PubMed]
108. Rochman, C.M.; Hoh, E.; Kurobe, T.; Teh, S.J. Ingested plastic transfers hazardous chemicals to fish and induces hepatic stress. *Sci. Rep.* **2013**, *3*, 3263. [CrossRef]
109. Güven, O.; Gökdağ, K.; Jovanović, B.; Kıdeyş, A.E. Microplastic litter composition of the Turkish territorial waters of the Mediterranean Sea, and its occurrence in the gastrointestinal tract of fish. *Environ. Pollut.* **2017**, *223*, 286–294. [CrossRef]
110. Lanctôt, C.M.; Al-Sid-Cheikh, M.; Catarino, A.I.; Cresswell, T.; Danis, B.; Karapanagioti, H.K.; Mincer, T.; Oberhänsli, F.; Swarzenski, P.; Tolosa, I.; et al. Application of nuclear techniques to environmental plastics research. *J. Environ. Radioact.* **2018**, *192*, 368–375. [CrossRef]
111. Yu, P.; Liu, Z.; Wu, D.; Chen, M.; Lv, W.; Zhao, Y. Accumulation of polystyrene microplastics in juvenile Eriocheir sinensis and oxidative stress effects in the liver. *Aquat. Toxicol.* **2018**, *200*, 28–36. [CrossRef]
112. Lee, W.S.; Cho, H.-J.; Kim, E.; Huh, Y.H.; Kim, H.-J.; Kim, B.; Kang, T.; Lee, J.-S.; Jeong, J. Bioaccumulation of polystyrene nanoplastics and their effect on the toxicity of Au ions in zebrafish embryos. *Nanoscale* **2019**, *11*, 3173–3185. [CrossRef]
113. Cheng, J.; Flahaut, E.; Cheng, S.H. Effect of carbon nanotubes on developing zebrafish (Danio rerio) embryos. *Environ. Toxicol. Chem. Int. J.* **2007**, *26*, 708–716. [CrossRef] [PubMed]
114. Kristofco, L.A.; Haddad, S.P.; Chambliss, C.K.; Brooks, B.W. Differential uptake of and sensitivity to diphenhydramine in embryonic and larval zebrafish. *Environ. Toxicol. Chem.* **2018**, *37*, 1175–1181. [CrossRef] [PubMed]
115. Brun, N.R.; van Hage, P.; Hunting, E.R.; Haramis, A.-P.G.; Vink, S.C.; Vijver, M.G.; Schaaf, M.J.M.; Tudorache, C. Polystyrene nanoplastics disrupt glucose metabolism and cortisol levels with a possible link to behavioural changes in larval zebrafish. *Commun. Biol.* **2019**, *2*, 382. [CrossRef]
116. Liegertová, M.; Wrobel, D.; Herma, R.; Müllerová, M.; Šťastná, L.Č.; Cuřínová, P.; Strašák, T.; Malý, M.; Čermák, J.; Smejkal, J. Evaluation of toxicological and teratogenic effects of carbosilane glucose glycodendrimers in zebrafish embryos and model rodent cell lines. *Nanotoxicology* **2018**, *12*, 797–818. [CrossRef]
117. Vranic, S.; Shimada, Y.; Ichihara, S.; Kimata, M.; Wu, W.; Tanaka, T.; Boland, S.; Tran, L.; Ichihara, G. Toxicological evaluation of SiO_2 nanoparticles by zebrafish embryo toxicity test. *Int. J. Mol. Sci.* **2019**, *20*, 882. [CrossRef] [PubMed]
118. Pitt, J. Developmental and Cross-Generational Distribution and Toxicity of Polystyrene Nanoparticles in Zebrafish (*Danio rerio*). Bachelor's Thesis, Duke University, Durham, NC, USA, 2018.
119. Duan, Z.; Duan, X.; Zhao, S.; Wang, X.; Wang, J.; Liu, Y.; Peng, Y.; Gong, Z.; Wang, L. Barrier function of zebrafish embryonic chorions against microplastics and nanoplastics and its impact on embryo development. *J. Hazard. Mater.* **2020**, *395*, 122621. [CrossRef] [PubMed]
120. Trevisan, R.; Voy, C.; Chen, S.; Di Giulio, R.T. Nanoplastics decrease the toxicity of a complex PAH mixture but impair mitochondrial energy production in developing zebrafish. *Environ. Sci. Technol.* **2019**, *53*, 8405–8415. [CrossRef]
121. Feng, Z.; Zhang, T.; Li, Y.; He, X.; Wang, R.; Xu, J.; Gao, G. The accumulation of microplastics in fish from an important fish farm and mariculture area, Haizhou Bay, China. *Sci. Total Environ.* **2019**, *696*, 133948. [CrossRef]
122. Zhu, J.; Zhang, Q.; Li, Y.; Tan, S.; Kang, Z.; Yu, X.; Lan, W.; Cai, L.; Wang, J.; Shi, H. Microplastic pollution in the Maowei Sea, a typical mariculture bay of China. *Sci. Total Environ.* **2019**, *658*, 62–68. [CrossRef]
123. Koongolla, J.B.; Lin, L.; Pan, Y.-F.; Yang, C.-P.; Sun, D.-R.; Liu, S.; Xu, X.-R.; Maharana, D.; Huang, J.-S.; Li, H.-X. Occurrence of microplastics in gastrointestinal tracts and gills of fish from Beibu Gulf, South China Sea. *Environ. Pollut.* **2020**, *258*, 113734. [CrossRef] [PubMed]
124. Park, T.-J.; Lee, S.-H.; Lee, M.-S.; Lee, J.-K.; Lee, S.-H.; Zoh, K.-D. Occurrence of microplastics in the Han River and riverine fish in South Korea. *Sci. Total Environ.* **2020**, *708*, 134535. [CrossRef]
125. Lu, Y.; Zhang, Y.; Deng, Y.; Jiang, W.; Zhao, Y.; Geng, J.; Ding, L.; Ren, H. Uptake and accumulation of polystyrene microplastics in zebrafish (*Danio rerio*) and toxic effects in liver. *Environ. Sci. Technol.* **2016**, *50*, 4054–4060. [CrossRef]
126. Jabeen, K.; Li, B.; Chen, Q.; Su, L.; Wu, C.; Hollert, H.; Shi, H. Effects of virgin microplastics on goldfish (*Carassius auratus*). *Chemosphere* **2018**, *213*, 323–332. [CrossRef]
127. Wang, W.; Gao, H.; Jin, S.; Li, R.; Na, G. The ecotoxicological effects of microplastics on aquatic food web, from primary producer to human: A review. *Ecotoxicol. Environ. Saf.* **2019**, *173*, 110–117. [CrossRef]
128. Huang, J.-N.; Wen, B.; Meng, L.-J.; Li, X.-X.; Wang, M.-H.; Gao, J.-Z.; Chen, Z.-Z. Integrated response of growth, antioxidant defense and isotopic composition to microplastics in juvenile guppy (*Poecilia reticulata*). *J. Hazard. Mater.* **2020**, *399*, 123044. [CrossRef] [PubMed]

129. Zitouni, N.; Bousserrhine, N.; Missawi, O.; Boughattas, I.; Chèvre, N.; Santos, R.; Belbekhouche, S.; Alphonse, V.; Tisserand, F.; Balmassiere, L.; et al. Uptake, tissue distribution and toxicological effects of environmental microplastics in early juvenile fish *Dicentrarchus labrax*. *J. Hazard. Mater.* **2021**, *403*, 124055. [CrossRef]

130. Yin, L.; Chen, B.; Xia, B.; Shi, X.; Qu, K. Polystyrene microplastics alter the behavior, energy reserve and nutritional composition of marine jacopever (*Sebastes schlegelii*). *J. Hazard. Mater.* **2018**, *360*, 97–105. [CrossRef] [PubMed]

131. Zhang, Y.; Kang, S.; Allen, S.; Allen, D.; Gao, T.; Sillanpää, M. Atmospheric microplastics: A review on current status and perspectives. *Earth-Sci. Rev.* **2020**, *203*, 103118. [CrossRef]

132. Hu, M.; Palić, D. Micro- and nano-plastics activation of oxidative and inflammatory adverse outcome pathways. *Redox Biol.* **2020**, *37*, 101620. [CrossRef] [PubMed]

133. Romano, N.; Renukdas, N.; Fischer, H.; Shrivastava, J.; Baruah, K.; Egnew, N.; Sinha, A.K. Differential modulation of oxidative stress, antioxidant defense, histomorphology, ion-regulation and growth marker gene expression in goldfish (*Carassius auratus*) following exposure to different dose of virgin microplastics. *Comp. Biochem. Physiol. Part C Toxicol. Pharmacol.* **2020**, *238*, 108862. [CrossRef]

134. Yang, H.; Xiong, H.; Mi, K.; Xue, W.; Wei, W.; Zhang, Y. Toxicity comparison of nano-sized and micron-sized microplastics to Goldfish *Carassius auratus* Larvae. *J. Hazard. Mater.* **2020**, *388*, 122058. [CrossRef]

135. Huang, J.-S.; Koongolla, J.B.; Li, H.-X.; Lin, L.; Pan, Y.-F.; Liu, S.; He, W.-H.; Maharana, D.; Xu, X.-R. Microplastic accumulation in fish from Zhanjiang mangrove wetland, South China. *Sci. Total Environ.* **2020**, *708*, 134839. [CrossRef]

136. Bussolaro, D.; Wright, S.L.; Schnell, S.; Schirmer, K.; Bury, N.R.; Arlt, V.M. Co-exposure to polystyrene plastic beads and polycyclic aromatic hydrocarbon contaminants in fish gill (RTgill-W1) and intestinal (RTgutGC) epithelial cells derived from rainbow trout (*Oncorhynchus mykiss*). *Environ. Pollut.* **2019**, *248*, 706–714. [CrossRef]

137. Lei, L.; Wu, S.; Lu, S.; Liu, M.; Song, Y.; Fu, Z.; Shi, H.; Raley-Susman, K.M.; He, D. Microplastic particles cause intestinal damage and other adverse effects in zebrafish *Danio rerio* and nematode *Caenorhabditis elegans*. *Sci. Total Environ.* **2018**, *619-620*, 1–8. [CrossRef]

138. Limonta, G.; Mancia, A.; Benkhalqui, A.; Bertolucci, C.; Abelli, L.; Fossi, M.C.; Panti, C. Microplastics induce transcriptional changes, immune response and behavioral alterations in adult zebrafish. *Sci. Rep.* **2019**, *9*, 15775. [CrossRef]

139. Yin, L.; Liu, H.; Cui, H.; Chen, B.; Li, L.; Wu, F. Impacts of polystyrene microplastics on the behavior and metabolism in a marine demersal teleost, black rockfish (*Sebastes schlegelii*). *J. Hazard. Mater.* **2019**, *380*, 120861. [CrossRef]

140. Xia, X.; Sun, M.; Zhou, M.; Chang, Z.; Li, L. Polyvinyl chloride microplastics induce growth inhibition and oxidative stress in *Cyprinus carpio* var. larvae. *Sci. Total Environ.* **2020**, *716*, 136479. [CrossRef] [PubMed]

141. van Pomeren, M.; Brun, N.R.; Peijnenburg, W.; Vijver, M.G. Exploring uptake and biodistribution of polystyrene (nano)particles in zebrafish embryos at different developmental stages. *Aquat. Toxicol.* **2017**, *190*, 40–45. [CrossRef] [PubMed]

142. Batel, A.; Borchert, F.; Reinwald, H.; Erdinger, L.; Braunbeck, T. Microplastic accumulation patterns and transfer of benzo[a]pyrene to adult zebrafish (*Danio rerio*) gills and zebrafish embryos. *Environ. Pollut.* **2018**, *235*, 918–930. [CrossRef]

143. Zhang, S.; Ding, J.; Razanajatovo, R.M.; Jiang, H.; Zou, H.; Zhu, W. Interactive effects of polystyrene microplastics and roxithromycin on bioaccumulation and biochemical status in the freshwater fish red tilapia (*Oreochromis niloticus*). *Sci. Total Environ.* **2019**, *648*, 1431–1439. [CrossRef] [PubMed]

144. Xu, K.; Zhang, Y.; Huang, Y.; Wang, J. Toxicological effects of microplastics and phenanthrene to zebrafish (*Danio rerio*). *Sci. Total Environ.* **2021**, *757*, 143730. [CrossRef] [PubMed]

145. Karami, A.; Romano, N.; Galloway, T.; Hamzah, H. Virgin microplastics cause toxicity and modulate the impacts of phenanthrene on biomarker responses in African catfish (*Clarias gariepinus*). *Environ. Res.* **2016**, *151*, 58–70. [CrossRef]

146. Grigorakis, S.; Mason, S.A.; Drouillard, K.G. Determination of the gut retention of plastic microbeads and microfibers in goldfish (*Carassius auratus*). *Chemosphere* **2017**, *169*, 233–238. [CrossRef]

147. Schirinzi, G.F.; Köck-Schulmeyer, M.; Cabrera, M.; González-Fernández, D.; Hanke, G.; Farré, M.; Barceló, D. Riverine anthropogenic litter load to the Mediterranean Sea near the metropolitan area of Barcelona, Spain. *Sci. Total Environ.* **2020**, *714*, 136807. [CrossRef] [PubMed]

148. Uddin, S.; Fowler, S.W.; Saeed, T. Microplastic particles in the Persian/Arabian Gulf – A review on sampling and identification. *Mar. Pollut. Bull.* **2020**, *154*, 111100. [CrossRef]

149. Miao, L.; Chi, S.; Wu, M.; Liu, Z.; Li, Y. Deregulation of phytoene-β-carotene synthase results in derepression of astaxanthin synthesis at high glucose concentration in *Phaffia rhodozyma* astaxanthin-overproducing strain MK19. *BMC Microbiol.* **2019**, *19*, 133. [CrossRef]

150. Jovanović, B.; Gökdağ, K.; Güven, O.; Emre, Y.; Whitley, E.M.; Kideys, A.E. Virgin microplastics are not causing imminent harm to fish after dietary exposure. *Mar. Pollut. Bull.* **2018**, *130*, 123–131. [CrossRef] [PubMed]

151. Wright, S.L.; Thompson, R.C.; Galloway, T.S. The physical impacts of microplastics on marine organisms: A review. *Environ. Pollut.* **2013**, *178*, 483–492. [CrossRef] [PubMed]

152. Von Moos, N.; Burkhardt-Holm, P.; Köhler, A. Uptake and effects of microplastics on cells and tissue of the blue mussel *Mytilus edulis* L. after an experimental exposure. *Environ. Sci. Technol.* **2012**, *46*, 11327–11335. [CrossRef] [PubMed]

153. Jeong, J.; Choi, J. Adverse outcome pathways potentially related to hazard identification of microplastics based on toxicity mechanisms. *Chemosphere* **2019**, *231*, 249–255. [CrossRef] [PubMed]

154. Qiao, R.; Sheng, C.; Lu, Y.; Zhang, Y.; Ren, H.; Lemos, B. Microplastics induce intestinal inflammation, oxidative stress, and disorders of metabolome and microbiome in zebrafish. *Sci. Total Environ.* **2019**, *662*, 246–253. [CrossRef]

155. Pedà, C.; Caccamo, L.; Fossi, M.C.; Gai, F.; Andaloro, F.; Genovese, L.; Perdichizzi, A.; Romeo, T.; Maricchiolo, G. Intestinal alterations in European sea bass *Dicentrarchus labrax* (Linnaeus, 1758) exposed to microplastics: Preliminary results. *Environ. Pollut.* **2016**, *212*, 251–256. [CrossRef] [PubMed]

156. Solomando, A.; Capó, X.; Alomar, C.; Álvarez, E.; Compa, M.; Valencia, J.M.; Pinya, S.; Deudero, S.; Sureda, A. Long-term exposure to microplastics induces oxidative stress and a pro-inflammatory response in the gut of *Sparus aurata* Linnaeus, 1758. *Environ. Pollut.* **2020**, *266*, 115295. [CrossRef]

157. Ohkubo, N.; Ito, M.; Hano, T.; Kono, K.; Mochida, K. Estimation of the uptake and gut retention of microplastics in juvenile marine fish: Mummichogs (*Fundulus heteroclitus*) and red seabreams (*Pagrus major*). *Mar. Pollut. Bull.* **2020**, *160*, 111630. [CrossRef]

158. Suman, S.; Upendra, P.H.S.; Vishnu, Y.S.S. Design and Analysis of Dynamic Comparator Implemented Using Stacking Technique. *J. Phys. Conf. Ser.* **2021**, *1804*, 012182. [CrossRef]

159. Romano, R.; Cristescu, S.M.; Risby, T.H.; Marczin, N. Lipid peroxidation in cardiac surgery: Towards consensus on biomonitoring, diagnostic tools and therapeutic implementation. *J. Breath Res.* **2018**, *12*, 027109. [CrossRef] [PubMed]

160. Strungaru, S.-A.; Jijie, R.; Nicoara, M.; Plavan, G.; Faggio, C. Micro-(nano) plastics in freshwater ecosystems: Abundance, toxicological impact and quantification methodology. *TrAC Trends Anal. Chem.* **2019**, *110*, 116–128. [CrossRef]

161. Cong, Y.; Jin, F.; Tian, M.; Wang, J.; Shi, H.; Wang, Y.; Mu, J. Ingestion, egestion and post-exposure effects of polystyrene microspheres on marine medaka (*Oryzias melastigma*). *Chemosphere* **2019**, *228*, 93–100. [CrossRef] [PubMed]

162. Karuppasamy, P.K.; Ravi, A.; Vasudevan, L.; Elangovan, M.P.; Dyana Mary, P.; Vincent, S.G.T.; Palanisami, T. Baseline survey of micro and mesoplastics in the gastro-intestinal tract of commercial fish from Southeast coast of the Bay of Bengal. *Mar. Pollut. Bull.* **2020**, *153*, 110974. [CrossRef] [PubMed]

163. McMinn, D.; Treacy, R.; Lin, K.; Pynsent, P. Metal on metal surface replacement of the hip: Experience of the McMinn prosthesis. *Clin. Orthop. Relat. Res.* **1996**, *329*, S89–S98. [CrossRef] [PubMed]

164. Greven, A.-C.; Merk, T.; Karagöz, F.; Mohr, K.; Klapper, M.; Jovanović, B.; Palić, D. Polycarbonate and polystyrene nanoplastic particles act as stressors to the innate immune system of fathead minnow (*Pimephales promelas*). *Environ. Toxicol. Chem.* **2016**, *35*, 3093–3100. [CrossRef]

165. Sendra, M.; Pereiro, P.; Yeste, M.P.; Mercado, L.; Figueras, A.; Novoa, B. Size matters: Zebrafish (*Danio rerio*) as a model to study toxicity of nanoplastics from cells to the whole organism. *Environ. Pollut.* **2021**, *268*, 115769. [CrossRef] [PubMed]

166. Zhao, J.; Rao, B.-Q.; Guo, X.-M.; Gao, J.-Y. Effects of Microplastics on Embryo Hatching and Intestinal Accumulation in Larval Zebrafish *Danio rerio*. *Huan Jing Ke Xue* **2021**, *42*, 485–491. [CrossRef]

167. Jeong, C.-B.; Kang, H.-M.; Lee, M.-C.; Kim, D.-H.; Han, J.; Hwang, D.-S.; Souissi, S.; Lee, S.-J.; Shin, K.-H.; Park, H.G.; et al. Adverse effects of microplastics and oxidative stress-induced MAPK/Nrf2 pathway-mediated defense mechanisms in the marine copepod *Paracyclopina nana*. *Sci. Rep.* **2017**, *7*, 41323. [CrossRef] [PubMed]

168. Sheng, C.; Zhang, S.; Zhang, Y. The influence of different polymer types of microplastics on adsorption, accumulation, and toxicity of triclosan in zebrafish. *J. Hazard. Mater.* **2021**, *402*, 123733. [CrossRef]

169. Fackelmann, G.; Sommer, S. Microplastics and the gut microbiome: How chronically exposed species may suffer from gut dysbiosis. *Mar. Pollut. Bull.* **2019**, *143*, 193–203. [CrossRef]

170. Welden, N.A.C.; Cowie, P.R. Long-term microplastic retention causes reduced body condition in the langoustine, *Nephrops norvegicus*. *Environ. Pollut.* **2016**, *218*, 895–900. [CrossRef] [PubMed]

171. Goyette, P.; Labbé, C.; Trinh, T.T.; Xavier, R.J.; Rioux, J.D. Molecular pathogenesis of inflammatory bowel disease: Genotypes, phenotypes and personalized medicine. *Ann. Med.* **2007**, *39*, 177–199. [CrossRef]

172. Furukawa, S.; Fujita, T.; Shimabukuro, M.; Iwaki, M.; Yamada, Y.; Nakajima, Y.; Nakayama, O.; Makishima, M.; Matsuda, M.; Shimomura, I. Increased oxidative stress in obesity and its impact on metabolic syndrome. *J. Clin. Investig.* **2017**, *114*, 1752–1761. [CrossRef] [PubMed]

173. Jovanović, B.; Bezirci, G.; Çağan, A.S.; Coppens, J.; Levi, E.E.; Oluz, Z.; Tuncel, E.; Duran, H.; Beklioğlu, M. Food web effects of titanium dioxide nanoparticles in an outdoor freshwater mesocosm experiment. *Nanotoxicology* **2016**, *10*, 902–912. [CrossRef]

174. Collard, F.; Gilbert, B.; Compère, P.; Eppe, G.; Das, K.; Jauniaux, T.; Parmentier, E. Microplastics in livers of European anchovies (*Engraulis encrasicolus*, L.). *Environ. Pollut.* **2017**, *229*, 1000–1005. [CrossRef]

175. Akhbarizadeh, R.; Dobaradaran, S.; Nabipour, I.; Tajbakhsh, S.; Darabi, A.H.; Spitz, J. Abundance, composition, and potential intake of microplastics in canned fish. *Mar. Pollut. Bull.* **2020**, *160*, 111633. [CrossRef]

176. Avio, C.G.; Gorbi, S.; Regoli, F. Experimental development of a new protocol for extraction and characterization of microplastics in fish tissues: First observations in commercial species from Adriatic Sea. *Mar. Environ. Res.* **2015**, *111*, 18–26. [CrossRef]

177. Mattsson, K.; Jocic, S.; Doverbratt, I.; Hansson, L.-A. Chapter 13—Nanoplastics in the Aquatic Environment. In *Microplastic Contamination in Aquatic Environments*; Zeng, E.Y., Ed.; Elsevier: Amsterdam, The Netherlands, 2018; pp. 379–399. [CrossRef]

178. Karbalaei, S.; Golieskardi, A.; Hamzah, H.B.; Abdulwahid, S.; Hanachi, P.; Walker, T.R.; Karami, A. Abundance and characteristics of microplastics in commercial marine fish from Malaysia. *Mar. Pollut. Bull.* **2019**, *148*, 5–15. [CrossRef] [PubMed]

179. Choi, J.S.; Jung, Y.-J.; Hong, N.-H.; Hong, S.H.; Park, J.-W. Toxicological effects of irregularly shaped and spherical microplastics in a marine teleost, the sheepshead minnow (*Cyprinodon variegatus*). *Mar. Pollut. Bull.* **2018**, *129*, 231–240. [CrossRef]

180. Zhu, M.; Chernick, M.; Rittschof, D.; Hinton, D.E. Chronic dietary exposure to polystyrene microplastics in maturing Japanese medaka (*Oryzias latipes*). *Aquat. Toxicol.* **2020**, *220*, 105396. [CrossRef]

181. Hu, L.; Chernick, M.; Lewis, A.M.; Ferguson, P.L.; Hinton, D.E. Chronic microfiber exposure in adult Japanese medaka (*Oryzias latipes*). *PLoS ONE* **2020**, *15*, e0229962. [CrossRef] [PubMed]

182. Espinosa, C.; Cuesta, A.; Esteban, M.Á. Effects of dietary polyvinylchloride microparticles on general health, immune status and expression of several genes related to stress in gilthead seabream (*Sparus aurata* L.). *Fish Shellfish Immunol.* **2017**, *68*, 251–259. [CrossRef]

183. Zhu, N.; Zhang, D.; Wang, W.; Li, X.; Yang, B.; Song, J.; Zhao, X.; Huang, B.; Shi, W.; Lu, R. A novel coronavirus from patients with pneumonia in China, 2019. *N. Engl. J. Med.* **2020**, *382*, 727–733. [CrossRef] [PubMed]

184. Hamed, M.; Soliman, H.A.M.; Osman, A.G.M.; Sayed, A.E.-D.H. Assessment the effect of exposure to microplastics in Nile Tilapia (*Oreochromis niloticus*) early juvenile: I. blood biomarkers. *Chemosphere* **2019**, *228*, 345–350. [CrossRef] [PubMed]

185. Brandts, I.; Balasch, J.; Tvarijonaviciute, A.; Barreto, A.; Martins, M.; Tort, L.; Oliveira, M.; Teles, M. The Role of Humic Acids on the Effects of Nanoplastics in Fish. In Proceedings of the 2nd International Conference on Microplastic Pollution in the Mediterranean Sea, Capri, Italy, 15–18 September 2019; pp. 164–169.

186. Espinosa, C.; Esteban, M.Á.; Cuesta, A. Dietary administration of PVC and PE microplastics produces histological damage, oxidative stress and immunoregulation in European sea bass (*Dicentrarchus labrax* L.). *Fish Shellfish Immunol.* **2019**, *95*, 574–583. [CrossRef]

187. Haghi, B.N.; Banaee, M. Effects of micro-plastic particles on paraquat toxicity to common carp (*Cyprinus carpio*): Biochemical changes. *Int. J. Environ. Sci. Technol.* **2017**, *14*, 521–530. [CrossRef]

188. Banaee, M.; Nemadoost Haghi, B.; Tahery, S.; Shahafve, S.; Vaziriyan, M. Effects of sub-lethal toxicity of paraquat on blood biochemical parameters of common carp, Cyprinus carpio (Linnaeus, 1758). *Iran. J. Toxicol.* **2016**, *10*, 1–5. [CrossRef]

189. Fazio, F.; Filiciotto, F.; Marafioti, S.; Di Stefano, V.; Assenza, A.; Placenti, F.; Buscaino, G.; Piccione, G.; Mazzola, S. Automatic analysis to assess haematological parameters in farmed gilthead sea bream (*Sparus aurata* Linnaeus, 1758). *Mar. Freshw. Behav. Physiol.* **2012**, *45*, 63–73. [CrossRef]

190. Fazio, F.; Saoca, C.; Sanfilippo, M.; Capillo, G.; Spanò, N.; Piccione, G. Response of vanadium bioaccumulation in tissues of *Mugil cephalus* (Linnaeus 1758). *Sci. Total Environ.* **2019**, *689*, 774–780. [CrossRef] [PubMed]

191. McCarthy, J.F.; Shugart, L.R. *Biological Markers of Environmental Contamination*; CRC Press: Boca Raton, FL, USA, 2018.

192. Sayed, A.E.-D.H.; Soliman, H.A.M. Modulatory effects of green tea extract against the hepatotoxic effects of 4-nonylphenol in catfish (*Clarias gariepinus*). *Ecotoxicol. Environ. Saf.* **2018**, *149*, 159–165. [CrossRef]

193. Satake, H.; Matsubara, S.; Shiraishi, A.; Yamamoto, T.; Osugi, T.; Sakai, T.; Kawada, T.J.C. Peptide receptors and immune-related proteins expressed in the digestive system of a urochordate. *Cell Tissue Res.* **2019**, *377*, 293–308. [CrossRef] [PubMed]

194. Fazio, F.; Arfuso, F.; Levanti, M.; Saoca, C.; Piccione, G. High stocking density and water salinity levels influence haematological and serum protein profiles in mullet *Mugil cephalus*, Linnaeus, 1758. *Cah. Biol. Mar.* **2017**, *58*, 331–339. [CrossRef]

195. Fazio, F.; Ferrantelli, V.; Saoca, C.; Giangrosso, G.; Piccione, G. Stability of haematological parameters in stored blood samples of rainbow trout *Oncorhynchus mykiss* (Walbaum, 1792). *Veterinární Med.* **2017**, *62*, 401–405. [CrossRef]

196. Fazio, F. Fish hematology analysis as an important tool of aquaculture: A review. *Aquaculture* **2019**, *500*, 237–242. [CrossRef]

197. Fazio, F.; Lanteri, G.; Saoca, C.; Iaria, C.; Piccione, G.; Orefice, T.; Calabrese, E.; Vazzana, I. Individual variability of blood parameters in striped bass Morone saxatilis: Possible differences related to weight and length. *Aquac. Int.* **2020**, *28*, 1665–1673. [CrossRef]

198. Fazio, F.; Saoca, C.; Costa, G.; Zumbo, A.; Piccione, G.; Parrino, V. Flow cytometry and automatic blood cell analysis in striped bass *Morone saxatilis* (Walbaum, 1792): A new hematological approach. *Aquaculture* **2019**, *513*, 734398. [CrossRef]

199. Fazio, F.; D'Iglio, C.; Capillo, G.; Saoca, C.; Peycheva, K.; Piccione, G.; Makedonski, L. Environmental Investigations and Tissue Bioaccumulation of Heavy Metals in Grey Mullet from the Black Sea (Bulgaria) and the Ionian Sea (Italy). *Animals* **2020**, *10*, 1739. [CrossRef]

200. Galgani, F.; Hanke, G.; Werner, S.; De Vrees, L. Marine litter within the European marine strategy framework directive. *ICES J. Mar. Sci.* **2013**, *70*, 1055–1064. [CrossRef]

201. Li, J.; Liu, H.; Paul Chen, J. Microplastics in freshwater systems: A review on occurrence, environmental effects, and methods for microplastics detection. *Water Res.* **2018**, *137*, 362–374. [CrossRef]

202. Sharma, S.; Chatterjee, S. Microplastic pollution, a threat to marine ecosystem and human health: A short review. *Environ. Sci. Pollut. Res.* **2017**, *24*, 21530–21547. [CrossRef] [PubMed]

203. Murray, F.; Cowie, P.R. Plastic contamination in the decapod crustacean *Nephrops norvegicus* (Linnaeus, 1758). *Mar. Pollut. Bull.* **2011**, *62*, 1207–1217. [CrossRef] [PubMed]

204. Van Cauwenberghe, L.; Janssen, C.R. Microplastics in bivalves cultured for human consumption. *Environ. Pollut.* **2014**, *193*, 65–70. [CrossRef] [PubMed]

205. Duis, K.; Coors, A. Microplastics in the aquatic and terrestrial environment: Sources (with a specific focus on personal care products), fate and effects. *Environ. Sci. Eur.* **2016**, *28*, 1–25. [CrossRef]

206. Teuten, E.L.; Saquing, J.M.; Knappe, D.R.; Barlaz, M.A.; Jonsson, S.; Björn, A.; Rowland, S.J.; Thompson, R.C.; Galloway, T.S.; Yamashita, R. Transport and release of chemicals from plastics to the environment and to wildlife. *Philos. Trans. R. Soc. B Biol. Sci.* **2009**, *364*, 2027–2045. [CrossRef]

207. Hahladakis, J.N.; Velis, C.A.; Weber, R.; Iacovidou, E.; Purnell, P. An overview of chemical additives present in plastics: Migration, release, fate and environmental impact during their use, disposal and recycling. *J. Hazard. Mater.* **2018**, *344*, 179–199. [CrossRef]
208. Osman, A.G.; AbouelFadl, K.Y.; Abd El Baset, M.; Mahmoud, U.M.; Kloas, W.; Moustafa, M.A. Blood biomarkers in Nile tilapia *Oreochromis niloticus* niloticus and African catfish *Clarias gariepinus* to evaluate water quality of the river Nile. *J. Fish.* **2018**, *12*, 1–15. [CrossRef]
209. Mukherjee, J.; Sinha, G. Cadmium toxicity on haematological and biochemical aspects in an Indian freshwater major carp, *Labeo rohita*(Hamilton). *J. Freshw. Biol.* **1993**, *5*, 245–251.
210. Tort, L. Stress and immune modulation in fish. *Dev. Comp. Immunol.* **2011**, *35*, 1366–1375. [CrossRef]
211. Dos Santos Norberto, R.S. Toxic effects of nickel alone and in combination with microplastics on early juveniles of the common goby (*Pomatoschistus microps*). Master's Thesis, School of Medicine and Biomedical Sciecnes, Buffalo, NY, USA, 2014.
212. Cedervall, T.; Hansson, L.-A.; Lard, M.; Frohm, B.; Linse, S. Food chain transport of nanoparticles affects behaviour and fat metabolism in fish. *PLoS ONE* **2012**, *7*, e32254. [CrossRef]
213. Ramos-Barron, M.; Bejarano, I.H.; Rodrigo, E.; Iglesias, E.C.; Hernandez, A.B.; Blanco, C.A.; Martin, A.M.; Vicente, M.P.; Hernandez, F.L.; Arias, M. Assessment of kidney graft function evolution measured by creatinine and cystatin c. *Transpl. Proc.* **2016**, *48*, 2913–2916. [CrossRef] [PubMed]
214. Manabe, I. Chronic inflammation links cardiovascular, metabolic and renal diseases. *Circ. J.* **2011**, *75*, 2739–2748. [CrossRef] [PubMed]
215. De Sá, L.C.; Luís, L.G.; Guilhermino, L. Effects of microplastics on juveniles of the common goby (*Pomatoschistus microps*): Confusion with prey, reduction of the predatory performance and efficiency, and possible influence of developmental conditions. *Environ. Pollut.* **2015**, *196*, 359–362. [CrossRef] [PubMed]
216. Rochman, C.M.; Browne, M.A.; Halpern, B.S.; Hentschel, B.T.; Hoh, E.; Karapanagioti, H.K.; Rios-Mendoza, L.M.; Takada, H.; Teh, S.; Thompson, R.C. Classify plastic waste as hazardous. *Nature* **2013**, *494*, 169–171. [CrossRef] [PubMed]
217. Ravit, B.; Cooper, K.; Moreno, G.; Buckley, B.; Yang, I.; Deshpande, A.; Meola, S.; Jones, D.; Hsieh, A. Microplastics in urban New Jersey freshwaters: Distribution, chemical identification, and biological affects. *AIMS Environ. Sci.* **2017**, *4*, 809–826. [CrossRef]
218. Pedersen, A.F.; Meyer, D.N.; Petriv, A.-M.V.; Soto, A.L.; Shields, J.N.; Akemann, C.; Baker, B.B.; Tsou, W.-L.; Zhang, Y.; Baker, T.R. Nanoplastics impact the zebrafish (*Danio rerio*) transcriptome: Associated developmental and neurobehavioral consequences. *Environ. Pollut.* **2020**, *266*, 115090. [CrossRef]
219. Pang, M.; Wang, Y.; Tang, Y.; Dai, J.; Tong, J.; Jin, G. Transcriptome sequencing and metabolite analysis reveal the toxic effects of nanoplastics on tilapia after exposure to polystyrene. *Environ. Pollut.* **2021**, *277*, 116860. [CrossRef]

Article

First Report and 3D Reconstruction of a Presumptive Microscopic Liver Lipoma in a Black Barbel (*Barbus balcanicus*) from the River Bregalnica in the Republic of North Macedonia

Katerina Rebok [1,*], Maja Jordanova [1], Júlia Azevedo [2,3] and Eduardo Rocha [2,3]

[1] Laboratory of Cytology, Histology and Embryology, Faculty of Natural Sciences and Mathematics-Skopje, Ss. Cyril and Methodius University in Skopje, 1000 Skopje, North Macedonia; majaj@pmf.ukim.mk

[2] Laboratory of Histology and Embryology, Department of Microscopy, Institute of Biomedical Sciences Abel Salazar, University of Porto, 4050-313 Porto, Portugal; j.azevedo122@gmail.com (J.A.); erocha@icbas.up.pt (E.R.)

[3] Team of Histomorphology, Physiopathology, and Applied Toxicology, Interdisciplinary Centre of Marine and Environmental Research, University of Porto, 4450-208 Matosinhos, Portugal

* Correspondence: katerinarebok@yahoo.com

Abstract: A lipoma is a benign tumour of mature adipocytes which may appear in various species, including marine and freshwater fish. It usually occurs in isolated locations, such as a superficial or deep mass, mainly in the skin and seldom in other organs. In non-mammalian vertebrates, there is no agreed minimal size for the mass to be considered a lipoma. This study histologically describes a case proposed to be a microlipoma in the liver of *Barbus balcanicus*. The structure was an oval-shaped mass of well-differentiated adipocytes, surrounded by hepatic parenchyma. The adipocyte cluster did not contact with major vascular or biliary tracts, the liver capsule, or the hilum. The cell mass reached a maximal linear length and width of ~0.5 mm and ~0.4 mm. A three-dimensional and software-assisted reconstruction of the adipocytic mass showed that it had the shape of a flattened prolate spheroid (~0.01 mm^3). Given the histological criteria currently used in the literature, we consider the mass as a lipoma, or, better, a microlipoma because it was tiny. We interpret this structure as an early growing lipoma. This work is the second description of a liver lipoma in a fish to the best of our knowledge.

Keywords: microlipoma; liver; *Barbus balcanicus*; 3D reconstruction

Citation: Rebok, K.; Jordanova, M.; Azevedo, J.; Rocha, E. First Report and 3D Reconstruction of a Presumptive Microscopic Liver Lipoma in a Black Barbel (*Barbus balcanicus*) from the River Bregalnica in the Republic of North Macedonia. *Appl. Sci.* **2021**, *11*, 8392. https://doi.org/10.3390/app11188392

Academic Editors: Panagiotis Berillis and Božidar Rašković

Received: 4 July 2021
Accepted: 7 September 2021
Published: 10 September 2021

Publisher's Note: MDPI stays neutral with regard to jurisdictional claims in published maps and institutional affiliations.

1. Introduction

Tumours of a neoplastic nature occur across vertebrates, and fish are no exception. The neoplasms in cold-blooded species were early recognised as fundamentally similar in structure and biologic behaviour to the matching tumours in warm-blooded animals [1]. The histological diagnostic criteria defined for benign/malignant neoplasms in mammals stand as guidelines. A lipoma is defined as a benign and slow-growing mesenchymal tumour consisting of well-differentiated mature white adipocytes, typically with minimal connective tissue stroma within [1–3]. In humans, this neoplasm is the most common soft tissue tumour, often subcutaneous in location, with an (under)estimated prevalence of 1% [2,3]. This neoplasm is rare and has only been diagnosed in a few fish species, typically appearing as isolated cases. The tumour was either encapsulated [4–8] or nonencapsulated and with the fat cells even merging with the surrounding tissue [9,10].

As it has been described for humans [11], and therefore suggesting phylogenetic conservation, most of the diagnosed lipomas in marine and freshwater fish were localised as dermal or hypodermal masses with visible skin protrusions [4–6,8–10,12–17], sometimes invading the underlying musculature [18,19]. Diagnoses have been grounded on the lesion's histological features. Contrary to major advances in diagnosing human lipomas, there is not one established histochemical, molecular or cytogenetic biomarker for any fish lipoma. In addition, there is no published data on the prevalence of lipomas in fish populations.

Despite the fact that lipomas rarely appear in other locations in fish, they were detected in the body cavity of yellow-eye mullet (*Aldrichetta forsteri*) [20], the mesentery of the Senegal seabream (*Diplodus bellottii*) [21], the musculature around the vertebral column of the southern bluefin tuna (*Thunnus maccoyuii*) [20] and northern bluefin tuna (*Thunnus thynnus*) [7], the stomach of the Indian oil sardine (*Sardinella longiceps*) [22] and the spleen of the largemouth bass (*Micropterus salmoides*) [23] and perch (*Perca fluviatilis*) [24]. Regarding liver lipoma, as far as we know, it has been described once, and with very limited imagery, in the marine coly (also known as the coalfish, saithe and Atlantic pollock) (*Pollachius virens*) almost a century ago [25].

When diagnosing a liver lipoma in fish, it is relevant to have in mind that there are species that naturally have non-hepatic cells inside the organ. Many fish have exocrine pancreatic acini within the liver (around blood vessels spreading into the organ from the hilum) [26]. Still, adipocytes are not seen intrahepatically, from the embryo (adipogenesis) to the adult, even though visceral adipose tissue is often found extrahepatically, at the hilum [26–29]. Therefore, accumulations of adipocytes inside the liver should be regarded a priori as abnormal features.

This study describes and presumptively diagnoses a microscopic liver lipoma in the black barbel (*Barbus balcanicus*) while creating the first 3D reconstruction of such nodule from serial histological sections. This finding seems to be the first report in this fish species and is discussed in the light of the literature, drawing attention to a rare and overlooked lesion.

2. Materials and Methods

The case black barbel was collected by electro-fishing in the upper part of the River Bregalnica, near the city of Berovo (22°51′27.9″ E, 41°41′59.8″ N), which does not receive effluents from urban, industrial or agricultural origins. This low anthropogenic-impacted area has been used as a reference location in ecotoxicological monitoring studies [30,31]. The animal with the unusual microanatomical feature was collected in summer. It was 1 out of 658 black barbels subjected to liver histopathology—with an intensive systematic sampling strategy—after being randomly caught over one year in the River Bregalnica.

The fish in consideration was measured and weighed, examined for external visual abnormalities and, after the necropsy, examined for internal abnormalities. Immediately after the gross visual inspection, the liver was removed and weighed. The liver was then sliced into a couple of slabs that were fixed in Bouin's fixative for 48 h, dehydrated in increasing concentrations of ethanol, cleared in xylene and embedded in paraffin wax blocks. All the blocks obtained from this particular fish were exhaustively cut in serial 5 µm-thick sections and routinely stained with haematoxylin and eosin (H&E). One section with the unusual feature was stained with periodic acid–Schiff (PAS) for carbohydrates.

All the serial sections containing the single intrahepatic agglomerate of adipocytes were chosen for the software-supported 3D reconstruction. In every section, the area of interest was photographed with a light microscope (BX50, Olympus, Tokyo, Japan) equipped with a digital camera (Camedia C5050, Olympus, Tokyo, Japan). The photographs for illustration were in TIFF format, and for reconstruction, they were high-resolution images (JPEG, 2560 × 1920 pixels). The latter photos were sequentially imported to a dedicated software (v3.0, BioVis3D, Montevideo, Uruguay), for building the baseline image stack that would support the 3D reconstruction. The software was calibrated for the magnification of the images. The final 3D reconstruction was subjected to digital rendering before being exported as TIFF files.

3. Results

One agglomerate of well-differentiated unilocular adipose cells was noticed when studying the serial histological sections from one black barbel's liver. Taking into account that 658 captured fish were analysed—in the context of a biomonitoring study—the estimated prevalence of the intrahepatic oddity in the Bregalnica black barbel population was 1.5‰.

The case fish was a two-year-old female with a body weight (BW) of 27.95 g, total length (TL) of 135 mm and liver weight (LW) of 0.68 g. The condition factor of the fish ($K = 100 \times BW/TL^3$) was 1.36, and the hepatosomatic index ($HSI = 100 \times LW/BW$) was 2.43.

During the necropsy, neither the external nor the internal examinations revealed gross changes or deformities that could signal pathological conditions (Figure 1). In particular, the liver had a normal appearance, with the typical reddish-brown colour and no gross lesions. However, the microscopic examination of the liver revealed the presence of one "island" made of distinctively empty-looking cells (Figure 2A). The cell mass was encircled by normal liver parenchyma. When sectioned, in the microscopic slides, the cell agglomerate was oval-shaped. It did not protrude the surface of the liver because it was buried deep in the organ, and it was tiny, reaching a maximum diameter of ~0.5 mm.

Figure 1. View of the necropsy procedure on the black barbel from the River Bregalnica. Gross lesions were not found on the body surface or in internal organs and cavities.

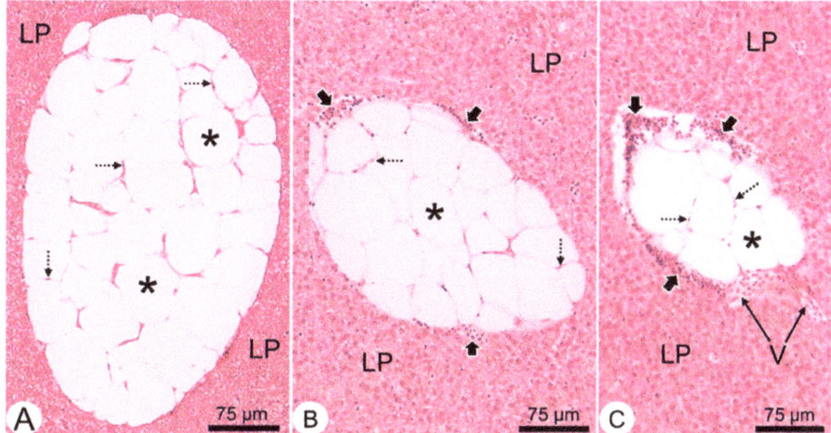

Figure 2. Representative histological sections (**A–C**) across the intrahepatic adipose cell agglomerate. There is a well-marked frontier between the adipocytes (*) and the liver parenchyma (LP). The flattened nuclei of the adipocytes are perceivable at the cell peripheries (dashed arrows). At specific locations around the adipose cell agglomerate, lymphocytic-like infiltrates are noted (block arrows) in the close vicinity of venules (V).

The nodular lesion was perfectly circumscribed and consisted of well-differentiated and large adipocytes. Their individual size varied little from one another. They contained an unambiguous single roundish empty space that filled most of the unilocular fat cell's cytoplasm. Indeed, it should be stressed that during the preparation of the tissue for histology, which includes dehydration in ethanol and clearing in xylene, free lipids are expected to dissolve. Therefore, the cytoplasmic location where the major lipid droplet stayed, in vivo, appeared empty in histological sections. The cytoplasm was negative for PAS, excluding the idea that carbohydrates could have caused the empty space. The nucleus was squeezed towards the cell's periphery. The intercellular limits were prominent. At times, an interstitial connective tissue matrix could be seen. Lipoblasts and mitotic figures were not observed. The internal vasculature was made of a few erratic thin-walled capillaries.

In specific smaller profiles of the adipose cell agglomerate, and at its margins, there were scattered groups of small roundish cells, with a morphology compatible with a leucocytic (mainly lymphocytic) infiltrate (Figure 2B,C). When the agglomerate profile was evidently smaller, we noted a relatively more extended infiltrate in the vicinity of a venule (Figure 2C). We saw no microparasites of any sort at the margin or inside the agglomerate.

The agglomerate of adipocytes was peripherally associated with scant small venules (Figure 2C), but not any major blood vessels or intrahepatic pancreatic tissue, and it was situated far from the liver hilum. Thus, the agglomerate was not a pocket made of an extension of the extrahepatic (peritoneal) fat. It was perfectly well demarked from the normal hepatic parenchyma but not encapsulated by a distinct connective tissue band. The surrounding hepatocytic elements caused no prominent peripheral compression of the agglomerate.

When looking at the 3D reconstructed agglomerate of adipose cells, we concluded that it was not lobulated, and that it approximately formed a flattened oblate spheroid, which looked like a sphere that was markedly squashed from top to bottom (Figure 3). The shorter "a" and longer "b" equatorial (horizontal) axes attained 375 μm and 520 μm, respectively, while the polar (vertical) "c" axis reached 65 μm. Applying the geometric formula for the volume (V) of a perfect oblate spheroid ($V = (4/3) \times \pi \times a/2 \times b/2 \times c/2$), the fat cell cluster had a V of approximately 6.7×10^6 μm^3 (~0.01 mm^3).

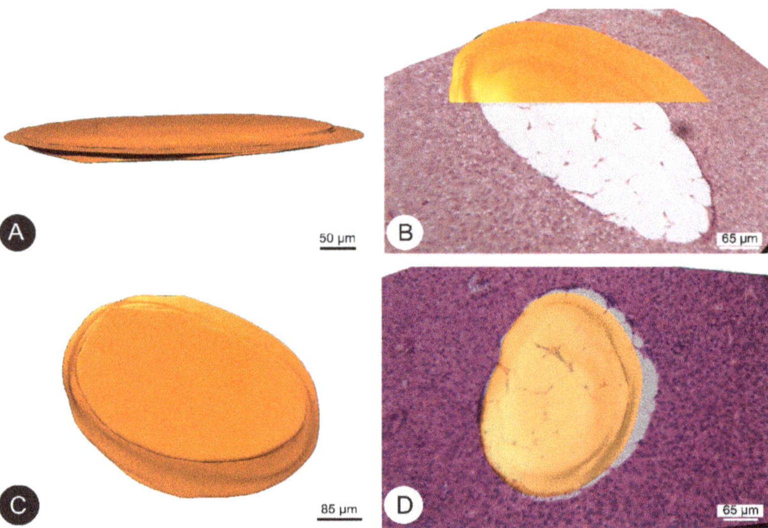

Figure 3. Three-dimensional (3D) reconstructions (**A,C**) of the intrahepatic adipose cell agglomerate, side by side with two "bi-dimensional" aspects (**B,D**). The last images have digital overlays (in orange) partially superimposed on the histological picture to explain how each histological image contributes to building the serial stack. The microlipoma flattened oval shape became evident when viewing the 3D reconstruction at different angles.

4. Discussion

This study histologically described and 3D reconstructed an intrahepatic collection of white adipocytes in a black barbel. Besides the structural evidence for such identification, the negativity for PAS staining indirectly supported that the major cytoplasm content in vivo should have been lipids; using paraffin, direct lipid staining was impossible. The oval-shaped and well-demarcated focus of unilocular fat cells was deeply nested into the liver. The histological characteristics of the adipocytic focus, including the absence of lipoblasts, increased mitosis, scarcity of vasculature and association with lymphocytes, nicely matched detailed lipoma descriptions [32]. Objectively, the focus fulfilled the basic structural definition of a lipoma [1] and therefore was presumptively diagnosed as such.

The good circumscription of the cell agglomerate favours a priori this diagnosis over another possibility: ectopic adipocytes. The presence of ectopic fat and ectopic adipocytes in unexpected locations is usually reported in human and experimental animal pathology in strict association with caloric intake (high-fat diets) exceeding expenditure, obesity, insulin resistance and type 2 diabetes mellitus [33,34]. Still, despite the fact the liver may accumulate ectopic fat (within steatotic hepatocytes), it is not among the organs where ectopic adipocytes typically appear under triggering conditions [35,36]. In the case revealed here, it is virtually impossible that just one wild barbel would have been consistently subjected to a high-fat diet, which, even so, would have sparked steatosis (not observed here) well in advance of ectopic adipocyte differentiation and proliferation. Therefore, the structural and circumstantial evidence favours the lipoma diagnosis over a pocket of ectopic adipocytes.

Anyway, it could be questioned if such a small focus of fat cells can be diagnosed as a lipoma in barbels. However, the literature does not specify a minimal size for an abnormal agglomerate of fat cells to be deemed as a lipoma. In humans, the size of (clinically detected) lipomas varies between 2 and 10 cm in diameter [36]. Yet, such benign neoplasms are very slow progressing. Therefore, the macroscopic tumours derive from earlier and much smaller (micro)lipomas, as illustrated well in intramucosal lipomas [37].

According to the predominant histological morphology, the World Health Organization classification of tumours of soft tissue considers the following lipoma types: conventional lipoma (the most common by far), fibrolipoma, angiolipoma, lipomatosis of nerve and lipoblastoma [38]. Since the liver lipoma in the studied black barbel consisted predominantly of mature adipocytes, without an excess of fibrous tissue or of substantial vascular involvement, it should be classified as a conventional lipoma.

Lipomas are rarely reported in fish, and, as far as we know, it seems that a lipoma has only been reported once in the liver [25]. Unlike the present findings in the black barbel, where the lipoma was noted only after histological examination, the liver lipoma found in the marine coly was visually apparent and noticed in the lower lobe of the liver, having a smooth surface and greenish colour [25]. Additionally, this liver lipoma was found in a seven-year-old specimen. Lipomas are generally noticed in older individuals. Their incidence increases with age in domestic animals [39] and in humans [40]. Contrarily, in the current case, the lipoma was noticed in a two-year-old female, an adult but not a particularly old fish. It is common to find specimens of *B. balcanicus* aged over two years [41,42], and, in general, barbels can live for more than 20 years [43].

The cause(s) of lipoma development in animals is (are) far from clear, but several theories were advanced, mostly from studies in humans. Some authors proposed genetic causes while favouring trauma as the aetiology [44], along with a diversity of risk factors, such as obesity, ethanol abuse, glucose intolerance and liver disease [40]. In farmed fish, dysmetabolic syndrome was suggested as the starting point for lipoma formation [19]. A link with lipid metabolism was reinforced when lipomas (also mentioned as lipoma-like fat tissue accumulations) were induced in a zebrafish model of obesity overexpressing the Akt1 "insulin signalling hub" gene [45]. In channel catfish (*Ictalurus punctatus*), subcutaneous lipoma was considered to be related to direct exposure to the carcinogenic and mutagenic *N*-Methyl-*N'*-Nitro-*N*-Nitrosoguanidine [14]. In the largemouth bass (*Micropterus salmoides*), a spleen lipoma was found in one specimen from a zone polluted by a range of contamination sources, but no toxicological aetiology was considered [23].

This microlipoma in a black barbel from a clean location was the only benign lesion found in all studied fish [31]. Facing the condition's rarity, a definitive causal explanation would be too speculative because the causes could reasonably range from purely genetic to toxicopathic. Although the latter aetiology is unlikely, it cannot be discarded in view of the sampling location, i.e., when knowing that liver lipomatous lesions resembling human hepatic lipomas evolved in rats from carcinogenic assays [46].

As with other lipomas, those in the human liver have no clear aetiology and are very rare [47]. A correlative study supported that, at least in humans, lipomas have a significantly greater association with steatosis when compared to non-lipomatous lesions [48]. In this black barbel single case, it is reiterated that no such association of lesions existed.

The use of 3D technology, including histopathologic reconstruction, to study lesions is a rapidly evolving practice in pathology, recognised as "useful in diagnostic, prognostic, and therapeutic decision-making for the medical and biomedical professions" [49]. Here, a 3D reconstruction of the microlipoma was created for the first time from serial histological sections, exposing a prolate spheroid that can only result from non-homogenous growth. If the fat cells expanded homogeneously in 3D, the focus would be spherical. Therefore, the 3D rebuild not only allowed insights on such kinetics but was also essential in confirming the lack of lobulation and in realistically knowing the mass's shape and size (volume).

In the end, what is the significance of a liver lipoma for a single fish or a fish population? Given its evident rarity, an impact on populations can be excluded. Regarding the individual, it can be deduced that the health impact will depend on the lipoma volume. In the black barbel, given the small size of the lipoma, it is inferred that it was in an early stage of development and growth. The mass minimally disturbed the nearby parenchyma, and therefore it unlikely altered the liver function. However, it may be theorised that severe morphofunctional impacts could appear if the lipoma expands over the years, finally leading to individual health effects. Notably, one grown liver lipoma in otherwise healthy human beings sparked abdominal pain, nausea and dyspepsia [50,51].

Irrespective of its cause and significance, the microlipoma described here seems to be the second report of a hepatic lipoma in any fish species and the first in a barbel species. The proposed diagnosis and detailed description of the finding in the black barbel can grab the attention of researchers conducting liver histopathology in fish and consequently raise awareness for this very likely underdiagnosed and underreported situation.

Author Contributions: Conceptualisation, K.R., M.J. and E.R.; data curation, K.R.; formal analysis, K.R., M.J., J.A. and E.R.; investigation, K.R., M.J., J.A. and E.R.; methodology, K.R., M.J., J.A. and E.R.; software, J.A.; supervision, M.J. and E.R.; writing—original draft, K.R. and J.A. All authors have read and agreed to the published version of the manuscript.

Funding: This study was partially supported by the Strategic Funding UID/Multi/04423/2019 provided by FCT and ERDF to CIIMAR/CIMAR, in the framework of the programme PT2020.

Institutional Review Board Statement: Not applicable. Ethical review and approval were waived for this study because it does not incur any actions falling within the "procedure" definition established by the Republic of North Macedonian and Portuguese law, in line with EU Directive No. 2010/63 (article 3, number 1). The latter excludes from the "procedure" the killing of animals solely for the use of their organs or tissues. This study used only organs and tissues of fish euthanised solely for the described purpose. In line with the EU Directive No. 2010/63, article 6, number 2, the killing was conducted by competent persons, trained and accredited by law to do so. The killing was at the site of capture.

Informed Consent Statement: Not applicable.

Data Availability Statement: The data are available on request from the corresponding author.

Conflicts of Interest: The authors declare no conflict of interest.

References

1. Schlumberger, H.H.; Lucké, B. Tumors of fishes, amphibians and reptiles. *Cancer Res.* **1948**, *8*, 657–754. [PubMed]
2. Johnson, C.N.; Ha, A.S.; Chen, E.; Davidson, D. Lipomatous soft-tissue tumors. *J. Am. Acad. Orthop. Surg.* **2018**, *26*, 779–788. [CrossRef] [PubMed]
3. Charifa, A.; Azmat, C.E.; Badri, T. Lipoma Pathology. In *StatPearls*; StatPearls Publishing: Treasure Island, FL, USA, 2020.
4. Stolk, A. Tumours of fishes 29. Lipoma in the eel *Anguilla anguilla*. *Beaufortia* **1959**, *7*, 193–198.
5. Tubiash, H.D.; Hendricks, J.D. A possible lipoma in the weakfish, *Cynoscion regalis* (Bloch and Schneider). *Chesap. Sci.* **1973**, *14*, 145–146. [CrossRef]
6. Bruno, D.W.; McVicar, A.H.; Fraser, C.O. Multiple lipoma in the common dab, *Limanda limanda* L. *J. Appl. Ichthyol.* **1991**, *7*, 238–243. [CrossRef]
7. Marino, F.; Monaco, S.; Salvaggio, A.; Macrì, B. Lipoma in a farmed northern bluefin tuna, *Thunnus thynnus* (L.). *J. Fish Dis.* **2006**, *29*, 697–699. [CrossRef] [PubMed]
8. Gómez, S. Multiple dermal lipomas in farmed striped seabream *Lithognathus mormyrus* on the Spanish Mediterranean coast. *Dis. Aquat. Organ.* **2009**, *85*, 77–79. [CrossRef]
9. McCoy, C.P.; Bowser, P.R.; Steeby, J.; Bleau, M.; Schwedler, T.E. Lipoma in channel catfish (*Ictalurus punctatus* Rafinesque). *J. Wild. Dis.* **1985**, *21*, 74–76. [CrossRef] [PubMed]
10. Easa, M.L.S.; Easa, M.; Harshbarger, J.C.; Hetrick, F.M. Hypodermal lipoma in a striped (grey) mullet *Mugil cephalus*. *Dis. Aquat. Organ.* **1989**, *6*, 157–160. [CrossRef]
11. Szewc, M.; Gawlik, P.; Żebrowski, R.; Sitarz, R. Giant lipoma in the fronto-temporo-parietal region in an adult man: Case report and literature review. *Clin. Cosmet. Investig. Dermatol.* **2020**, *13*, 1015–1020. [CrossRef]
12. Schwartz, F.J.; Màrquez, R. A lipoma in the striped mojarm, *Diapterus plumieri* (Pisces: Gerridae) from Mexico. *J. Elisha Mitchell Sci. Soc.* **1971**, *87*, 87–90.
13. Hard, G.C.; Williams, R.; Lee, J. Survey of Demersal Fish in Port Phillip Bay for Incidence of Neoplasia. *Aust. J. Mar. Freshw. Res.* **1979**, *30*, 73–79. [CrossRef]
14. Chen, H.C.; Pan, I.J.; Tu, W.J.; Lin, W.H.; Hong, C.C.; Brittelli, M.R. Neoplastic Response in Japanese Medaka and Channel Catfish Exposed to *N*-Methyl-*N'*-Nitro-*N*-Nitrosoguanidine. *Toxicol. Pathol.* **1996**, *24*, 696–706. [CrossRef] [PubMed]
15. Abowei, J.F.N.; Briyai, O.F.; Bassey, S.E. A Review of Some Viral, Neoplastic, Environmental and Nutritional Diseases of African Fish. *Br. J. Pharmacol. Toxicol.* **2011**, *2*, 227–235.
16. De Stefano, C.; Bonfiglio, R.; Montalbano, G.; Giorgianni, P.; Lanteri, G. Multicentric lipoma in a molly (*Poecilia velifera*). *Bull. Eur. Assoc. Fish Pathol.* **2012**, *32*, 220–224.
17. Sood, N.; Swaminathan, T.R.; Yadav, M.K.; Pradhan, P.K.; Kumar, R.; Sood, N.K. First report of cutaneous infiltrative lipoma in goldfish *Carassius auratus*. *Dis. Aquat. Org.* **2017**, *125*, 243–247. [CrossRef]
18. Johnston, C.J.; Deveney, M.R.; Bayly, T.; Nowak, B.F. Gross and histopathological characteristics of two lipomas and a neurofi-brosarcoma detected in aquacultured southern bluefin tuna, *Thunnus maccoyii* (Castelnau), in South Australia. *J. Fish Dis.* **2008**, *31*, 241–247. [CrossRef]
19. Marino, F.; Chiofalo, B.; Mazzullo, G.; Panebianco, A. Multicentric infiltrative lipoma in a farmed Mediterranean seabass *Dicentrarchus labrax*: A pathological and biochemical case study. *Dis. Aquat. Organ.* **2011**, *96*, 259–264. [CrossRef] [PubMed]
20. Lester, R.J.G.; Kelly, W.R. Tumour-like growths from southern Australian marine fish. *Tasmanian Fish Res.* **1983**, *25*, 27–32.
21. Ramos, P.; Faisca, P.; Carneiro, M.; Rosa, F. Dermal Melanoma and Mesenteric Lipoma in a Senegal Seabream, *Diplodus bellottii* (Steindachner, 1882) from the Portuguese Coast. *Exp. Pathol. Health Sci.* **2016**, *8*, 67–68.
22. Singaravel, V.; Gopalakrishnan, A.; Vijayakumar, R.; Raja, K. Prevalence and pathology of gastric tumours in Indian oil sardine (*Sardinella longiceps*) from Parangipettai coastal waters, southeast coast of India. *J. Coast. Life Med.* **2015**, *3*, 592–595. [CrossRef]
23. Hinck, J.E.; Blazer, V.S.; Denslow, N.D.; Echols, K.R.; Gale, R.W.; Wieser, C.; May, T.W.; Ellersieck, M.; Coyle, J.J.; Tillitt, D.E. Chemical contaminants, health indicators, and reproductive biomarker responses in fish from rivers in the Southeastern United States. *Sci. Total Environ.* **2008**, *390*, 538–557. [CrossRef] [PubMed]
24. Stejskal, V.; Kouřil, J.; Policar, T.; Svobodová, Z. Splenic lipidosis in intensively cultured perch, *Perca fluviatilis* L. *J. Fish Dis.* **2015**, *39*, 87–93. [CrossRef]
25. Thomas, L. Sur un lipome abdominal chez un colin. *Bull. Ass. Fr. Etude Cancer* **1933**, *22*, 419–435.
26. Figueiredo-Fernandes, A.M.; Fontaínhas-Fernandes, A.A.; Monteiro, R.A.F.; Reis-Henriques, M.A.; Rocha, E. Spatial relationships of the intrahepatic vascular-biliary tracts and associated pancreatic acini of Nile tilapia, *Oreochromis niloticus* (Teleostei, Cichlidae): A serial section study by light microscopy. *Ann. Anat.* **2007**, *189*, 17–30. [CrossRef] [PubMed]
27. Imrie, D.; Sadler, K.C. White adipose tissue development in zebrafish is regulated by both developmental time and fish size. *Dev. Dyn.* **2010**, *239*, 3013–3023. [CrossRef]
28. Salmerón, C. Adipogenesis in fish. *J. Exp. Biol.* **2018**, *221*, jeb161588. [CrossRef] [PubMed]
29. Sousa, S.; Rocha, M.J.; Rocha, E. Characterization and spatial relationships of the hepatic vascular–biliary tracts, and their associated pancreocytes and macrophages, in the model fish guppy (*Poecilia reticulata*): A study of serial sections by light microscopy. *Tissue Cell* **2018**, *50*, 104–113. [CrossRef]

30. Jordanova, M.; Rebok, K.; Naskovska, M.; Kostov, V.; Rocha, E. Splenic pigmented macrophage aggregates in barbel (*Barbus peloponnesius*, Valenciennes, 1844) from River Bregalnica—Influences of age, sex and season on a pollution biomarker. *Turkish J. Fish Aquat. Sci.* **2016**, *16*, 881–890. [CrossRef]

31. Rebok, K.; Jordanova, M.; Slavevska-Stamenković, V.; Ivanova, L.; Kostov, V.; Stafilov, T.; Rocha, E. Frequencies of erythrocyte nuclear abnormalities and of leucocytes in the fish *Barbus peloponnesius* correlate with a pollution gradient in the River Bregalnica (Macedonia). *Environ. Sci. Pollut. Res.* **2017**, *24*, 10493–10509. [CrossRef]

32. McTighe, S.; Chernev, I. Intramuscular lipoma: A review of the literature. *Orthop. Rev. Pavia* **2014**, *6*, 5618. [CrossRef] [PubMed]

33. Church, C.; Horowitz, M.; Rodeheffer, M. WAT is a functional adipocyte? *Adipocyte* **2012**, *1*, 38–45. [CrossRef]

34. Snel, M.; Jonker, J.T.; Schoones, J.; Lamb, H.; de Roos, A.; Pijl, H.; Smit, J.W.A.; Meinders, A.E.; Jazet, I.M. Ectopic fat and insulin resistance: Pathophysiology and effect of diet and lifestyle interventions. *Int. J. Endocrinol.* **2012**, *2012*, 983814. [CrossRef]

35. De Munck, T.J.I.; Soeters, P.B.; Koek, G.H. The role of ectopic adipose tissue: Benefit or deleterious overflow? *Eur. J. Clin. Nutr.* **2021**, *75*, 38–48. [CrossRef] [PubMed]

36. Kosztyuova, T.; Shim, T.N. Rapidly enlarging lipoma. *BMJ Case Rep.* **2017**, *2017*, bcr2017221272. [CrossRef] [PubMed]

37. Caliskan, A.; Kohlmann, W.; Affolter, K.; Downs-Kelly, E.; Kanth, P.; Bronner, M.P. Intramucosal lipomas of the colon implicate Cowden syndrome. *Mod. Pathol.* **2018**, *31*, 643–651. [CrossRef] [PubMed]

38. WHO Classification of Tumours Editorial Board. *WHO Classification of Tumours of Soft Tissue and Bone*, 4th ed.; International Agency for Research on Cancer: Lyon, France, 2013; p. 427.

39. Moulton, J.E. (Ed.) *Tumors in Domestic Animals*; University of California Press: Berkeley, CA, USA, 1990.

40. Kolb, L.; Yarrarapu, S.N.S.; Ameer, M.A.; Rosario-Collazo, J.A. Lipoma. In *StatPearls*; StatPearls Publishing: Treasure Island, FL, USA, 2020. Available online: https://www.ncbi.nlm.nih.gov/books/NBK507906/ (accessed on 6 October 2020).

41. Bertoli, M.; Pizzul, E.; Devescovi, V.; Franz, F.; Pastorino, P.; Giulianini, P.G.; Ferrari, C.; Nonnis Marzano, F. Biology and distribution of Danube barbel (*Barbus balcanicus*) (Osteichthyes: Cyprinidae) at the Northwestern limit of its range. *Eur. Zool. J.* **2019**, *86*, 280–293. [CrossRef]

42. Žutinić, P.; Jelić, D.; Jelić, M.; Buj, I. A contribution to understanding the ecology of the large spot barbel-sexual dimorphism, growth and population structure of *Barbus balcanicus* (Actinopterygii; Cyprinidae) in Central Croatia. *N.-West. J. Zool.* **2013**, *10*, 158–166.

43. Amat Trigo, F.; Roberts, C.G.; Britton, J.R. Spatial variability in the growth of invasive European barbel *Barbus barbus* in the River Severn basin, revealed using anglers as citizen scientists. *Knowl. Manag. Aquat. Ecosyst.* **2017**, *418*, 17. [CrossRef]

44. Aust, M.C.; Spies, M.; Kall, S.; Jokuszies, A.; Gohritz, A.; Vogt, P. Posttraumatic lipoma: Fact or fiction? *Skinmed* **2007**, *6*, 266–270. [CrossRef]

45. Chu, C.-Y.; Chen, C.-F.; Rajendran, R.S.; Shen, C.-N.; Chen, T.-H.; Yen, C.-C.; Chuang, C.-K.; Lin, D.-S.; Hsiao, C.-D. Overexpression of Akt1 enhances adipogenesis and leads to lipoma formation in zebrafish. *PLoS ONE* **2012**, *7*, e36474. [CrossRef] [PubMed]

46. Dixon, D.; Yoshitomi, K.; Boorman, G.A.; Maronpot, R.R. "Lipomatous" lesions of unknown cellular origin in the liver of B6C3F1 mice. *Vet. Pathol.* **1994**, *31*, 173–182. [CrossRef] [PubMed]

47. Puljiz, Ž.; Petričević, M.; Bratanić, A.; Barišić, I.; Puljiz, M.; Karin, Ž. An unusually large liver lipoma. Case report. *Med. Glas. Zenica* **2012**, *9*, 411–414.

48. Martin-Benitez, G.; Marti-Bonmati, L.; Barber, C.; Vila, R. Hepatic lipomas and steatosis: An association beyond chance. *Eur. J. Radiol.* **2011**, *81*, 491–494. [CrossRef]

49. Farahani, N.; Braun, A.; Jutt, D.; Huffman, T.; Reder, N.; Liu, Z.; Yagi, Y.; Pantanowitz, L. Three-dimensional imaging and scanning: Current and future applications for pathology. *J. Pathol. Inform.* **2017**, *8*, 36. [CrossRef]

50. Manenti, G.; Picchi, E.; Castrignanò, A.; Muto, M.; Nezzo, M.; Floris, R. Liver lipoma: A case report. *BJR Case Rep.* **2016**, *3*, 20150467. [CrossRef] [PubMed]

51. Reddy, O.J.; Gafoor, J.A.; Suresh, B.; Prasad, P.O. Lipoma in liver: A rare presentation. *J. Dr. NTR Univ. Health Sci.* **2015**, *4*, 185–187. [CrossRef]

Article

Effects of Dietary Fishmeal Replacement by Poultry By-Product Meal and Hydrolyzed Feather Meal on Liver and Intestinal Histomorphology and on Intestinal Microbiota of Gilthead Seabream (*Sparus aurata*)

Pier Psofakis [1], Alexandra Meziti [1], Panagiotis Berillis [1,*], Eleni Mente [1,2], Konstantinos A. Kormas [1] and Ioannis T. Karapanagiotidis [1,*]

[1] Department of Ichthyology and Aquatic Environment, School of Agricultural Sciences, University of Thessaly, Fytoko Street, 38446 Volos, Greece; psofakis@uth.gr (P.P.); ameziti@gmail.com (A.M.); emente@uth.gr (E.M.); kkormas@uth.gr (K.A.K.)

[2] Laboratory of Ichthyology—Culture and Pathology of Aquatic Animals, School of Veterinary Medicine, Aristotle University of Thessaloniki, University Campus, 54006 Thessaloniki, Greece

* Correspondence: pveril@uth.gr (P.B.); ikarapan@uth.gr (I.T.K.)

Featured Application: Poultry by-product meal and hydrolyzed feather meal can successfully replace fishmeal at low dietary levels in feeds for farmed gilthead seabream and thus enhance the environmental and economic sustainability of its production.

Abstract: The effects on liver and intestinal histomorphology and on intestinal microbiota in gilthead seabream (*Sparus aurata*) fed diets that contained poultry by-product meal (PBM) and hydrolyzed feather meal (HFM) as fishmeal replacements were studied. Fish fed on a series of isonitrogenous and isoenergetic diets, where fishmeal protein of the control diet (FM diet) was replaced by either PBM or by HFM at 25%, 50% and 100% without amino acid supplementation (PBM25, PBM50, PBM100, HFM25, HFM50 and HFM100 diets) or supplemented with lysine and methionine (PBM25+, PBM50+, HFM25+ and HFM50+ diets). The use of PBM and HFM at 25% fishmeal replacement generated a similar hepatic histomorphology to FM-fed fish, indicating that both land animal proteins are highly digestible at low FM replacement levels. However, 50% and 100% FM replacement levels by either PBM or HFM resulted in pronounced hepatic alterations in fish with the latter causing more severe degradation of the liver. Dietary amino acid supplementation delivered an improved tissue histology signifying their importance at high FM replacement levels. Intestinal microbiota was dominated by *Proteobacteria* (58.8%) and *Actinobacteria* (32.4%) in all dietary groups, but no specific pattern was observed among them at any taxonomic level. This finding was probably driven by the high inter-individual variability observed.

Keywords: nutrition; aquaculture; fishmeal replacement; land animal proteins; histology; intestinal microbiota; *Sparus aurata*

Citation: Psofakis, P.; Meziti, A.; Berillis, P.; Mente, E.; Kormas, K.A.; Karapanagiotidis, I.T. Effects of Dietary Fishmeal Replacement by Poultry By-Product Meal and Hydrolyzed Feather Meal on Liver and Intestinal Histomorphology and on Intestinal Microbiota of Gilthead Seabream (*Sparus aurata*). *Appl. Sci.* **2021**, *11*, 8806. https://doi.org/10.3390/app11198806

Academic Editor: Simona Picchietti

Received: 28 July 2021
Accepted: 16 September 2021
Published: 22 September 2021

Publisher's Note: MDPI stays neutral with regard to jurisdictional claims in published maps and institutional affiliations.

1. Introduction

Gilthead seabream (*Sparus aurata*) is one of the most important carnivorous farmed fish species in European aquaculture with an annual production of approximately 186,000 mt [1]. As aquaculture is becoming the major fish-food production sector [2] there is a search for suitable protein sources in aquafeeds that are alternatives to fishmeal to enhance its environmental and economic sustainability. Fishmeal was, and in many cases remains the primary protein source for the nutrition of farmed fish. However, it has become necessary to use low fishmeal diets because the global availability of fishmeal is stagnant, especially for those sourced from the wild, and its price has increased [3]. Land animal proteins, such as hydrolyzed feather meal (HFM) and poultry by-product meal (PBM) are currently incorporated in European aquafeeds. After their re-approval in 2013, proved to be valuable feedstuffs for dietary fishmeal replacement in the diet of most fish species [4–6], including gilthead seabream [7,8]. Although the poultry sector is responsible for a substantial proportion of greenhouse gases emissions [9], these feedstuffs provide a valuable mean of animal by-products utilization and upgrade the ecological efficiency of the whole poultry production process [10]. Thus, the use of land animal proteins could enhance aquaculture's sustainability and eco-efficiency, as these have a more favorable carbon footprint and a higher environmental efficiency when compared with fishmeal and plant alternatives [11,12].

Dietary protein manipulations, however, are known to affect the functionality of the digestive system [13,14]. A functional digestive system is a prerequisite for the optimal growth of fish with the liver being the main organ of nutrient deposition and metabolism and the intestine being the primary site of nutrient digestion and absorption. Therefore, studying any possible effects and alterations in the histomorphology of these tissues is fundamental for the evaluation of the use of land animal proteins as fishmeal substitutes. Most studies have focused on the effects of plant proteins as fish meal replacements [15] with high substitution levels resulting in marked changes in hepatic and intestinal tissues, such a reduced number of goblet cells, lipid accumulation in hepatocytes, shorter and thinner mucosal folds and villi, steatosis, submucosal layer hypertrophy and impaired structural integrity of the gut [13,15–17]. These alterations are mainly due to the presence of various anti-nutritional factors [18] which in turn cause pathophysiological changes in the gastrointestinal tract and reduce nutrient digestibility.

On the other hand, very little is known about the effects of land animal proteins as fishmeal replacements on the liver and intestinal histology. Findings from the limited studies that have been reported up to date have revealed that high inclusion levels of land animal proteins may induce hepatic steatosis and increase hepatic lipid vacuolization in *Lateolabrax japonicus* [19], in hybrid grouper [20,21] and in *Lates calcarifer* [22]. In addition, negative effects on the intestinal histology have also been reported with the fishmeal replacement by land animal proteins [22,23].

Gut microbes are essential for host nutrition and immunity [24] and changes in their community composition are related to stress and dysbiosis. Fish gut microbes are linked to the diet since different microbiota persist under different nutritional conditions along with the different enzymes produced (proteases, lipases, esterases, cellulases) that contribute to better food digestion by the host [25,26]. It has been shown that the use of alternative protein sources can alter the gut microbiome of the host having a beneficial impact on growth and immunity by triggering lactic acid bacteria (LAB) and cytokines respectively [27].

The present study addressed the effects of poultry by-product meal and hydrolyzed feather meal on liver and intestinal histomorphology and on the intestinal microbiota of gilthead seabream (*Sparus aurata*).

2. Materials and Methods

All experimental procedures were conducted according to the guidelines of the EU Directive 2010/63/EU regarding the protection of animals used for scientific purposes. The experiments were performed at the registered experimental facility (EL-43BIO/exp-01) of the Laboratory of Aquaculture, Department of Ichthyology and Aquatic Environment, University of Thessaly by FELASA accredited scientists (functions A–D).

2.1. Feeding Trials and Experimental Diets

Two feeding trials were conducted in which the growth data were not the object of the present study and are described in detail elsewhere [7,8]. Briefly, gilthead sea bream (*S. aurata*) juveniles were raised in glass tanks (125 L) with recirculating seawater of standard water quality (21.0 ± 1.0 °C, pH at 8.0 ± 0.4, salinity at 33 ± 0.5 g/L, dissolved oxygen at >6.5 mg/L, total ammonia nitrogen at <0.1 mg/L). In feeding trial I, juveniles with an initial mean weight of 2.5 ± 0.2 g were raised in quadruplicate groups (25 fish/tank, 4 tanks/dietary group). For 100 days they were fed to satiation with one of the five isonitrogenous (50%) and isoenergetic (21 KJ/Kg) experimental diets [7,8], where the FM protein of the control diet (FM diet) was replaced by either poultry by-product meal (PBM) at 50% (PBM50 diet) and 100% (PBM100 diet) or by hydrolyzed feather meal (HFM) at 50% (HFM50 diet) and 100% (HFM100 diet). In feeding trial II, juveniles with an initial mean weight of 2.9 ± 0.3 g were raised in triplicate groups (25 fish/tank, 3 tanks/dietary group). For 110 days they were fed to satiation with one of the seven isonitrogenous (50%) and isoenergetic (21 KJ/Kg) experimental diets [7,8]. These diets used the same FM control diet as before, but the FM protein was now replaced by PBM and HFM at lower levels: 25% without amino acid supplementation (PBM25 and HFM25 diets), 25% supplemented with lysine and methionine ((PBM25+ and HFM25+ diets), and 50% supplemented with lysine and methionine (PBM50+ and HFM50+ diets). At the end of each feeding trial, fish were weighed individually and euthanized with an overdose of tricaine methanesulfonate (MS 222, 300+ mg/L) according to the Directive 2010/63/EU and FELASA guidelines.

2.2. Histological Analysis and Measurements

For the feeding trial I, two fish per tank were randomly sampled (eight fish per dietary group). The liver of each fish was removed quickly and weighed for the determination of hepatosomatic index. Liver and midgut samples were collected from each fish, fixated into 10% formalin in filtered seawater for 24 h at 4 °C and then were immediately dehydrated in graded series of ethanol, immersed in xylol, and embedded in paraffin wax. Intestinal samples from the HFM100 group were not taken because the fish intestine was too thin. A part of the liver of fish was also collected for the determination of its fat content by exhaustive Soxhlet extraction using petroleum ether on a Soxtherm Multistat/SX PC (Sox-416 Macro, Gerhard, Germany). Liver and intestine sections of 4–7 μm were taken and stained with hematoxylin and eosin. All sections were examined under a microscope (Bresser Science TRM 301, Bresser GmbH, Rhede, Germany) and any histological abnormalities were recorded. A digital camera (Bresser MikroCam 5.0 MP, Bresser GmbH, Rhede, Germany) adjusted to the microscope was used for acquiring histological section images. For feeding trial II, 2 fish per tank were randomly sampled (six fish per dietary group). The same procedures as in the feeding trial I were followed for histological examination.

A semi-quantitative grading system was used in order to quantify the histopathological alterations of the examined tissues [28]. Severity grading used the following system: Grade 0 (not remarkable), Grade 1 (minimal), Grade 2 (mild), Grade 3 (moderate), Grade 4 (severe).

2.3. Microbiota Analysis-DNA Extraction, Bioinformatics and Data Analysis

In the present study, the effect of PBM and HFM on the intestinal microbiota of juvenile *S. aurata* was investigated at the 50% FM replacement level that negatively affected the fish growth performance see [7,8]. Thus, for the microbiota analysis fish from PBM50 and HFM50 groups of the feeding trial I were used. Two fish per tank (eight fish per dietary group) were randomly sampled and dissected using sterile lancets and forceps. The midgut was transferred in sterile particle-free (<0.2 mL) sea water (SPFSW). The gut's contents were extruded by mechanical force with forceps, as we targeted the resident gut microorganisms and not the ones associated with the ingested food. DNA extraction and 454 tag-pyrosequencing were performed as shown at Nikouli et al. [29].

Processing of the resulting sequences, i.e., trimming and quality control, was performed with the MOTHUR software (v 1.35.0 open access, University of Michigan, MI, USA) [30]. Only sequences with ≥250 bp and no ambiguous or no homopolymers ≥8 bp were considered for further analysis. These sequences were aligned and classified using the SILVA SSU database (release 119) [31]. All sequences were binned into Operational Taxonomic Units (OTUs) and were clustered (average neighbor algorithm) at 97% sequence similarity. Coverage values were calculated with MOTHUR (v 1.35.0). The batch of sequences from this study has been submitted in NCBI Short Read Archive under accession number SRS1839183. The heatmaps of the dominant OTUs and orders were implemented by the pheatmap function in the pheatmap package in R version 3.0.2). For the prediction of abundant metabolic pathways the Piphillin algorithm [http://secondgenome.com/Piphillin (accessed on 1 October 2020)] was used with support of KEGG database [32].

2.4. Statistical Analysis

For the microbiota analysis, canonical correspondence analysis (CCA) was performed using the R package vegan [33]. Similarly, the significance of morphological parameters and diversity indices for the ordination of the samples was calculated using the function envfit of the same package. Differentially abundant categories (taxa or subsystems) between samples were identified with DESeq package version 1.14.0 [34] using the binomial test and false discovery rate ($p < 0.05$). For liver fat data, percentages were subjected to one-way analysis of variance (ANOVA) followed by Tukey's post-hoc test to rank the groups using SPSS 18.0 (SPSS, Chicago, IL, USA).

3. Results

3.1. Liver Histology

In fish fed the control FM diets only minimal alterations (grade 1) were detected in their hepatic tissues (Table 1, Figures 1A and 2A). In general, the liver had normal structure with central hepatocytes' nuclei and a small amount of lipid droplets in their hepatocytes cytoplasm. In some of the hepatocytes the nuclei were not central but pressed against the periphery of the cells. In the cytoplasm of the exocrine pancreas' pyramidal cells many large eosinophilic zymogen granules were observed. Fish fed the diets with low inclusion levels of PBM (PBM25 and PBM25+ diets) showed a similar histomorphology to that of the control FM group (Figure 1B,C) and only two fish of the PBM25+ group showed large lipid droplets around pancreatic islets (Figure 1D). Fish fed the diets with a higher inclusion level of PBM, had mild (PBM50 fish, grade 2, Figure 1E) to moderate (PBM100 fish, grade 3, Figure 1F) alterations. The latter group had also increased signs of degeneration (Figure 1F). In some of the hepatocytes, the nuclei were not central but pressed against the periphery of the cells (Figure 1E). Within the hepatocytes, medium size lipid droplets were observed, but no steatosis or liver hemorrhage signs were detected to any fish. The liver histomorphology of fish fed the PBM50+ diet that was supplemented with lysine and methionine was slightly better than that of PBM50 fish.

Table 1. Severity score (0–4) for the observed histopathological alterations and liver fat (% of dry weight) in *S. aurata* fed the experimental diets.

	FM	PBM25	PBM25+	PBM50	PBM50+	PBM100	HFM25	HFM25+	HFM50	HFM50+	HFM100
	Severity score										
Liver	1	1	1	2	2	3	1	1	3	2	4
Intestine	0	0	0	0	0	0	0	0	0	0	0
Liver fat (%)	38.0	36.2	40.6	42.0	43.2	42.5	35.7	36.9	42.7	40.8	7.8

As far as the effect of dietary hydrolyzed feather meal is concerned, fish fed with high inclusion levels of HFM showed moderate (grade 3, HFM50 fish) to severe (grade 4, HFM100 fish) alterations in their liver tissue (Figure 2C,D). These fish showed enlarged lipid droplets, signs of pancreatic islets necrosis and hemorrhage, which were more intense in the HFM100 fish. Moreover, the latter fish showed signs of liver cirrhosis (Figure 2D) with the regenerative nodules of hepatocytes to be surrounded by fibrous connective tissue. The supplementation of lysine and methionine at the HFM50+ diet resulted in less hepatic alterations (grade 2) and a normal hepatic structure compared to the HFM50 fish, but still large lipid droplets and more hepatocytes with no central nuclei were detected in these fish (Figure 2F). However, the replacement of fishmeal by HFM at lower levels (HFM25,

HFM25+) resulted in a normal liver histomorphology that was similar to that of the control FM group (Figure 2B,E).

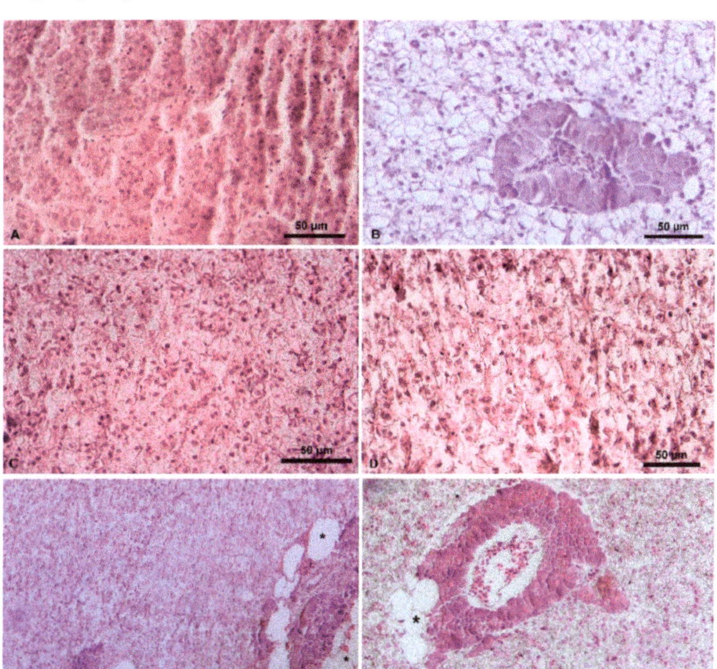

Figure 1. Liver histopathological examination of *S. aurata* fed on PBM diets. (**A**) fish fed FM diet—normal liver structure; (**B**) fish fed PBM25 diet—normal liver structure; (**C**) fish fed PBM25+ diet—normal liver structure; (**D**) fish fed PBM25+ diet—large lipid droplets (*) around pancreatic islets in some fish; (**E**) fish fed PBM50 diet—medium size lipid droplets with some nuclei pressed against the periphery of the cells; (**F**) fish fed PBM100 diet—liver degeneration.

3.2. Intestinal Histology

All the experimental groups of fish of both feeding trials revealed a normal intestinal histology and none of them showed any signs of inflammation (Figures 3 and 4). Enterocytes were distinct, while goblet cells and apical epithelial vacuoles were normally present (Figure 3A). In addition, abundant eosinophils cells were normally observed within the submucosa layer of all fish (Figure 3B).

3.3. Intestinal Microbiota

The gut bacterial diversity of fish was studied using 454-pyrosequencing. From the 24 samples analyzed in total only eight provided a satisfactory number of sequences (>100) combined with coverage >90% (Table 2). Taxonomic and potential species habitat origin was further studied, as well as similarities between the bacterial community composition (BCC) of the different dietary groups. A total of 125 Operational Taxonomic Units (Figure 5A) were identified from all 454 datasets, containing 5876 rRNA sequences in total (Table 2). Coverage was above 95% for all samples, while diversity was low (Shannon < 3) in all samples with the lowest values (<2) being observed in the FM-fed fish (Table 2).

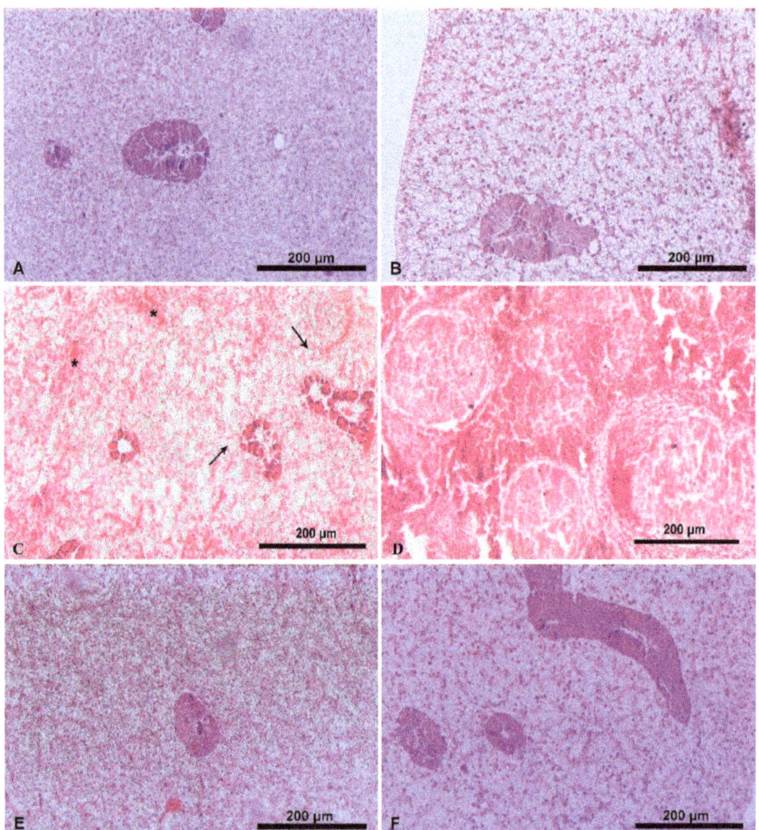

Figure 2. Liver histopathological examination of *S. aurata* fed on HFM diets. (**A**) fish fed FM diet—normal liver structure; (**B**) fish fed HFM25 diet—normal liver structure; (**C**) fish fed HFM100 diet—hemorrhage signs (*) and large lipid droplets (arrows); (**D**) fish fed HFM100 diet—signs of liver cirrhosis; (**E**) fish fed HFM25+ diet—normal liver structure; (**F**) fish fed HFM50+ diet—hepatocytes with no central nuclei were detected.

Table 2. Sequencing results, diversity indices and coverage values of fish fed the FM, PBM50 and HFM50 diets.

	FMa	FMb	HFMa	HFMb	HFMc	HFMd	PBMa	PBMb
Richness	14	14	48	29	14	28	15	16
Sequences	2283	132	2206	325	196	181	234	319
Shannon	1.32	1.71	2.27	2.97	2.05	2.87	2.10	2.23
Cumulative abundance >1%	99.21	96.21	92.84	97.54	97.45	97.24	98.72	98.12
Coverage	1.00	0.95	0.99	0.99	0.97	0.97	0.99	0.99

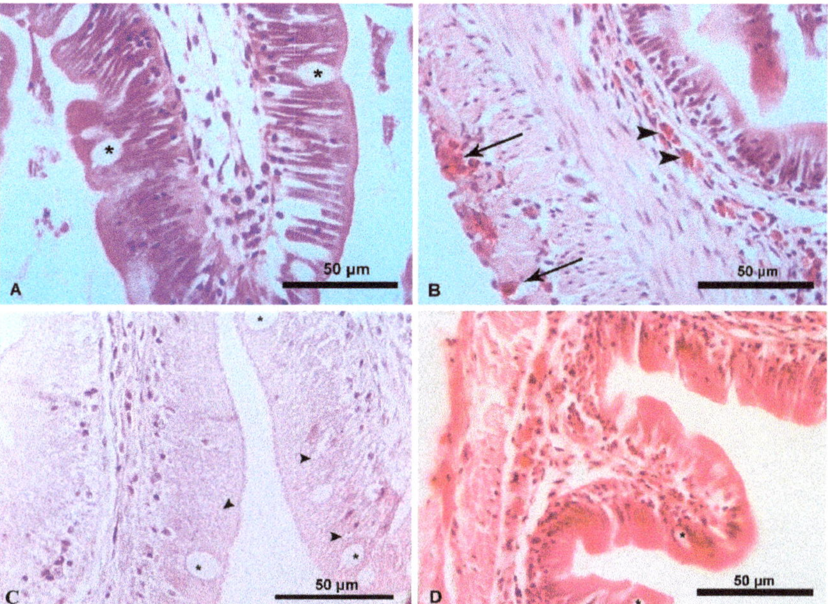

Figure 3. Midgut histopathological examination of *S. aurata* fed on PBM diets. (**A**) fish fed FM diet—normal gut structure with goblet cells (*) present; (**B**) fish fed PBM50 diet—eosinophil cells (arrow) accumulation within the muscularis layer. Abundant eosinophils cells were normally observed within the submucosa layer (arrowhead); (**C**) fish fed PBM25 diet—normal structure; (**D**) fish fed PBM100 diet—normal structure with goblet cells (*) present.

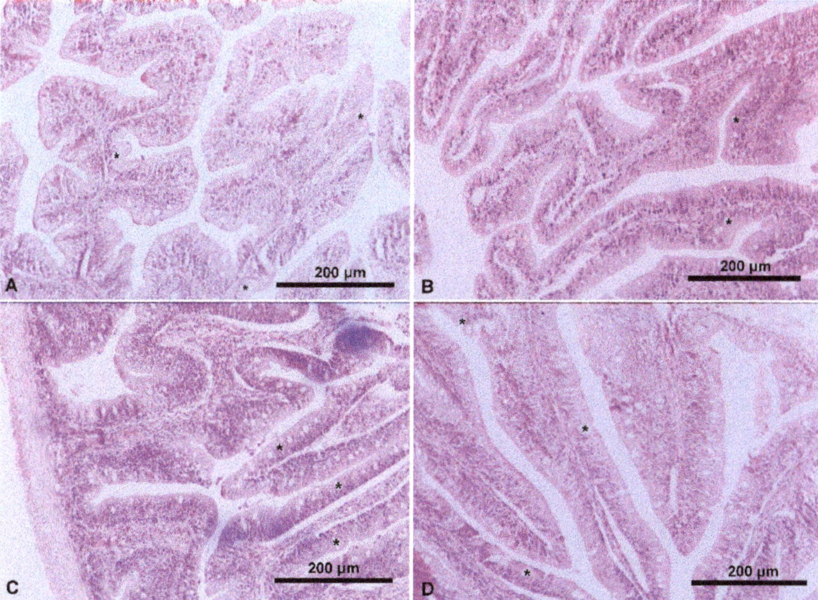

Figure 4. Midgut histopathological examination of *S. aurata* fed on HFM diets. (**A**) fish fed FM diet; (**B**) fish fed HFM50 diet; (**C**) fish fed HFM25 diet; (**D**) fish fed HFM50+ diet. In all images, the gut structure appeared normal with goblet cells (*) present.

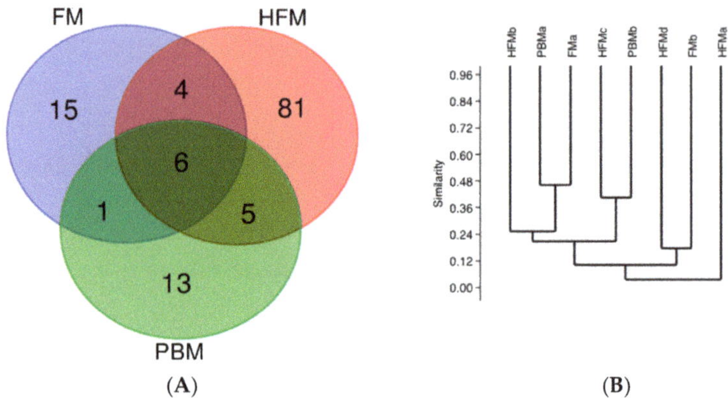

Figure 5. (**A**) Venn diagram showing shared and unique OTUs between the dietary groups (FM, PBM50 and HFM50); (**B**) Morisita similarities between fish fed the FM, PBM50 and HFM50 diets.

Morisita similarities between samples were very low ($<50\%$) showing no specific pattern according to the different diet fed (Figure 5B). Canonical correspondence analysis also exhibited no pattern among fish of either the same or of a different dietary group but revealed the importance of body weight for the ordination of the samples ($p < 0.05$) (Figure 6). At the phylum level, all samples were characterized by *Proteobacteria* and *Actinobacteria*, which were the most abundant in almost all fish, as well as by *Bacteroidetes* and *Firmicutes* (Figure 7A). At the OTU level, a total number of 64 OTUs were detected in relative abundances $>1\%$ in at least one fish. Overall, these OTUs accounted for more than 97% of the total diversity in all samples (Table 2). Similarly, OTUs with relative abundance $>10\%$ clearly representing persistent members of BCC reached cumulative abundances $>50\%$ in all samples (Figure 7B) and represented different species of *Alpha-*, *Beta-*, *Gamma-proteobacteria*, *Actinobacteria*, *Flavobacteria*, *Bacilli* and *Clostridia* (Figure 7B). In total, 6 OTUs were shared amongst all dietary groups (Figure 5) and belonged to the genera *Staphylococcus*, *Pseudomonas*, *Delftia*, *Cutibacterium* and *Hydrogenophaga* (Figure 7B). Most of them (5) belonged to the abundant species that dominated ($>10\%$) at least in one sample (Figure 7B). The cumulative abundances of this 'core' microbiome that was identified from habitat ranged from 6.1% (HFMa) to 99.2% (FMa) (Figure 7B). The lowest values for core microbiome relative abundances were observed in the HFM fish with an average of 20.15% contrary to 63.34% and 43.81% in the FM and PBM fish, respectively. This was attributed to the unique abundant species that were detected in the HFM fish, belonging to different *Actinobacteria*, such as *Propiomicromonospora* and to other species, such as *Roseomonas* and *Sphingomonas*.

Differences in predicted functional pathways based on the bacterial abundance did not exhibit any significant grouping of the samples based on the different diet fed. However, some pathways were found to be significantly different ($p < 0.05$) among the different dietary groups (Figure 8). Overall, pathways for renin-angiotensin system, retinol metabolism and cAMP signaling were decreased in fish fed the FM diet compared to those fed either the HFM or the PBM diet. This suggests gut dysbiosis in the two latter groups and possibly an effort to use alternative carbon sources. Steroid degradation pathways showed a statistically significant increase in the HFM fed fish, indicating that microbial communities were using alternative carbon sources.

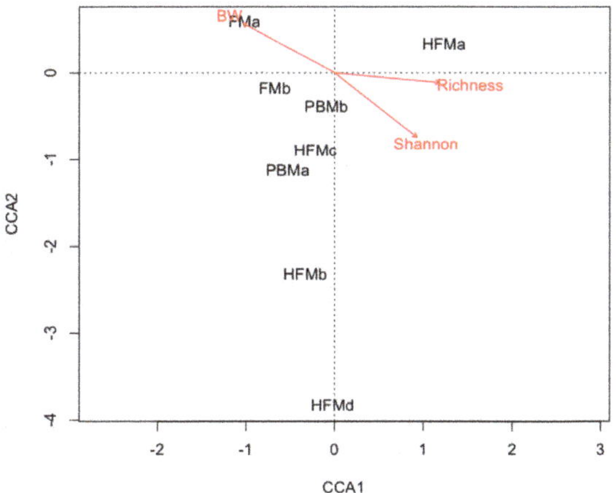

Figure 6. Canonical correspondence analysis for fish fed the FM, PBM50 and HFM50 diets. Only factors with significant values ($p < 0.05$) are presented.

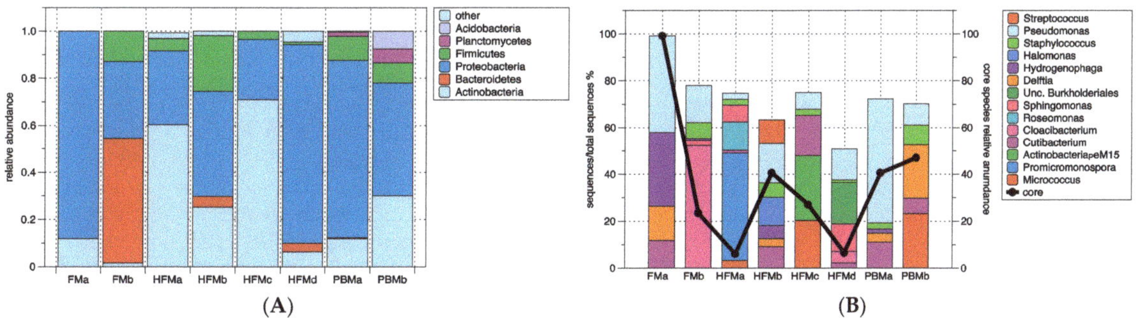

Figure 7. (**A**) Relative abundances of different bacterial phyla in fish fed the FM, PBM50 and HFM50 diets; (**B**) Relative abundances of abundant (>10%; left axis) and shared (between treatments) OTUs (right *y*-axis) in fish fed the FM, PBM50 and HFM50 diets.

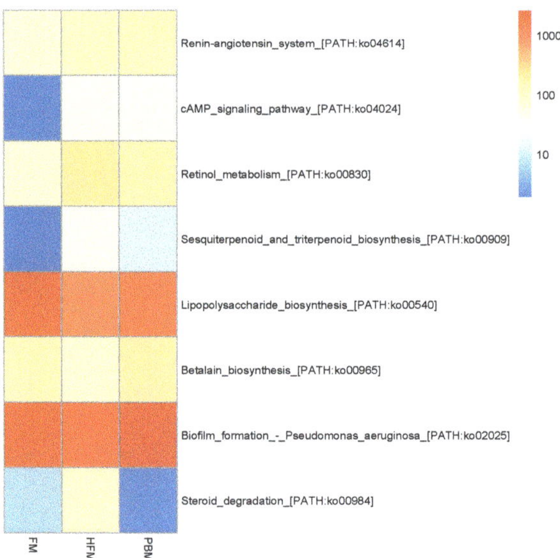

Figure 8. Significantly different ($p < 0.05$) predicted functional pathways of the bacterial relative abundances between fish fed the FM, PBM50 and HFM50 diets.

4. Discussion

Non-ruminant processed animal proteins, such as hydrolyzed feather meal (HFM) and poultry by-product meal (PBM) have been used successfully to replace fishmeal protein in the diets of several farmed fish and crustacean species [5–8,35–39]. However, knowledge of their effects on the histology of digestive organs, the intestinal microbiota and digestive physiology is extremely limited. Histology is a valuable tool that is used to describe tissue alterations and to detect any possible pathological signs in fish that may be caused by dietary protein modifications. In addition, intestinal microbiota profiling may assess fish intestinal function, health and nutrition [40,41].

4.1. Liver Histology

In the present study, the inclusion of poultry by-product meal or hydrolyzed feather meal caused no to severe alterations in the hepatic tissue of *S. aurata* and these alterations were dependent on the level of fishmeal protein replacement. Neither PBM nor HFM altered the liver histomorphology of seabream when these animal proteins replaced fishmeal at 25%. However, at higher replacement levels more lipid droplets and increased hepatic vacuolization were observed, and these changes were more pronounced in fish fed HFM diets. In general, high inclusion levels of PBM caused mild to moderate hepatic alterations compared to the high inclusion levels of HFM that caused severe alterations, particularly in the case of total fishmeal replacement. Although there were no signs of steatosis, which may be caused by the increased lipid vacuolization, the total FM replacement by HFM, contrary to PBM, led to haemorrhage, pancreatic islet necrosis and cirrhosis in a substantial number of fish that were examined. The dietary supplementation of PBM and HFM with essential amino acids, such as lysine and methionine, seemed to improve the digestive physiology. Fish fed these diets showed fewer hepatic alterations and abnormalities compared to fish fed diets of a similar replacement level but without amino acid supplementation.

The present findings contradict with those reported by Sabbagh et al. [42] in which the 100% replacement of FM by PBM did not cause any liver alteration in *S. aurata*. Although in both studies the total FM replacement by PBM did not lead to any clear signs of steatosis, the mild increase in lipid vacuolization with the increase of PBM dietary inclusion observed in the current study may indicate a lower lipid digestibility of the PBM fat. This indication is supported by the fact that fish fed diets with high inclusion levels of PBM had increased fat in their livers (Table 1), although this cannot be clearly said for the HFM fed fish. It has been suggested that that a hepatic lipid accumulation may occur because the excessive dietary intake of lipids that surpasses the physiological capability of the liver to β-oxidize them, thus leading to larger lipid droplets and subsequent steatosis [43]. However, in our study the dietary lipid intake of PBM and HFM fed fish was lower than that of FM fed fish see [7,8], which indicates that dietary lipids, specifically of PBM and HFM lipids, inducehepatic lipid accumulation in *S. aurata*.

Increased lipid droplets and hepatic vacuolization with signs of inflammation have also been reported in *Lates calcarifer* [22,44] and in *Tinca tinca* [45] when the FM in the fish fed diets were completely replaced with PBM. Moreover, Zhou et al. [21] reported an induced steatosis in the hepatocytes of hybrid grouper even when fish were fed diets with 50–70% FM replacement levels by PBM. On the other hand, the total replacement of FM by PBM did not cause any increased vacuolization or alterations in the hepatocytes of *Salmo salar* [14] and of *Oreochromis niloticus* [46]. The effects of dietary HFM as a sole FM replacement on fish liver histomorphology are not well studied. Hartviksen et al. [14] stated that a diet with total FM replacement by HFM did not reveal signs of steatosis in *Salmo salar* with hepatocytes having even a lower fat accumulation than to those fed on FM. Studies using HFM in a blend with other animal proteins, including PBM, for FM replacement have reported that high substitution levels induce hepatic lipidosis and steatosis in *Lateolabrax japonicus* [19] and in hybrid grouper [20].

4.2. Intestinal Histology

The inclusion of poultry by-product meal or hydrolyzed feather meal did not cause any intestinal histological alterations in seabream compared to fish fed the FM-based diet. All fish showed distinct enterocytes with abundant eosinophils cells, with goblet cells and apical epithelial vacuoles being present along the entire intestine of fish. Goblet cells assist fish health and nutrition as the mucus secreted by them acts as a protection medium to the epithelium, while also lubricates undigested materials for onward progression into the rectum [47,48]. Moreover, apical epithelial vacuoles consist integral structural components of the intestine that are responsible for nutrient absorption [49]. Although the present findings do not provide a sufficient evidence for the absorption of PBM and HFM, it can be claimed that these land animal proteins did not result in signs of malnutrition or inflammation, such as enteritis. This applies for all tested FM replacement levels, except for total replacement by HFM which was not feasible to examine. Similarly, an unaffected intestinal histology was reported in Atlantic salmon fed on high levels of PBM [50] and in Nile tilapia fed on high levels of HFM [51] replacing dietary FM. Hartviksen et al. [14] working with Atlantic salmon reported no severe signs of enteritis and unaffected numbers of eosinophilic granular cells in fish fed either with PBM or HFM replacing dietary fishmeal at a level close to 50%. However, the authors reported that PBM led to a decreased submucosa width, while HFM led to a decreased presence of goblet cells and an increased presence of apical epithelial vacuoles in the intestine. Moreover, Chaklader et al. [22] reported a dysregulated intestinal morphology with smaller microvilli of shorter diameter in juvenile barramundi fed on PBM totally replacing dietary fishmeal. Furthermore, Yu et al. [23] working with Pengze crucian carp reported shortened microvillus and enterocytes and thinner muscular thickness when HFM replaced more than 30% of dietary fishmeal protein.

4.3. Intestinal Microbiota

Regarding the dietary effects on the intestinal microbiota, it can be argued that no clear effects were detected from the use of HFM or PBM. The gut Bacterial Community Composition (BCC) of *S. aurata* fed either FM, PBM or HFM was characterized by groups that are commonly found in the fish gut microbiome. The core genera *Pseudomonas*, *Cutibacterium*, *Staphylococcus* and *Delftia* have been previously reported as core microbiome of *S. aurata* farmed in several geographical sites [29,52], suggesting that certain bacterial genera are capable of colonizing the seabream gut independently of the diet and location. Similar findings have been reported for *Salmo salar* [41] with authors stating that the role of fish-hosts in selecting or promoting core microbes is unclear. *Actinobacteria* species were dominant and unique in the HFM fed fish and their dominance could be related to their antimicrobial functions (i.e., antibiotics) that protect the host [53,54]. Additionally, the prevalence of terpenoid biosynthesis genes that were prevalent in the HFM fed fish is mostly detected in *Actinobacteria* [55] and was in accordance with the taxonomic BCC (Figure 7A,B). The *Sphingomonas* species in the gut have been found to be negatively correlated with weight gain [56] which also agrees with our CCA data (Figure 6).

The predicted pathways that were enriched in the HFM and PBM fed fish compared to FM fed fish were mostly related to functions that imply gut dysbiosis. For instance, increased renin-angiotensin system (RAS) genes have been related to RAS system activation which implies malnutrition [57,58]. Gut dysbiosis that promotes RAS activation is mostly related to decreased abundances of fermenting bacteria, which was the case in the HFM and PBM groups of fish. Similarly, bacterial retinol metabolism predicted genes indicate a potential need for vitamin A production that enhances mucosal immunity. Thus, bacteria able to participate in retinol metabolism also assist in avoiding pathogen invasion [59].

Knowledge of the effects of fishmeal replacement by land animal proteins on the intestinal microbiota is extremely limited. In the present study, as stated above, there was no specific pattern in the bacterial communities among fish fed either fishmeal or the tested land animal proteins. Gajardo et al. [41], working with *Salmo salar*, reported a significant effect of the dietary PBM on the distal intestine digesta and mucosa. This study found significantly higher and lower abundances of specific genera in fish fed PBM than in fish fed on a FM-based diet. Hartviksen et al. [14], also working with *Salmo salar*, reported that the use of HFM and PBM as fishmeal replacements increased the total allochthonous and total autochthonous bacteria in the distal intestine, but the total autochthonous bacteria in the proximal intestine remained unaffected. The authors also reported that the supplementation of HFM caused significant increases in specific genera (*Corynebacteriaceae, Lactobacillaceae, Streptococcaceae, Pseudomonadaceae, Xanthomonadaceae*) and decreases in another (*Vibrionaceae*). Furthermore, PBM caused increases in *Corynebacteriaceae* and decreases in *b-Proteobacteria, Bacilli-like, Peptostreptococcaceae and Vibrionaceae*. Certainly, a better understanding of the functional roles of the intestinal microbiota communities of fish and to what extent these are affected by the use of land animal proteins is needed.

5. Conclusions

In conclusion, fishmeal replacement by either poultry by-product meal or hydrolyzed feather meal did not cause any intestinal histological alterations. Thus, these results indicate normal digestion and absorption in the midgut of *S. aurata* even when their dietary fishmeal protein is completely replaced. Neither land animal proteins altered the liver histomorphology of gilthead seabream when fishmeal was replaced at 25%. However, at higher replacement levels increased lipid droplets and hepatic vacuolization were observed to be more pronounced in fish fed HFM diets. Moreover, the dietary supplementation of PBM and HFM with essential lysine and methionine seemed to improve the digestive physiology, as fish fed these diets showed fewer hepatic alterations and abnormalities compared to diets of a similar replacement level but without amino acid supplementation. The dominant phyla in the intestinal microbiota were *Proteobacteria* (58.8%) and *Actinobacteria* (32.4%), but no specific pattern was observed among the different dietary fish groups at any taxonomic level. This finding was probably driven by the high inter-individual variability observed.

Author Contributions: Conceptualization, I.T.K., E.M. and K.A.K.; Methodology, P.P., A.M. and P.B.; Software, P.P., A.M. and P.B.; Validation, I.T.K., E.M. and K.A.K.; Formal Analysis, P.P., A.M. and P.B.; Investigation, P.P., A.M. and P.B.; Resources, I.T.K., E.M., P.B. and K.A.K.; Data Curation, I.T.K., K.A.K. and E.M.; Writing—Original Draft Preparation, P.P. and A.M.; Writing—Review & Editing,

I.T.K., P.B., K.A.K., E.M.; Supervision, I.T.K., E.M. and K.A.K.; Project Administration, I.T.K.; Funding Acquisition, I.T.K. All authors have read and agreed to the published version of the manuscript.

Funding: This research was funded by the Operational Programme "Fisheries 2007–2013" (Ministry of Rural Development and Food of the Hellenic Republic/European Fisheries Fund) under the project title "The use of Processed Animal Proteins in the feeds of gilt-head seabream (*Sparus aurata*)", Grant Number: 2014ΣE086800090000.

Institutional Review Board Statement: The study was conducted according to the guidelines of EU legal frameworks related to the welfare and protection of animals for scientific purposes (Directive 2010/63/EU) and in strict accordance with the University of Thessaly's Ethics Committee on the Use of Animals in Research (protocol approval 20 December 2016).

Informed Consent Statement: Not applicable.

Data Availability Statement: The data presented in this study are available on request from the corresponding author.

Conflicts of Interest: The authors declare no conflict of interest.

References

1. FAO. *The State of World Fishery and Aquaculture 2020 (SOFIA)*; Food and Agriculture Organization of the United Nations: Rome, Italy, 2020. [CrossRef]
2. FAO. Cultured Aquatic Species Information Programme. *Sparus aurata* (Linnaeus, 1758). Available online: http://www.fao.org/fishery/culturedspecies/Sparus_aurata/en (accessed on 24 July 2021).
3. Jannathulla, R.; Rajaram, V.; Kalanjiam, R.; Ambasankar, K.; Muralidhar, M.; Dayal, J.S. Fishmeal availability in the scenarios of climate change: Inevitability of fishmeal replacement in aquafeeds and approaches for the utilization of plant protein sources. *Aquac. Res.* **2019**, *50*, 3493–3506. [CrossRef]
4. Yu, H.-R.; Zhang, Q.; Cao, H.; Wang, X.-Z.; Huang, G.-Q.; Zhang, B.-R.; Fan, J.-J.; Liu, S.-W.; Li, W.-Z.; Cui, Y. Apparent digestibility coefficients of selected feed ingredients for juvenile snakehead, *Ophiocephalus argus*. *Aquac. Nutr.* **2012**, *19*, 139–147. [CrossRef]
5. González-Rodríguez, Á.; Celada, J.D.; Carral, J.M.; Sáez-Royuela, M.; García, V.; Fuertes, J.B. Evaluation of poultry by-product meal as partial replacement of fish meal in practical diets for juvenile tench (*Tinca tinca* L.). *Aquac. Res.* **2014**, *47*, 1612–1621. [CrossRef]
6. Campos, I.; Matos, E.; Marques, A.; Valente, L.M. Hydrolyzed feather meal as a partial fishmeal replacement in diets for European seabass (*Dicentrarchus labrax*) juveniles. *Aquaculture* **2017**, *476*, 152–159. [CrossRef]
7. Karapanagiotidis, I.T.; Psofakis, P.; Mente, E.; Malandrakis, E.; Golomazou, E. Effect of fishmeal replacement by poultry by-product meal on growth performance, proximate composition, digestive enzyme activity, haematological parameters and gene expression of gilthead seabream (*Sparus aurata*). *Aquac. Nutr.* **2019**, *25*, 3–14. [CrossRef]
8. Psofakis, P.; Karapanagiotidis, I.; Malandrakis, E.; Golomazou, E.; Exadactylos, A.; Mente, E. Effect of fishmeal replacement by hydrolyzed feather meal on growth performance, proximate composition, digestive enzyme activity, haematological parameters and growth-related gene expression of gilthead seabream (*Sparus aurata*). *Aquaculture* **2020**, *521*, 735006. [CrossRef]
9. Enzing, C.; Ploeg, M.; Barbosa, M.; Sijtsma, L. *Microalgae-Based Products for the Food and Feed Sector: An Outlook for Europe*; Report EUR 26255; Publications Office of the European Union: Luxembourg, 2014.
10. Maiolo, S.; Parisi, G.; Biondi, N.; Lunelli, F.; Tibaldi, E.; Pastres, R. Fishmeal partial substitution within aquafeed formulations: Life cycle assessment of four alternative protein sources. *Int. J. Life Cycle Assess.* **2020**, *25*, 1455–1471. [CrossRef]
11. den Hartog, L.A.; Sijtsma, S.R. Sustainable feed ingredients. In Proceedings of the 12th International Symposium of Australian Renderers Association "Rendering for Sustainability", Victoria, Australia, 23–26 July 2013; pp. 18–26.
12. Maiolo, S.; Cristiano, S.; Gonella, F.; Pastres, R. Ecological sustainability of aquafeed: An emergy assessment of novel or underexploited ingredients. *J. Clean. Prod.* **2021**, *294*, 126266. [CrossRef]
13. Martínez-Llorens, S.; Baeza-Ariño, R.; Nogales-Mérida, S.; Jover-Cerdá, M.; Tomás-Vidal, A. Carob seed germ meal as a partial substitute in gilthead sea bream (*Sparus aurata*) diets: Amino acid retention, digestibility, gut and liver histology. *Aquaculture* **2012**, *338-341*, 124–133. [CrossRef]
14. Hartviksen, M.A.B.; Vecino, J.L.G.; Ringo, E.; Bakke, A.M.; Wadsworth, S.; Krogdahl, A.; Ruohonen, K.; Kettunen, A. Alternative dietary protein sources for Atlantic salmon (*Salmo salar* L.) effect on intestinal microbiota, intestinal and liver histology and growth. *Aquac. Nutr.* **2014**, *20*, 381–398. [CrossRef]
15. Baeza-Ariño, R.; Martínez-Llorens, S.; Nogales-Mérida, S.; Jover-Cerdá, M.; Tomás-Vidal, A. Study of liver and gut alterations in sea bream, *Sparus aurata* L., fed a mixture of vegetable protein concentrates. *Aquac. Res.* **2014**, *47*, 460–471. [CrossRef]
16. Sitjà-Bobadilla, A.; Peña-Llopis, S.; Requeni, P.G.; Médale, F.; Kaushik, S.; Sánchez, J.P. Effect of fish meal replacement by plant protein sources on non-specific defence mechanisms and oxidative stress in gilthead sea bream (*Sparus aurata*). *Aquaculture* **2005**, *249*, 387–400. [CrossRef]

17. Kokou, F.; Sarropoulou, E.; Cotou, E.; Rigos, G.; Henry, M.; Alexis, M.; Kentouri, M. Effects of Fish Meal Replacement by a Soybean Protein on Growth, Histology, Selected Immune and Oxidative Status Markers of Gilthead Sea Bream, *Sparus aurata*. *J. World Aquac. Soc.* **2015**, *46*, 115–128. [CrossRef]

18. Krogdahl, Å.; Penn, M.; Thorsen, J.; Refstie, S.; Bakke, A.M. Important antinutrients in plant feedstuffs for aquaculture: An up-date on recent findings regarding responses in salmonids. *Aquac. Res.* **2010**, *41*, 333–344. [CrossRef]

19. Hu, L.; Yun, B.; Xue, M.; Wang, J.; Wu, X.; Zheng, Y.; Han, F. Effects of fish meal quality and fish meal substitution by animal protein blend on growth performance, flesh quality and liver histology of Japanese seabass (*Lateolabrax japonicus*). *Aquaculture* **2013**, *372–375*, 52–61. [CrossRef]

20. Ye, H.; Zhou, Y.; Su, N.; Wang, A.; Tan, X.; Sun, Z.; Zou, C.; Liu, Q.; Ye, C. Effects of replacing fish meal with rendered animal protein blend on growth performance, hepatic steatosis and immune status in hybrid grouper (*Epinephelus fuscoguttatus*♀ × *Epinephelus lanceolatus*♂). *Aquaculture* **2019**, *511*, 734203. [CrossRef]

21. Zhou, Z.; Yao, W.; Ye, B.; Wu, X.; Li, X.; Dong, Y. Effects of replacing fishmeal protein with poultry by-product meal protein and soybean meal protein on growth, feed intake, feed utilization, gut and liver histology of hybrid grouper (*Epinephelus fuscoguttatus* ♀× *Epinephelus lanceolatus* ♂) juveniles. *Aquaculture* **2020**, *516*, 734503. [CrossRef]

22. Chaklader, R.; Siddik, M.A.B.; Fotedar, R. Total replacement of fishmeal with poultry by-product meal affected the growth, muscle quality, histological structure, antioxidant capacity and immune response of juvenile barramundi, *Lates calcarifer*. *PLoS ONE* **2020**, *15*, e0242079. [CrossRef]

23. Yu, R.; Cao, H.; Huang, Y.; Peng, M.; Kajbaf, K.; Kumar, V.; Tao, Z.; Yang, G.; Wen, C. The effects of partial replacement of fishmeal protein by hydrolysed feather meal protein in the diet with high inclusion of plant protein on growth performance, fillet quality and physiological parameters of Pengze crucian carp (*Carassius auratus* var. Pengze). *Aquac. Res.* **2019**, *51*, 636–647. [CrossRef]

24. Perry, W.B.; Lindsay, E.; Payne, C.J.; Brodie, C.; Kazlauskaite, R. The role of the gut microbiome in sustainable teleost aquaculture. *Proc. R. Soc. B* **2020**, *287*, 20200184. [CrossRef] [PubMed]

25. Ray, A.; Ghosh, K.; Ringø, E. Enzyme-producing bacteria isolated from fish gut: A review. *Aquac. Nutr.* **2012**, *18*, 465–492. [CrossRef]

26. Wu, S.; Ren, Y.; Peng, C.; Hao, Y.; Xiong, F.; Wang, G.; Li, W.; Zou, H.; Angert, E.R. Metatranscriptomic discovery of plant biomass-degrading capacity from grass carp intestinal microbiomes. *FEMS Microbiol. Ecol.* **2015**, *91*, fiv107. [CrossRef] [PubMed]

27. Foysal, J.; Fotedar, R.; Tay, C.-Y.; Gupta, S.K. Dietary supplementation of black soldier fly (*Hermetica illucens*) meal modulates gut microbiota, innate immune response and health status of marron (*Cherax cainii*, Austin 2002) fed poultry-by-product and fishmeal based diets. *PeerJ* **2019**, *7*, e6891. [CrossRef] [PubMed]

28. Johnson, R.; Wolf, J.; Braunbeck, T. *OECD Guidance Document for the Diagnosis of Endocrine-Related Histopathology of Fish Gonads*; Organization for Economic Co-operation and Development 2009; p. 96. Available online: https://www.oecd.org/chemicalsafety/testing/42140701.pdf (accessed on 14 April 2020).

29. Nikouli, E.; Meziti, A.; Antonopoulou, E.; Mente, E.; Kormas, K.A. Gut Bacterial Communities in Geographically Distant Populations of Farmed Sea Bream (*Sparus aurata*) and Sea Bass (*Dicentrarchus labrax*). *Microorganisms* **2018**, *6*, 92. [CrossRef] [PubMed]

30. Schloss, P.D.; Westcott, S.L.; Ryabin, T.; Hall, J.R.; Hartmann, M.; Hollister, E.B.; Lesniewski, R.A.; Oakley, B.B.; Parks, D.H.; Robinson, C.J.; et al. Introducing mothur: Open-Source, Platform-Independent, Community-Supported Software for Describing and Comparing Microbial Communities. *Appl. Environ. Microbiol.* **2009**, *75*, 7537–7541. [CrossRef]

31. Pruesse, E.; Quast, C.; Knittel, K.; Fuchs, B.M.; Ludwig, W.; Peplies, J.; Glöckner, F.O. SILVA: A comprehensive online resource for quality checked and aligned ribosomal RNA sequence data compatible with ARB. *Nucleic Acids Res.* **2007**, *35*, 7188–7196. [CrossRef]

32. Iwai, S.; Weinmaier, T.; Schmidt, B.L.; Albertson, D.G.; Poloso, N.J.; Dabbagh, K.; DeSantis, T.Z. Piphillin: Improved Prediction of Metagenomic Content by Direct Inference from Human Microbiomes. *PLoS ONE* **2016**, *11*, e0166104. [CrossRef] [PubMed]

33. Oksanen, J.; Blanchet, F.G.; Friendly, M.; Kindt, R.; Legendre, P.; McGlinn, D.; Minchin, P.R.; O'Hara, R.B.; Simpson, G.L.; Solymos, P.; et al. Vegan: Community Ecology Package. Ordination methods, Diversity Analysis and Other Functions for Community and Vegetation Ecologists. 2021. Available online: http://outputs.worldagroforestry.org/cgi-bin/koha/opac-detail.pl?biblionumber=39504 (accessed on 27 July 2021).

34. Anders, S.; Huber, W. Differential expression analysis for sequence count data. *Genome Biol.* **2010**, *11*, R106. [CrossRef]

35. Li, K.; Wang, Y.; Zheng, Z.X.; Jiang, R.L.; Xie, N.X. Replacing fishmeal with rendered animal protein ingredients in diets for Malabar grouper, *Epinephelus malabaricus*, reared in net pens. *J. World Aquac. Soc.* **2009**, *40*, 67–75. [CrossRef]

36. Hernández, C.; Olvera-Novoa, M.; Hardy, R.; Hermosillo, A.; Reyes, C.; González, B. Complete replacement of fish meal by porcine and poultry by-product meals in practical diets for fingerling Nile tilapia *Oreochromis niloticus*: Digestibility and growth performance. *Aquac. Nutr.* **2010**, *16*, 44–53. [CrossRef]

37. Hernández, C.; Sanchez-Gutierrez, Y.; Hardy, R.W.; Benitez-Hernández, A.; Domínguez-Jimenez, P.; González-Rodríguez, B.; Osuna-Osuna, L.; Tortoledo, O. The potential of pet-grade poultry by-product meal to replace fish meal in the diet of the juvenile spotted rose snapper *Lutjanus guttatus* (Steindachner, 1869). *Aquac. Nutr.* **2014**, *20*, 623–631. [CrossRef]

38. Fuertes, J.; Celada, J.D.; Carral, J.M.; Sáez-Royuela, M.; Gonzalez-Rodriguez, A. Effects of fishmeal replacement by feather meal in practical diets for juvenile crayfish (*Pacifastacus leniusculus* Dana, Astacidae). *Aquac. Nutr.* **2013**, *20*, 36–43. [CrossRef]

39. Zapata, D.B.; Lazo, J.P.; Herzka, S.Z.; Viana, M.T. The effect of substituting fishmeal with poultry by-product meal in diets for *Totoaba macdonaldi* juveniles. *Aquac. Res.* **2016**, *47*, 1778–1789. [CrossRef]

40. Ringø, E.; Zhou, Z.; Vecino, J.L.G.; Wadsworth, S.; Romero, J.; Krogdahl, Å.; Olsen, R.E.; Dimitroglou, A.; Foey, A.; Davies, S.; et al. Effect of dietary components on the gut microbiota of aquatic animals. A never-ending story? *Aquac. Nutr.* **2016**, *22*, 219–282. [CrossRef]
41. Gajardo, K.; Jaramillo-Torres, A.; Kortner, T.M.; Merrifield, D.L.; Tinsley, J.; Bakke, A.M.; Krogdahl, Å. Alternative Protein Sources in the Diet Modulate Microbiota and Functionality in the Distal Intestine of Atlantic Salmon (*Salmo salar*). *Appl. Environ. Microbiol.* **2017**, *83*, 83. [CrossRef]
42. Sabbagh, M.; Schiavone, R.; Brizzi, G.; Sicuro, B.; Zilli, L.; Vilella, S. Poultry by-product meal as an alternative to fish meal in the juvenile gilthead seabream (*Sparus aurata*) diet. *Aquaculture* **2019**, *511*, 734220. [CrossRef]
43. Spisni, E.; Tugnoli, M.; Ponticelli, A.; Mordenti, T.; Tomasi, V. Hepatic steatosis in artificially fed marine teleosts. *J. Fish Dis.* **1998**, *21*, 177–184. [CrossRef] [PubMed]
44. Siddik, M.A.B.; Chungu, P.; Fotedar, R.; Howieson, J. Bioprocessed poultry by-product meals on growth, gut health and fatty acid synthesis of juvenile barramundi, *Lates calcarifer* (Bloch). *PLoS ONE* **2019**, *14*, e0215025. [CrossRef]
45. Panicz, R.; Żochowska-Kujawska, J.; Sadowski, J.; Sobczak, M. Effect of feeding various levels of poultry by-product meal on the blood parameters, filet composition and structure of female tenches (*Tinca tinca*). *Aquac. Res.* **2017**, *48*, 5373–5384. [CrossRef]
46. Aydín, B.; Gümüş, E.; Balci, B.A. Effect of dietary fish meal replacement by poultry by-product meal on muscle fatty acid composition and liver histology of fry of Nile tilapia, *Oreochromis niloticus* (Actinopterygii: Perciformes: Cichlidae). *Acta Ichthyol. Piscat.* **2015**, *45*, 343–351. [CrossRef]
47. Deplancke, B.; Gaskins, H.R. Microbial modulation of innate defense: Goblet cells and the intestinal mucus layer. *Am. J. Clin. Nutr.* **2001**, *73*, 1131S–1141S. [CrossRef]
48. Berillis, P.; Mente, E. Histology of Goblet Cells in the Intestine of the Rainbow Trout Can Lead to Improvement of the Feeding Management. *J. Fish.* **2017**, *11*, 32–33. [CrossRef]
49. Ferguson, H.W. Systemic pathology of fish. In *A Text and Atlas of Comparative Tissue Responses in Diseases of Teleosts*, 2nd ed.; Scotian Press: London, UK, 2006.
50. Hu, H.; Kortner, T.M.; Gajardo, K.; Chikwati, E.M.; Tinsley, J.; Krogdahl, Å. Intestinal Fluid Permeability in Atlantic Salmon (*Salmo salar* L.) Is Affected by Dietary Protein Source. *PLoS ONE* **2016**, *11*, e0167515. [CrossRef]
51. Tran-Ngoc, K.T.; Haidar, M.N.; Roem, A.J.; Sendão, J.; Verreth, J.; Schrama, J.W. Effects of feed ingredients on nutrient digestibility, nitrogen/energy balance and morphology changes in the intestine of Nile tilapia (*Oreochromis niloticus*). *Aquac. Res.* **2019**, *50*, 2577–2590. [CrossRef]
52. Nikouli, E.; Meziti, A.; Smeti, E.; Antonopoulou, E.; Mente, E.; Kormas, K.A. Gut Microbiota of Five Sympatrically Farmed Marine Fish Species in the Aegean Sea. *Microb. Ecol.* **2021**, *81*, 460–470. [CrossRef]
53. El-Rhman, A.M.A.; Khattab, Y.A.; Shalaby, A.M. Micrococcus luteus and Pseudomonas species as probiotics for promoting the growth performance and health of Nile tilapia, *Oreochromis niloticus*. *Fish Shellfish. Immunol.* **2009**, *27*, 175–180. [CrossRef]
54. Akayli, T.; Albayrak, G.; Ürkü, Ç.; Çanak, Ö.; Yörük, E. Characterization of *Micrococcus luteus* and *Bacillus marisflavi* Recovered from Common Dentex (*Dentex dentex*) Larviculture System. *Mediterr. Mar. Sci.* **2015**, *17*, 163. [CrossRef]
55. Yamada, Y.; Kuzuyama, T.; Komatsu, M.; Shin-Ya, K.; Omura, S.; Cane, D.-E.; Ikeda, H. Terpene synthases are widely distributed in bacteria. *Proc. Natl. Acad. Sci. USA* **2015**, *112*, 857–862. [CrossRef]
56. Dvergedal, H.; Sandve, S.R.; Angell, I.L.; Klemetsdal, G.; Rudi, K. Association of gut microbiota with metabolism in juvenile Atlantic salmon. *Microbiome* **2020**, *8*, 1–8. [CrossRef]
57. Perlot, T.; Penninger, J.M. ACE2—From the renin–angiotensin system to gut microbiota and malnutrition. *Microbes Infect.* **2013**, *15*, 866–873. [CrossRef] [PubMed]
58. Lu, C.C.; Ma, K.L.; Ruan, X.Z.; Liu, B.C. Intestinal dysbiosis activates renal renin-angiotensin system contributing to incipient diabetic nephropathy. *Int. J. Med. Sci.* **2018**, *15*, 816–822. [CrossRef] [PubMed]
59. Iyer, N.; Vaishnava, S. Vitamin A at the interface of host-commensal-pathogen interactions. *PLoS Pathog.* **2019**, *15*, e1007750. [CrossRef] [PubMed]

applied sciences

Review

Effects of Glyphosate and Its Metabolite AMPA on Aquatic Organisms

Nikola Tresnakova *, Alzbeta Stara and Josef Velisek

Research Institute of Fish Culture and Hydrobiology, South Bohemian Research Center of Aquaculture and Biodiversity of Hydrocenoses, Faculty of Fisheries and Protection of Waters, The University of South Bohemia in Ceske Budejovice, Zatisi 728/II, 389 25 Vodnany, Czech Republic; staraa01@frov.jcu.cz (A.S.); velisek@frov.jcu.cz (J.V.)
* Correspondence: tresnakova@frov.jcu.cz; Tel.: +420-387-774-625

Abstract: Glyphosate (N-(phosphonomethyl)glycine) was developed in the early 1970s and at present is used as a herbicide to kill broadleaf weeds and grass. The widely occurring degradation product aminomethylphosphonic acid (AMPA) is a result of glyphosate and amino-polyphosphonate degradation. The massive use of the parent compound leads to the ubiquity of AMPA in the environment, and particularly in water. Considering this, it can be assumed that glyphosate and its major metabolites could pose a potential risk to aquatic organisms. This review summarizes current knowledge about residual glyphosate and its major metabolite AMPA in the aquatic environment, including its status and toxic effects in aquatic organisms, mainly fish. Based on the above, we identify major gaps in the current knowledge and some directions for future research knowledge about the effects of worldwide use of herbicide glyphosate and its major metabolite AMPA. The toxic effect of glyphosate and its major metabolite AMPA has mainly influenced growth, early development, oxidative stress biomarkers, antioxidant enzymes, haematological, and biochemical plasma indices and also caused histopathological changes in aquatic organisms.

Keywords: toxicity; effect; fish; invertebrate; mussels

Citation: Tresnakova, N.; Stara, A.; Velisek, J. Effects of Glyphosate and Its Metabolite AMPA on Aquatic Organisms. *Appl. Sci.* **2021**, *11*, 9004. https://doi.org/10.3390/app11199004

Academic Editors: Panagiotis Berillis and Božidar Rašković

Received: 31 August 2021
Accepted: 24 September 2021
Published: 27 September 2021

1. Introduction

Over the last few years, the importance of knowledge about pesticide's persistence, mobility, and ecotoxicity has increased. Using pesticides and other agrochemicals is the most cost-effective way to maintain economic viability in the increasing human population [1,2]. On the other hand, the intensive application and repeated use of pesticides in fields in order to increase the crop yield lead to long-term risk for humans, fauna, flora, and the whole ecosystem (soil, air, and water) [1–3]. The extensive use of pesticides is not only a problem in agricultural areas but also in urban settings where pesticides are applied for horticultural purposes. Therefore, it is challenging to control the source of diffuse chemical pollution and its consequences [4]. In particular, the presence of pesticides and their metabolites occurring in residual concentratiosn in drinking, ground, and surface waters poses a global problem [1,3].

Before World War II, natural and organic pesticides were used. However, after the war, it was necessary to increase crop production to prevent starvation and malnutrition, which was an opportunity for industrial companies to produce new synthetic agrochemicals and disseminate worldwide use of them [5]. A considerable amount of pesticide-based chemicals with different uses and modes of action have been successfully brought to market thanks to the fact that the chemical structures, the way, or period of modes of action had the desired effect on the target organisms (according to the United States Environmental Protection Agency 40% herbicides, following insecticides and fungicides) [6]. Later after World War II, in 1970, John Franz discovered the glyphosate-based herbicide effect working in Monsanto (USA). The herbicide was registered in 1974 under the trade name

"Roundup". [7,8]. Due to initial toxicity tests, which showed relatively low risks to nontarget organisms, including mammals, the exposure limits of glyphosate were set relatively high worldwide. In a short time, use of this popular herbicide increased dramatically due to genetically modified crops (soybean, canola, alfalfa, maize, cotton, and corn) which proved to be glyphosate-tolerant. This high frequency of use in agronomy and urban areas caused the general public to perceive this herbicide as having low toxicity and not being very mobile in the environment [9–11]. However, ecotoxicology and epidemiology studies published in the last decade indicate the need for further intensive glyphosate toxicity testing. [11]. Furthermore, the World Health Organization's International Agency for Research on Cancer concluded recently that glyphosate is "probably carcinogenic to humans" [12–14].

Generally, the mobility and concentration of glyphosate and aminomethylphosphonic acid (AMPA) are mainly influenced by their bioavailability, bioaccumulation, persistence, ecotoxicity, and transfer into the aquatic environment (Figure 1) [1,8,11].

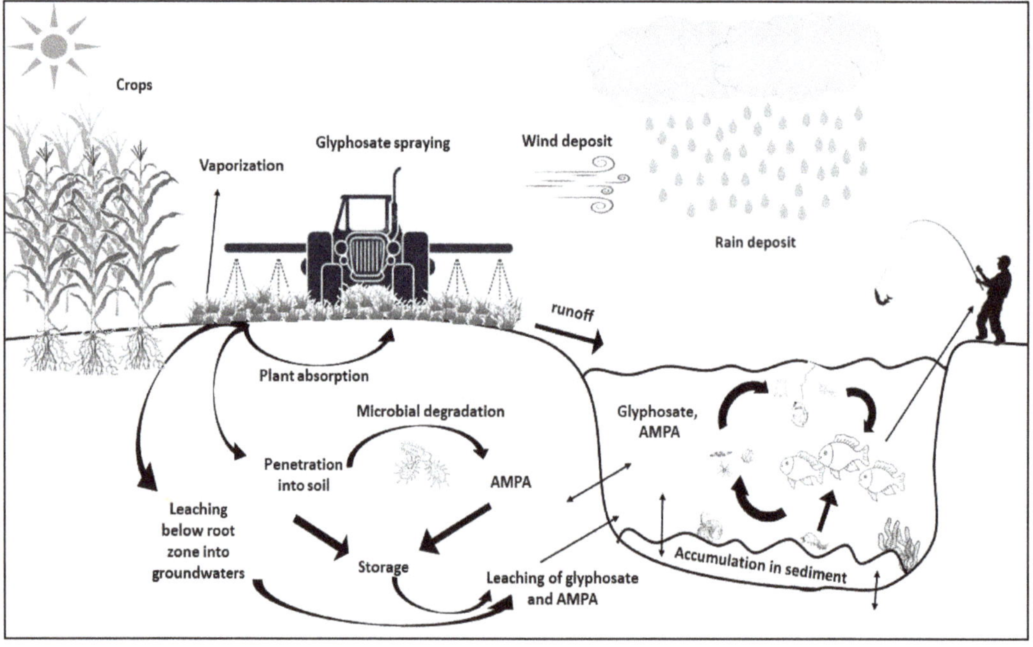

Figure 1. Distribution and transport of glyphosate and its major metabolite AMPA into the aquatic environment.

Directly after spraying herbicide in agriculture or in urban areas, glyphosate is absorbed by crops or weeds and penetrates the soil simultaneously. The glyphosate degradation pathway in bacterial strains is the cleavage of the C-N bond and conversion to AMPA, which is either further decomposed or excreted into the environment [15,16]. AMPA is a primary product of the degradation process of glyphosate and the following nontoxic products are sarcosine and glycine. Unlike AMPA, which is 3–6-fold times more toxic and persistent than glyphosate [17], sarcosine is barely detected in the natural environment [18], except under experimental conditions in a laboratory [16]. On the one hand, the soil has functioned as storage; on the other hand, these contaminants leach below the root zone into groundwater. Glyphosate is also transported by runoff into surface water and consequently accumulated in sediment where glyphosate can be highly mobile [10,17]. The residual concentrations of glyphosate and AMPA in waters contaminate aquatic organisms via the food web (Figure 1) [11,15].

Indeed, pesticides have the ability to dissolve themselves to some extent in the environment. However, there is also the potential risk of residues from the biodegradation process [4,19]. As a result of the extensive use of pesticides, residual concentrations of pesticides and their metabolites are commonly found ubiquitously through different environmental constituents ranging from 1 ng/L to 1 mg/L or higher concentrations [3,4]. There is also a potential risk of banned pesticides. They were excluded because of their long-term persistence and toxicity in the ecosystem. For example, organochlorine insecticides were still detectable in water after 20 years [20] and Acetochlor ESA, the major metabolite of prohibited Acetochlor in the European Union in 2012 no. 1372/2011 [21], was found in the waters of the Czech Republic in recent years [22,23]. Although these pesticides are usually detected only in low concentrations in the environment, they may be present as complex mixtures. The metabolites may be as toxic as their parental compound or even moreso. Therefore, the presence of these substances is of great concern to ecotoxicologists, e.g., [24–26].

Due to repeated application of pesticides, the physical and chemical changes in water properties rise considerably, which is reflected in the modification of the cellular and biochemical biology of aquatic communities, leading to significant changes in their tissues, physiology, and behaviour [27,28]. Therefore, it may affect the daily or seasonal rhythm of aquatic organisms and also their reproduction ability. The environmental stress from xenobiotics may cause loss of habitats and consequently loss of freshwater biodiversity [29,30], which implies that the use of pesticides, despite their advantage in controlling pests, diseases, fungi, etc., has adversely impacted their ubiquity in the environment, e.g., [16,17,31,32].

As far as is known, several studies and reports about the occurrence and toxic effects of different types of pesticides are available in the literature; nevertheless, their global extent and spatial extent of exposure remain largely unknown [2,33]. Considering this information, we decided to write a review to summarise the toxic effects of the often used herbicide glyphosate and its metabolite AMPA on aquatic organisms.

2. Glyphosate (N-(phosphonomethyl)glycine)

Glyphosate (GLY) belongs to the phosphonoamino acid class of pesticides. Glyphosate is an acid that can be associated with different counter cations to form salts [15]. This herbicide is a crop desiccant, broad-spectrum, nonselective, postemergency herbicide that affects all annual and multiannual plants and aquatic weed control in ponds, lakes, canals, etc. [34,35].

Unlike GLY, whose small molecule consists of a linear chain with weak bonds, the molecules of other herbicides are usually arranged in aromatic circular structures. This difference reduces the persistence of glyphosate in the environment [36]. For higher water solubility, GLY is formulated as potassium salts or isopropylamine salts and a surfactant, poly-oxyethylene amine (POEA), is added to enhance the efficacy of the herbicide. Another formulation, Rodeo, contains the isopropylamine salt (IPA) of GLY without the surfactant and is primarily used for controlling aquatic weeds [35,37] or Roundup Transorb, which contains a mix of 15% POEA and additional surfactants [38]. Roundup includes 48% of active agent IPA [34] or potassium salts in the range 167–480 g L^{-1}. The exact amount depends on the type of area where the Roundup is applied [39].

2.1. Environmental Fate

Although a strong bond to the soil amount of GLY leaching up or runoff into surface or ground water is low [40], the aerial applications of glyphosate spray drifts from the ground and may enter into aquatic ecosystems (Figure 1) [41]. Height application rates, rainfall, and a flow route that does not include transportation of GLY through the soil from watersheds pose the highest risk to offsite transport of GLY [9]. For example, the United States Environmental Protection Agency [15] reports predicted GLY concentration from direct applications into a standard pond in 103.8–221.5 µg/L for daily peak,

101.8–217.5 µg/L for 21-day average, and 98.4–210 µg/L for 60-day average. In water bodies, the glyphosate-based herbicide is usually detectable as glyphosate acid equivalent at the range level from 0.01 mg/L to 0.7 mg/L and has the worst impact on surface waters with the value of 1.7 mg/L [42–44]. Coupe et al. [9] reported concentration of GLY for Mississippi, Iowa, and France ranged from 0.03 to 73 µg/L, 0.02 to1.6 µg/L, and 1.9 to 4.7 µg/L, approximately.

This herbicide is unique for its ability to transform itself to the major metabolite AMPA due to microbial degradation [16,40], and its physiochemical properties: water solubility 11.6 g/L at 25 °C, low lipophilicity LogP <-3.2 at 20 °C, dissociation constant of 2.3 at 25 °C [40]. Under aerobic conditions, the halflife of GLY ranges from 1.8 to 109 days in soil and 14–518 days in water-sediment systems; however, in anaerobic water-sediment systems it ranges from 199 to 208 days [15]. Nevertheless, according to the published data the halflife of GLY ranges from 7 to 14 days [40].

GLY contamination has emerged as a pressing issue their high-water solubility and extensive usage in the environment (especially in shallow water systems). Therefore, the exposure of nontarget aquatic organisms to these herbicides is a concern of ecotoxicologists [16,37]. Many objects of ecotoxicologists studies are the toxicity of GLY to different species of fauna and flora and as the final food chain link human. For example, the recent review by Matozzo [45] summarizes only the impact on marine invertebrates but compares glyphosate and its commercial formulations. On the other hand, we report published data about adverse effects of GLY on more aquatic species as a summarisation of harmful impact on aquatic biota.

2.2. Acute Toxicity

It has been already mentioned that the initial testing of GLY did not fully demonstrate its toxic effects, and therefore the amount for use was not strictly regulated. U.S. The EPA divided the toxicity of GLY into slight toxicity with concentrations ranging from 10 to 100 mg/L and almost nontoxicity with concentration higher than 100 mg/L to fish species with acute LC50 values from >10 to >1000 mg/L [15]. Lethal concentrations are various for 24, 48, and 96 h ranging from 0.295 to 645 mg/L for fish species (Table 1); from 6.5 to 115 mg/L for amphibian's species (Table 2); and from 35 to 461.54 mg/L for invertebrate species (Table 3).

Table 1. Acute toxicity values (LC50) of glyphosate and its commercial products on fish.

Species	Formulation	Exposure (Hours)	Concentration (mg/L)	References
Rainbow trout (*Oncorhynchus mykiss*)	GLY	96	140	[41]
	Roundup [1]	96	52–55	[46]
Common carp (*Cyprinus carpio*)	GLY	48 96	645 620	[47]
	Roundup [1]	96	22.19	[48]
	GLY	48 96	602.61 520.77	[49]
Blackhead minnow (*Pimephales promelas*)	GLY	96	97	[41]
Channel catfish (*Ictalurus punctatus*)		96	130	
Bluegills (*Lepomis macrochirus*)	GLY	24 96	150 140	[41]
Guppy (*Poecilia reticulata*)	GLY	96	69.83	[50]

Table 1. *Cont.*

Species	Formulation	Exposure (Hours)	Concentration (mg/L)	References
Rhamdia quelen	GLY	96	7.30	[51]
North African catfish (*Clarias gariepinus*)	GLY	96	0.295	[52]
Zebrafish (*Danio rerio*)	Atnor 48 [2]	96	76.50	[53]
Ten spotted live-bearer (*Cnesterodon decemmaculatus*)	Glyfoglex [3]	96	41.40	[54]

[1] Roundup (active substance glyphosate, 41%), [2] Atnor 48 (active substance glyphosate, 48%), [3] Glyfoglex (active substance glyphosate, 48%).

Table 2. Acute toxicity values (LC50) of glyphosate and its commercial products on amphibians.

Species	Formulation	Exposure (Hours)	Concentration (mg/L)	References
Boana pardalis	GLY	96	106	[55]
Physalaemus cuvieri		96	115	
Green frog (*Lithobates clamitans*)	Roundup [1]	24 / 96	6.6 / 6.5	[56]
Northern leopard frog (*Lithobates pipiens*)		24 / 96	11.9 / 9.2	
Wood frog (*Lithobates sylvaticus*)		24 / 96	18.1 / 16.5	
Dwarf American toad (*Anaxyrus americanus*)		24 / 96	13.5 / <12.9	
Rhinella arenarum	Roundup Ultra-Max [2]	48	2.42 / 77.52	[57]

[1] Roundup (active substance glyphosate, 41%), [2] Roundup Ultra-Max (active substance glyphosate, 36%).

Table 3. Acute toxicity values (LC50) of glyphosate and its commercial products on invertebrate species.

Species	Formulation	Exposure (Hours)	Concentration (mg/L)	References
Midge larvae (*Chironomus plumosus*)	GLY	48	55	[41]
Ceriodaphnia dubia	Roundup [1]	48	147	[37]
Acartia tonsa		48	35	
Chinese mitten crab (*Eriocheir sinensis*)	GLY	24 / 96	461.54 / 97.89	[58]

[1] Roundup (active substance glyphosate, 41%).

2.3. Toxic Effects

2.3.1. Fish

In recent years, GLY toxicity has been studied on various kinds of aquatic organisms. The exposure to GLY may cause several changes in fish (Table 4), such as haematologic and biochemical processes in tissues [38], genotoxicity [53,59], histopathological damage, immunotoxicity [50,60], or cardiotoxicity [61].

Table 4. Toxic effects of glyphosate and its commercial products on fish.

Species	Concentration	Exposure	Effects	References
Common carp (*Cyprinus carpio*)	2.5, 5, 10 mg/L (GLY)	96 h	↑ ALP in liver, heart, GOT in liver and kidney, GPT in kidney;Subepithelial edema and epithelial hyperplasia in gills, focal fibrosis in liver	[47]
	3.5, 7, 14 mg/L (Roundup [1])	16 days	↑ MCV, MCH; ↓ AChE in muscle, brain and liver, Hb, HCT, RBC, WBC, AST, ALT, LDH	[48]
	52.08, 104.15 mg/L (GLY)	7 days	Vacuolization of the renal parenchyma and intumescence of the renal tubule in kidney, immunotoxicity	[49]
			↑ AST, ALT, MDA, PC; ↓ GSH, inhibition of NA^+/K^+ -ATPase, SOD, CAT, GPx, GR, T-AOC, induce inflammatory response in gills	[60]
European eel (*Anguilla Anguilla*)	58, 116 μg/L (Roundup [1])	1, 3 days	↑ TBARS, LPO, GDI, ENA	[42]
			↑ GDI, damaged nucleoids, EndoIII	[59]
Curimbata (*Prochilodus lineatus*)	10 mg/L (Roundup [1])	24 h	↑ GSH, GST, LPO; ↓ SOD, GPx, inhibition of AChE in muscle	[38]
		96 h	↑ GST, LPO;inhibition of AChE in muscle in brain and muscle	
Spotted snakehead (*Channa punctatus*)	32.54 mg/L (Roundup [1])	1, 7, 14, 21, 28, 35 days	↑ TBARS, DNA damage, LPO, ROS; ↓ CAT, SOD, GR in gill and blood	[62]
Ten spotted live-bearer (*Cnesterodon decemmaculatus*)	1, 1.75, 35 mg/L (GLY)	96 h	↓ AChE	[63]
Megaleporinus obtusidens	3, 6, 10, 20 mg/L (Roundup [1])	96 h	↑ hepatic GL, GLU, NH_3 in liver and muscle, PCV, Hb, RBC, WBC, P; ↓ AChE in brain, LACT, P in liver, muscle GL, GLU	[64]
	5 mg/L (Roundup [1])	90 days	↑ LACT in liver and muscle, P in liver; ↓ AChE, GL in liver, P in muscle, PCV, Hb, RBC, WBC	[65]
Rhamdia quelen	0.2, 0.4 mg/L (Roundup [1])	96 h	↑ hepatic GL, LACT in liver and muscle, P in liver and muscle, NH_3 in liver and muscle, TBARS in muscle; ↓ muscle GL, GLU in liver and muscle, AChE in brain	[66]
	0.730 mg/L (GLY)	24, 96 h, 10 days	↑ immature circulating cells; ↓ RBC, THR, WBC, phagocytic activity, agglutination activity, lysozyme activity	[67]
Rhamdia quelen	18, 36, 72 μg/L (Roundup [1])	7 days	↑ TP in liver, ↑ GL in muscle; ↓ TP, GL, TL in gills, liver, and kidney	[68]
Goldfish (*Carassius auratus*)	2.5–20 mg/L (Roundup [1])	2 months	↑ CAT in liver and kidney; ↓ GR in kidney, liver, and brain, G6PDH in kidney, liver and brain, SOD in kidney, liver and brain	[69]
	0.22, 0.44, 0.88 mmol/L (GLY)	96 h	Behaviour abnormalities (observed depression, erratic swimming, partial loss of equilibrium), liver tissue damage (cellular swelling, inflammatory cell infiltration, hydropic degeneration, loose cytoplasm, ↑ brown particles), kidney tissue damage (edema in the epithelial cells of renal tubules, ↑ cell volume, loose cytoplasm, slight staining), changes in plasma (↑ CK, UN, ↓ LDH)	[70]
	0.2 mmol/L (Nongteshi [2])	90 days	Hyaline cast in kidney, ↑ CRE, BUN, ALT, AST, LDH, MDA, ↑ 3-hydroxybutyrate, LACT, alanine, acetamide, glutamate, glycine, histidine, inosine, GLU;↓ SOD, GSH-Px, GR, lysine, NAA, citrate, choline, phosphocholine, myo-inosine, nicotinamide,	[71]
North African catfish (*Clarias gariepinus*)	0, 19, 42, 94, 207, 455mg/L (GLY)	96 h	Cellular infiltration in gills; fatty degeneration, fat vacuolation, diffuse hepatic necrosis, infiltration of leukocytes in liver; hematopoietic necrosis, pyknotic nuclei in kidney; mononuclear infiltration, neuronal degeneration, spongiosis in brain; respiratory stress, erratic swimming	[52]

Table 4. *Cont.*

Species	Concentration	Exposure	Effects	References
Hybrid fish jundiara (*Leiarius marmoratus* × *Psedoplatystoma reticulatum*)	1.357 mg/L (Roundup [1])	6, 24, 48,96 h	↑ LACT in liver, P level in liver, ALT, AST, CHOL, TAG in plasma; ↓ GL in liver and muscle, plasma GLU, Hb, PCV, RBC, WBC	[28]
Zebrafish (*Danio rerio*)	50 µg/mL (GLY)	24 h	↓ gene expression in eye, fore, and midbrain delineated brain ventricles and cephalic regions	[72]
	32.5, 65, 130 µg/L (Transorb [3])	48 h	↓ integrity of plasma membrane of hepatocytes, viability of cells, mitochondrial activity in the cell, lysosomal integrity, inhibition in ABC transporter activity	[73]
	10, 50, 100, 200, 400 µg/L (GLY)	48 h	↓ heartbeat, NO generation, downregulation of Cacana1C and ryr2a genes, upregulation of hspb11	[61]
Climbing bass (*Anabas testudineus*)	17.20 mg/L (Excel Mera 71 [4])	30 days	↑ AChE, LPO, CAT; ↓ TP, GST	[74]
Heteropneustes fossilis				

[1] Roundup (active substance glyphosate, 41%), [2] Nongteshi (active substance glyphosate, 30%), [3] Transorb (active substance glyphosate, 48%), [4] Excel Mera 71 (active substance glyphosate, 71%).

There are just a few data about the chronic effects of glyphosate on nontarget organisms. For example, Le Du-Carrée [75] have studied chronic exposure to glyphosate with a concentration of 1 µg/L on rainbow trout for 10 months. No significant changes in reproduction, metabolism, nor even oxidative response were observed. However, some occasional impacts on immune response have occurred. Other chronic effects have been studied with different concentrations of glyphosate (0.2, 0.8, 4 and 16 mg/L) in *Oreochromis niloticus* for 80 days [76]. It was found that glyphosate exposure reduces antioxidative ability, disturbs liver metabolism, promote inflammation, and suppresses immunity.

2.3.2. Invertebrate Species

The exposure to GLY may cause several changes in invertebrate species (Table 5), such as biochemical processes in tissues, development, or behaviour; changes in haemolymph [77], changes in the reproduction system, and 50% inhibition of cholinesterase activity [78] in mussels.

Table 5. Toxic effects of glyphosate and its commercial products on invertebrate species.

Species	Concentration	Exposure	Effects	References
Mediterranean mussel (*Mytilus galloprovincialis*)	100 µg/L (GLY)	7 days	↑ THC, haemocyte proliferation; ↓ Haemocyte diameter, AChE in gills	[77]
		14 days	↑ AChE in gills, CAT in digestive gland; ↓ CAT in gills	
		21 days	↑ CAT in gills; ↓ THC, haemocyte diameter, haemocyte volume, HL, AChE in gills	
	10, 100, 1000 µg/L (GLY)	7, 14, 21 days	↑ cell volume of haemocyte, haemolymph pH; ↓ HL, haemolymph acid phosphatase activity; AChE in gills; SOD in digestive gland, THC,	[79]
Limnoperna fortunei	1, 3, 6 mg/L (GLY)	26 days	↑ TBARS, GST, ALP; ↓ CES, SOD	[80]
	10, 20, 40 mg/L (GLY)	3 weeks	↓ presence of large mussel by 40%, presence empty shell by 25%	[81]
Pacific oyster (*Crassostrea gigas*)	0.1, 1, 100 µg/L (Roundup Expres [1])	35 days	↑ GST; ↓ growth; LPO, MDA	[82]
California blackworm (*Lumbriculus variegatus*)	0.05–5 mg/L (GLY)	4 days	↑ SOD; ↓ GST, membrane bound GST	[83]

Table 5. *Cont.*

Species	Concentration	Exposure	Effects	References
Chinese mitten crab (*Eriocheir sinensis*)	4.4, 9.8, 44, 98 mg/L(GLY)	96 h	↑ % DNA in tail, SOD, POD, β-GD;↓ THC, granulocytes, phagocytic activity, ACP, AKP	[58]
American bullfrog (*Lithobates catesbeianus*)	1 mg/L (Roundup [2])	48 h	↑ swimming activity, CPM; SOD, CAT and LPO in liver; LPO in muscle; ↓ SOD, CAT in muscle, TtHR	[84]
Rhinella arenarum	1.85, 3.75, 7.5, 15, 30, 60, 120, 240 mg/L (Roundup Ultra-Max [3])	48 h	↓ AChE, BChE, CbE, GST	[57]
Northern leopard frog (*Rana pipiens*)	0.6, 1.8 mg/L (Roundup [2])	166 days	↑ TRβ mRNA; Late metamorphic climax, developmental delay, abnormal gonads, necrosis of the tail tip, fin damage, abnormal growth on the tail tip, blistering on the tail fin	[56]
Snail(*Biomphalaria alexandrina*)	3.15 mg/L (Roundup [2])	6 weeks	↑ mortality, stopped egg lying, abnormal laid eggs, ↑ GLU, LACT, FAC; ↓ egg hatchability, GL, TP, pyruvate, nucleic acids levels	[85]
	10 mg/L (Roundup [2])	7 days	↑ in vitro phagocytic activity, DNA damage in haemocytes	[86]

[1] Roundup Expres (active substance glyphosate, 15%), [2] Roundup (active substance glyphosate, 41%), [3] Roundup Ultra-Max (active substance glyphosate, 36%).

3. AMPA (Aminomethylphosphonic Acid)

AMPA belongs to the aminomethylenephosphonates chemical group. It is the primary metabolite of the GLY degradation process (Figure 1) with a significant measured concentration in the environment. Additional sources of AMPA originate from organic phosphonates used in water treatment [87], from the degradation of phosphonic acids used in Europe in detergent and industrial boilers, and cooling (EDTMA, DTMP, ATMP, and HDTMP) [15,87]. Due to phosphonate and amine functional groups, AMPA will form metal complexes with Ca^{2+}, Mg^{2+}, Mn^{2+}, and Zn^{2+}. AMPA is adsorbed firmly to soil [88].

3.1. Environmental Fate

AMPA has a lower water solubility and longer soil halflife than glyphosate. The presence of AMPA in freshwater, sediment, and suspended particulate is commonly measured in significant quantities [10,89], and even more frequently (67.5%) than glyphosate (17.5%) [15,90,91]. The Water Framework Directive [92] provides a procedure to set Environmental Quality Standards for AMPA at level 450 mg/L. Coupe et al. [9] reported concentrations of AMPA in freshwater environments for Mississippi and Iowa were 2.6 μg/L and 0.02–5.7 μg/L. In France, AMPA was detected with the highest concentration at a level of 44 μg/L.

A concentration of AMPA in soil was found from 299 to 2256 $\mu g\ kg^{-1}$ by Aparicio et al. [93]. This study pointed out the difficulty of establishing specific conditions for the presence of AMPA in soils. It depends on complex and multifactorial processes, including agronomic conditions, local agrometeorological conditions, mineralogy, and soil conditions. Moreover, AMPA was detected in soil with no exposure time to glyphosate. It could be caused by surface runoff. This movement of soil particles can end up in surface water, and next, the substance can be desorbed, biodegraded, and accumulated in the bottom sediment.

AMPA, like glyphosate, also degrades in water and soil but significantly slower. Because its adsorption to particulates is possibly stronger, its penetrability of cell membranes is lower. The concentration of AMPA in the sediment can fluctuate depending on its degradation rate relative to GLY (Figure 1) [93].

3.2. Acute Toxicity

AMPA toxicity has been already studied in recent years on various kinds of organisms. Although de Brito Rodrigues et al. [53] observed no acute toxic effect of AMPA on fish species, other studies showed acute toxicity values ranging from 27 to 452 mg/L (Table 6).

Table 6. Toxicity values of AMPA for aquatic organisms.

Species	Value	Concentration (mg/L)	References
Fish			
Guppy (*Poecilia reticulata*)	96hLC50	180 for male / 164.32 for female	[50]
Invertebrate			
Pacific oyster (*Crassostrea gigas*)	36hEC10	38.55	[94]
	36hEC20	42.68	
	36hEC50	50.78	
	24hEC10	27.08	
	24hEC20	39.80	
	24hEC50	76.90	
Daphnia magna	48hEC10	$>100^5$	
	48hEC20		
	48hEC50		
Algae			
Pseudokirchneriella subcapitata	72hEC10	85.05	[94]
	72hEC20	>100	
	72hEC50		
Desmodesmus subspicatus	72hIC50	117.8	[95]
	72hEC50	89.8 [1] / 452 [2]	[96]

[1] biomass test, [2] algal growth inhibition tests.

3.3. Toxic Effects

Although AMPA has been studied less than glyphosate, Reddy et al. [97] pointed to affecting chlorophyll biosynthesis which leads to plant growth reduction. That means that AMPA can also be translocated to diverse plant tissue. AMPA is also known as a phytotoxin, which can amplify the indirect effects of glyphosate on physiological processes. On the other hand, due to its chemical similarity, AMPA can compete with glycine in biological sites and pathways, affecting chlorophyll biosynthesis and therefore the photosynthetic process as well [98]. Plants treated with AMPA showed a decreased glycine, serine, and glutamate [99]. According to published data, AMPA seems to be highly toxic on aquatic organisms (Table 7).

Table 7. Toxic effects of AMPA on aquatic organisms.

Species	Concentration	Exposure	Effects	References
European eel (*Anguilla Anguilla*)	11.8, 23.6 µg/L	1, 3 days	↑ GDI, FPG, EndoIII	[100]
Zebrafish (*Danio rerio*)	1.7, 5, 10, 23, 50, 100 mg/L	24, 48, 72, 96 h	Genotoxicity with LOEC 1.7 mg/L, induce primary DNA lesions,	[53]
Guppy (*Poecilia reticulata*)	82 mg/L	96 h	Proliferation of the interlamellar epithelium, fusion of secondary lamellae in gill, steatosis, pyknotic nuclei in liver, degeneration of hepatocytes	[50]
Mediterranean mussel (*Mytilus galloprovincialis*)	100 µg/L	7 days	↑ haemocyte diameter, haemocyte volume, haemocyte proliferation, LDH in haemolymph, HL; ↓ THC, AChE in gills	[77]
Mediterranean mussel (*Mytilus galloprovincialis*)	100 µg/L	14 days	↑ THC, haemocyte diameter, haemocyte volume, haemocyte proliferation, AChE in gills, CAT in digestive gland; ↓ HL	[101]
		21 days	↑ haemocyte volume, LDH in haemolymph; ↓ THC, haemocyte proliferation, HL, AChE in gills	
	1, 10, 100 µg/L	7 days	↓ THC	
		14 days	↑ THC, haemocyte diameter and volume, lysosome activity, acid phosphatase; ↓ haemocyte proliferation, SOD in gill and digestive gland	
		21 days	↑ haemocyte proliferation, lysosome activity, acid phosphatase, LDH; ↓ THC, haemocyte diameter and volume	
Bufo spinosus	0.07, 0.32, 3.57 µg/L	16 days	↓ embryonic survival, development delay, short tail length	[102]

4. Conclusions

There is a large amount of information about the benefits, environmental fate, effects, and risks of using glyphosate throughout the scientific literature. Nevertheless, evaluating chronic exposures of nontarget aquatic organisms is missing. The results of the studies that have been summarized in this review indicate that GLY, as an individual compound or as a component of commercial products used in agriculture, and its main metabolite AMPA may have adverse effects on freshwater and marine organisms at different levels of biological organization. GLY mainly caused oxidative stress, and affected antioxidant enzymes, blood parameters, and caused several histopathologic changes in the gills, liver, and kidneys, and not least genotoxicity, immunotoxicity, and cardiotoxicity in fish and oxidative stress, antioxidant enzymes, and haemocyte parameters in mussels. In comparison to AMPA, there are many gaps in the scientific literature regarding the knowledge of its toxicity on aquatic organisms. AMPA may cause genotoxicity and immunotoxicity in fish, adverse changes in haemolymph parameters, effects on mussels' antioxidant enzymes, and developmental delay and survival of tadpoles.

There are also concerns about potential bioconcentration effects and breeding in organisms of these compounds. Considering the increasing consumption of herbicides and their repeated application worldwide, the European Commission implemented a regulation (EU 2017/2324 [103] that GLY can be used as an active substance until 15 December 2022 on the condition that national authorities have to authorize each product. Therefore, we assume that the presence of GLY and AMPA in the aquatic environment requires a stricter control and further studies of the potentially toxic effects of these substances on nontarget organisms. As the lethal concentrations indicate glyphosate, its commercial products, and AMPA at very high levels impact long-term exposure at real environmental concentrations, more detailed information about the ecotoxicity needs to be evaluated.

Author Contributions: Conceptualization, data curation, writing—original draft preparation, N.T.; writing—review and editing A.S.; supervision J.V. All authors have read and agreed to the published version of the manuscript.

Funding: The research was funded by the Ministry of Agriculture of the Czech Republic—project No. QK1910282.

Institutional Review Board Statement: Not applicable.

Informed Consent Statement: Not applicable.

Data Availability Statement: Not report data.

Conflicts of Interest: The authors declare no conflict of interest regarding the publication of this review paper.

Abbreviations

ABC transporter activity: adenosine triphosphate-binding cassette transporters constitute; AChE: acetylcholinesterase; ACP: acid phosphatase; AKP/ALP: alkaline phosphatase;; ALT: alanine aminotransferase; AMPA: aminomethylphosphonic acid; AST: aspartate aminotransferase; ATMP: amino tris(methylenephospohonate); β-GD: β- glucuronidase; BChE: butyrylcholinesterase; BCF: bioaccumulation factor; BUN: blood urea nitrogen; C-N bond: carbon–nitrogen bond; Ca^{2+}: calcium ion; Cacanal1C: L-type calcium chanel; CAT: catalase; CbE: carboxylesterase, CES: carboxylesterases; ChOL: cholesterol; CK: creatinine; CPM: cardiac pumping capacity; CRE: creatine; DNA: deoxyribonucleic acid; DTPMP: diethylenetriamine penta(methylenephosphonate); EC10: equivalent to the No observed effect concentration; EC20: equivalent to the Low observed effect concentration; EC50: effective concentration that affects 50% of the population; EDTMA: ethylenediamine tetra(methylenephosphonate); ENA: erytrocytic nuclear abnormalities; EndoIII: endonuclease III; FAC: free amino acid levels; FPG: formamidopyrimidine DNA glycosylase; G6PDH: glucose-6-phosphate dehydrogenase; GDI: total DNA damage; GL: glycogen; GLU: glucose; GLY: glyphosate; GOT: glutamic–oxaloacetic transaminases; GPx: glutathione peroxidase; GPT: glutamic–pyruvic transaminases; GR: glutathione reductase; GSH: glutathione; GSH-Px: glutathione peroxidase; GST: glutation-S-transferase; Hb: hemoglobin; HCT: hematocrit; HDTMP: hexamethylenediamine tetra(methylenephosphonate); HL: haemocyte lysate; hspb11: heat shock protein; IC: inhibition concentration; IPA: isopropylamine salt; LACT: lactate; LC: lethal concentration; LDH: lactate dehydrogenase; LPO: lipid peroxidation; Na^+/K^+-ATPase: sodium–potassium adenosine triphosphatase; NOEC: no observed effect concentration; MCH: mean cell hemoglobin; MCV: mean cell volume; MDA: methanedicarboxylic aldehyde; Mg^{2+}: magnesium ion; mg ae/L: miligrams active ingredient per liter; Mn^{2+}: manganese ion; NAA: N-Acetyl aspartate; NH3: ammonia; NO: nitric oxide; P: protein; PC: protein carbonyl; PCV: hematocrit; POD: peroxidase; RBC: erythrocytes; ROS: reactive oxygen species; ryr2a: Ryanodine receptor; SOD: superoxide dismutase; T-AOC: total antioxidant activity; TAG: triacylglycerides; TBARS: thiobarbituric acid reactive substances; THC: total hemocyte count; THR: thrombocytes; TL: Total lipids; TP: total protein; TRβ mRNA: TRβ mRNA: Thyroid hormone receptor beta of messenger ribonucleic acid; TtHR: time to half relaxation; UN: urine nitrogen; U.S. EPA: United States Environmental Protection Agency; WBC: leukocytes; Zn^+: zinc ion.

References

1. Arias-Estévez, M.; López-Periago, E.; Martínez-Carballo, E.; Simal-Gándara, J.; Mejuto, J.-C.; García-Río, L. The mobility and degradation of pesticides in soils and the pollution of groundwater resources. *Agric. Ecosyst. Environ.* **2008**, *4*, 247–260. [CrossRef]
2. Bilal, M.; Igbal, H.M.N.; Barceló, D. Persistence of pesticides-based contaminants in the environment and their effective degradation using laccase-assisted biocatalytic systems. *Sci. Total Environ.* **2019**, *695*, 133896. [CrossRef]
3. Riahi, B.; Rafatpanah, H.; Mahmoudi, M.; Memar, B.; Brook, A.; Tabasi, N.; Karimia, G. Immunotoxicity of paraquat after subacute exposure to mice. *Food Chem. Toxicol.* **2010**, *48*, 1627–1631. [CrossRef]
4. Fenner, K.; Canonica, S.; Wackett, L.P.; Elsner, M. Evaluating Pesticide Degradation in the Environment: Blind Spots and Emerging Opportunities. *Science* **2013**, *341*, 752–758. [CrossRef]
5. Gill, J.P.K.; Sethi, N.; Mohan, A.; Datta, S.; Girdhar, M. Glyphosate toxicity for animals. *Environ. Chem. Lett.* **2018**, *16*, 401–426. [CrossRef]
6. Grube, A.; Donaldson, D.; Kiely, T.; Wu, L. *Pesticides Industry Sales and Usage, 2006 and 2007 Market Estimates*; U.S. Environmental Protection Agency: Washington, DC, USA, 2011; p. 41.
7. Franz, J. *N-Phosphonomethyl-Glycine Phytotoxicant Compositions*; Monsanto CO, US: St. Louis, MO, USA, 1974; p. 3799758. Available online: https://www.freepatentsonline.com/3799758.html (accessed on 12 February 2021).
8. Henderson, A.M.; Gervais, J.A.; Luukinen, B.; Buhl, K.; Stone, D.; Strid, A.; Cross, A.; Jenkins, J. Glyphosate Technical Fact Sheet; National Pesticide Information Center, Oregon State University Extension Services. 2010. Available online: http://npic.orst.edu/factsheets/archive/glyphotech.html (accessed on 10 February 2021).
9. Coupe, R.H.; Kalkhoff, S.J.; Capel, P.D.; Gregoire, C. Fate and transport of glyphosate and Aminomethylphosphonic acid in surface waters of agricultural basins. *Pest. Manag. Sci.* **2012**, *68*, 16–30. [CrossRef]
10. Battaglin, W.A.; Myer, M.T.; Kuivila, K.M.; Dietze, J.E. Glyphosate and its gedradation product AMPA occur frequently and widely in U.S. soils, surface water, groundwater, and precipitation. *J. Am. Water Resour. Assoc.* **2014**, *50*, 275–290. [CrossRef]
11. Myers, J.P.; Antoniou, M.N.; Blumberg, B.; Carroll, L.; Colborn, T.; Everett, L.G.; Hansen, M.; Landrigan, P.J.; Lanphear, B.P.; Mesnage, R.; et al. Concerns over use of glyphosate-based herbicides and risks associated with exposures: A consensus statement. *Environ. Health* **2016**, *15*, 19. [CrossRef]
12. Guyton, K.Z.; Loomis, D.; Grosse, Y.; El Ghissassi, F.; Benbrahim-Tallaa, L.; Guha, N.; Scoccianti, C.; Mattock, H.; Straif, K. International Agency for Research on Cancer Monograph Working Group, IARC, Lyon, France. Carcinogenicity of tetrachlorvinphos, parathion, malathion, diazinon, and glyphosate. *Lancet Oncol.* **2015**, *5*, 490–491. [CrossRef]
13. IARC (International Agency for Research on Cancer). IARC Rejecets False Claims in Reuters Article ("In Glyphosate Review, WHO Cancer Agency Edited Out "Non-Carcinogenic" Findings"). 2017. Available online: https://www.iarc.who.int/wp-content/uploads/2018/07/IARC_Response_Reuters_October2017.pdf (accessed on 14 February 2021).
14. IARC (International Agency for Research on Cancer). IRAC Response to Critisms of the Monographs and the Glyphosate Evaluation. 2018. Available online: https://www.iarc.who.int/wp-content/uploads/2018/07/IARC_response_to_criticisms_of_the_Monographs_and_the_glyphosate_evaluation.pdf (accessed on 14 February 2021).
15. U.S. EPA (United States Environmental Protection Agency). *Preliminary Ecological Risk Assessment in Support of the Registration Review of Glyphosate and Its Salts*; U.S. EPA: Washington, DC, USA, 2015; p. 318.
16. Zhan, H.; Feng, Y.; Fan, X.; Chen, S. Recent advances in glyphosate biodegradation. *Appl. Microbiol. Biotechnol.* **2018**, *102*, 5033–5043. [CrossRef]
17. Sun, M.; Li, H.; Jaisi, D.P. Degradation of glyphosate and bioavailability of phosphorus derived from glyphosate in a soil-water system. *Water Res.* **2019**, *163*, 114840. [CrossRef]
18. Wang, S.; Seiwert, B.; Kästner, M.; Miltner, A.; Schäffer, A.; Reemtsma, T.; Yang, Q.; Nowak, K.M. (Bio)degradation of glyphosate in water-sediment microcosms–A stable isotope co-labeling approach. *Water Res.* **2016**, *99*, 91–100. [CrossRef]
19. Al-Mamun, A. Pesticide Degradations Resides and Environmental Concerns. In *Pesticide Residue in Foods: Sources, Management, and Control*; Khan, M.S., Rahman, M.S., Eds.; Springer International Publishing AG: Cham, Switzerland, 2017; pp. 87–102. [CrossRef]
20. Larson, S.J.; Capel, P.D.; Majewski, M.S. Pesticide in surface waters—Distribution, trends, and governing factors. In *Series of Pesticides in Hydrologic System*; Gilliom, R.J., Ed.; Ann Arbor Press: Chelsea, MI, USA, 1997; Volume 3, p. 392.
21. Commission Implementing Regulation (EU) no 1372/2011. Concerning the Non-Approval of the Active Substance Acetochlor, in Accordance with Regulation (EC) No 1107/2009 of the European Parliament and of the Council Concerning the Placing of Plant Protection Products on the Market, and Amending Commission Decision 2008/934/EC. OJEU, L 341/45. Available online: https://eur-lex.europa.eu/legal-content/EN/TXT/PDF/?uri=CELEX:32011R1372 (accessed on 25 November 2020).
22. Moulisová, A.; Bendakovská, L.; Kožíšek, F.; Vavrouš, A.; Jeligová, H.; Kotal, F. Pesticidy a jejich metabolity v pitné vodě. Jaký je současný stav v České Republice? *Vodní Hospodářství* **2017**, *68*, 4–10.
23. CHMI (Czech Hydrometeorological Institute). On-Line Water Quality Database. 2020. Available online: http://hydro.chmi.cz/ (accessed on 10 December 2020).
24. Kolpin, D.W.; Thurman, E.M.; Linhart, S.M. The environmental occurrence of herbicides: The importance of degradates in ground water. *Arch. Environ. Contam. Toxicol.* **1998**, *35*, 385–390. [CrossRef]
25. Schwarzenbach, R.P.; Escher, B.I.; Fenner, K.; Hofstetter, T.B.; Johnson, C.A.; Von Gunten, U.; Wehrli, B. The challenge of micropollutants in aquatic systems. *Science* **2006**, *313*, 1072–1077. [CrossRef]

26. Ceyhun, S.B.; Şentürk, M.; Ekinci, D.; Erdoğan, O.; Çiltaş, A.; Kocaman, E.M. Deltamethrin attenuates antioxidant defense system and induces the expression of heat shock protein 70 in rainbow trout. *Comp. Biochem. Physiol. C* **2010**, *152*, 215–222. [CrossRef]

27. Abrantes, N.; Pereira, R.; Gonçalves, F. Occurrence of pesticides in water, sediments, and fish tissues in a Lake Surrounded by agricultural lands: Concerning risks to humans and ecological receptors. *Water Air Pollut.* **2010**, *212*, 77–88. [CrossRef]

28. De Moura, F.R.; da Silva Lima, R.R.; da Cunha, A.P.S.; da Costa Marisco, P.; Aguiar, D.H.; Sugui, M.M.; Sinhorin, A.P.; Sinhorin, V.D.G. Effects of glyphosate-based herbicide on pintado da Amazônia: Hematology, histological aspects, metabolic parameters and genotoxic potential. *Environ. Toxicol. Pharmacol.* **2017**, *56*, 241–248. [CrossRef]

29. Geist, J. Integrative freshwater ecology and biodiversity conservation. *Ecol. Indic.* **2011**, *11*, 1507–1516. [CrossRef]

30. Malaj, E.; Peter, C.; Grote, M.; Kühne, R.; Mondy, C.P.; Usseglio-Polatera, P.; Schäfer, R.B. Organic chemicals jeopardize the health of freshwater ecosystems on the continental scale. *Proc. Nat. Acad. Sci. USA* **2014**, *111*, 9549–9554. [CrossRef]

31. Qui, Y.W.; Zeng, E.Y.; Qiu, H.; Yu, K.; Cai, S. Bioconcentration of polybrominated diphenyl ethers and organochlorine pesticides in algae is an important contaminant route to higher trophic levels. *Sci. Total. Environ.* **2017**, *579*, 1885–1893. [CrossRef]

32. Eddleston, M. Poisoning by pesticides. *Medicine* **2020**, *48*, 214–217. [CrossRef]

33. Ippolito, A.; Kattwinkel, M.; Rasmussen, J.J.; Schäfer, R.B.; Fornaroli, R.; Liess, M. Modeling global distribution of agricultural insecticides in surface waters. *Environ. Pollut.* **2015**, *198*, 54–60. [CrossRef]

34. WHO (World Health Organization). Glyphosate/Published under the Joint Sponsorship of the United Nations Environment Programme, the International Labour Organisation, and the World Health Organization. World Health Organization & International Programme on Chemical Safety. 1994. Available online: https://apps.who.int/iris/handle/10665/40044 (accessed on 3 December 2020).

35. Franz, J.E.; Mao, M.K.; Sikorski, J.A. *Glyphosate: A Unique Global Herbicide*; American Chemical Society: Washington, DC, USA, 1997; p. 615.

36. Wallace, J.; Lingenfelter, D. Glyphosate (Roundup): Understanding Risks to Human Health. PennState Extension: College of Agricultural Sciences. 1–3. 2019. Available online: https://extension.psu.edu/glyphosate-roundup-understanding-risks-to-human-health#:~{}:text=Glyphosate%20fate%20in%20environment.%20The%20chemical%20properties%20of,water.%20Glyphosate%20does%20not%20degrade%20quickly%20in%20plants (accessed on 10 July 2020).

37. Tsui, M.T.K.; Chu, L.M. Aquatic toxicity of glyphosate-based formulations: Comparison between different organisms and the effects of environmental factors. *Chemosphere* **2003**, *52*, 1189–1197. [CrossRef]

38. Modesto, K.A.; Martinez, C.B.R. Roundup causes oxidative stress in liver and inhibits acetylcholinesterase in muscle and brain of the fish *Prochilodus lineatus*. *Chemosphere* **2010**, *78*, 294–299. [CrossRef] [PubMed]

39. Roundup.cz. RoundupBioaktiv: Okolo Vodních Toků a Nádrží, Fakta. Monsanto ČR s.r.o. Available online: https://www.roundup.cz/roundup-biaktiv/fakta (accessed on 12 December 2020).

40. Blake, R.J.; Pallet, K. The environmental fate and ecotoxicity of glyphosate. *Outlooks Pest. Manag.* **2018**, *29*, 266–269. [CrossRef]

41. Folmar, L.C.; Sanders, H.O.; Julin, A.M. Toxicity of the herbicide glyphosate and several of its formulations to fish and aquatic invertebrates. *Arch. Environ. Contam. Toxicol.* **1979**, *8*, 269–278. [CrossRef] [PubMed]

42. Guilherme, S.; Gaivao, I.; Santos, M.A.; Pacheco, M. European eel (*Anguilla Anguilla* genotoxic and prooxidant responses following short-term exposure to Roundup®—A glyphosate-based herbicide. *Mutagenesis* **2010**, *25*, 523–530. [CrossRef]

43. Wagner, N.; Reichenbecher, W.; Teichmann, H.; Tappeser, B.; Lotters, S. Questions concerning the potential impact of glyphosate-base herbicides on amphibians. *Environ. Toxicol. Chem.* **2013**, *32*, 1688–1700. [CrossRef]

44. Rodrigues, N.R.; de Souza, A.P.F. Occurrence of glyphosate and AMPA residue in soy-based infant formula sold in Brazil. *Food Addit. Contam. A* **2018**, *35*, 724–731. [CrossRef]

45. Matozzo, V.; Fabrello, J.; Marin, M.G. The effects of glyphosate and its commercial formulations to marine invertebrates: A review. *J. Mar. Sci. Eng.* **2020**, *8*, 399. [CrossRef]

46. Hildebrand, L.D.; Sullivan, D.S.; Sullivan, T.P. Experimental studies of rainbow trout populations exposed to field applications of Roundup®herbicide. *Arch. Environ. Contam. Toxicol.* **1982**, *11*, 93–98. [CrossRef]

47. Neskovic, N.K.; Poleksic, V.; Elezovic, I.; Karan, V.; Budimir, M. Biochemical and histopathological effects of glyphosate on carp, *Cyprinus carpio* L. *Bull. Environ. Contam. Toxicol.* **1996**, *56*, 295–302. [CrossRef] [PubMed]

48. Gholami-Seyedkolaei, S.; Mirvaghefi, A.; Farahmand, H.; Kosari, A.A. Effect of a glyphosate-based herbicide in *Cyprinus carpio*: Assessment of acetylcholinesterase activity, hematologigal responses and serum biochemical parameters. *Ecotoxicol. Environ. Saf.* **2013**, *98*, 135–141. [CrossRef] [PubMed]

49. Ma, J.; Bu, Y.; Li, X. Immunological and histopathological responses of the kidney of common carp (*Cyprinus carpio* L.) sublethally exposed to glyphosate. *Environ. Toxicol. Pharmacol.* **2015**, *39*, 1–8. [CrossRef] [PubMed]

50. Antunes, A.M.; Rocha, T.L.; Pires, F.S.; de Freitas, M.A.; Leite, V.R.M.C.; Arana, S.; Moreira, P.C.; Sabóia-Morais, S.M.T. Gender-specific histopathological response in guppies *Poecilia reticulata* exposed to glyphosate or its metabolite Aminomethylphosphonic acid. *J. Appl. Toxicol.* **2017**, *37*, 1098–1107. [CrossRef] [PubMed]

51. Kreutz, L.C.; Barcellos, L.J.G.; Silva, T.O.; Anziliero, D.; Martins, D.; Lorenson, M.; Martheninghe, A.; da Silva, L.B. Acute toxicity test of agricultural pesticides on silver catfish (*Rhamdia quelen*). *Cienc. Rural* **2008**, *38*, 1050–1055. [CrossRef]

52. Ayoola, S.O. Histopathological effects of glyphosate on juvenile African catfish (*Clarius gariepinus*). *Am-Euras. J. Agric. Environ. Sci.* **2008**, *4*, 362–367.

53. de Brito Rodrigues, L.; Costa, G.G.; Thá, E.L.; da Silva, L.R.; de Oliveira, R.; Leme, D.M.; Cestari, M.M.; Grisolia, C.K.; Valadares, M.C.; de Oliveira, G.A.R. Impact of the glyphosate-based commercial herbicide, its components and its metabolite AMPA on non-target aquatic organisms. *Mut. Res.-Genet. Toxicol. Environ. Mut.* **2019**, *842*, 94–101. [CrossRef]
54. Brodeur, J.C.; Malpel, S.; Anglesio, A.B.; Cristos, D.; D'Andrea, M.F.; Poliserpi, M.B. Toxicities of glyphosate- and cypermethrin-based pesticides are antagonic in the tenspotted livebearer fish (*Cnesterodon decemmaculatus*). *Chemosphere* **2016**, *155*, 429–435. [CrossRef]
55. Daam, M.A.; Moutinho, M.F.; Espíndola, E.L.G.; Schiesari, L. Lethal toxicity of the herbicides acetochlor, ametryn, glyphosate and metribuzin to tropical frog larvae. *Ecology* **2019**, *28*, 707–712. [CrossRef] [PubMed]
56. Howe, C.M.; Berrill, M.; Pauli, B.D.; Helbing, C.C.; Werry, K.; Veldhoen, N. Toxicity of Glyphosate-based pesticides to four north American frog species. *Environ. Toxicol. Chem.* **2004**, *23*, 1928–1938. [CrossRef] [PubMed]
57. Lajmanovich, R.C.; Attademo, A.M.; Peltzer, P.M.; Junges, C.M.; Cabagna, M.C. Toxicity of four herbicide formulations with glyphosate on *Rhinella arenarum* (Anura: Bufonidae) tadpoles: B-esterases and gluthation S-transferase inhibitors. *Arch. Environ. Contam. Toxicol.* **2011**, *60*, 681–689. [CrossRef] [PubMed]
58. Hong, Y.; Yang, X.; Yan, G.; Huang, Y.; Zuo, F.; Shen, Y.; Ding, Y.; Cheng, Y. Effects of glyphosate on immune responses and haemocyte DNA damage of Chinese mitten crab, *Eriocheir sinensis*. *Fish Shellfish Immunol.* **2017**, *71*, 19–27. [CrossRef] [PubMed]
59. Guilherme, S.; Santos, M.; Barroso, C.; Gaivão, I.; Mário, P. Differential genotoxicity of Roundup®formulation and its constituents in blood cells of fish (*Anguilla anguilla*): Considerations on chemical interactions and DNA damaging mechanisms. *Ecotoxicology* **2012**, *21*, 1381–1390. [CrossRef]
60. Ma, J.; Zhu, J.; Wang, W.; Ruan, P.; Rajeshkumar, S.; Li, X. Biochemical and molecular impacts of glyphosate-based herbicide on the gills of common carp. *Environm. Pollut.* **2019**, *252*, 1288–1300. [CrossRef]
61. Gaur, H.; Bhargava, A. Glyphosate induces toxicity and modulates calcium and NO signaling in zebrafish embryos. *Biochem. Bioph. Res. Comm.* **2019**, *513*, 1070–1075. [CrossRef]
62. Nwani, C.D.; Nagpure, N.S.; Kumar, R.; Kushwaha, B.; Lakra, W.S. DNA damage and oxidative stress modulatory effects of glyphosate-based herbicide in freshwater fish, *Channa punctatus*. *Environ. Toxicol. Pharmacol.* **2013**, *36*, 539–547. [CrossRef]
63. Menéndez-Helman, R.; Ferreyroa, G.V.; dos Santos Afonso, M.; Salibán, A. Glyphosate as an Acetylcholinesterase inhibitor in *Cnesterodon decemmaculatus*. *Bull. Environ. Contam. Toxicol.* **2012**, *88*, 6–9. [CrossRef]
64. Glusczak, L.; dos Santos Miron, D.; Crestani, M.; da Fonseca, M.B.; de Araújo Pedron, F.; Duarte, M.F.; Pimentel Vieira, V.L. Effect of glyphosate herbicide on acetylcholinesterase activity and metabolic and hematological parameters in piava (*Leporinus obtusidens*). *Ecotoxicol. Environ. Saf.* **2006**, *65*, 237–241. [CrossRef]
65. Salbego, J.; Pretto, A.; Gioda, C.R.; de Menezes, C.C.; Lazzari, R.; Neto, J.R.; Baldisserotto, B.; Loro, V.L. Herbicide formulation with glyphosate affects growth, acetylcholinesterase activity, and metabolic and hematological parameters in Piava (*Leporinus obtusidens*). *Arch. Environ. Contam. Toxicol.* **2010**, *58*, 740–745. [CrossRef]
66. Glusczak, L.; dos Santos Miron, D.; Moares, B.S.; Simoes, R.R.; Chitolina Schetinger, M.R.; Morsch, V.M.; Loro, V.L. Acute effects of glyphosate herbicide on metabolic and enzymatic parameters of silver catfish. *Comp. Biochem. Physiol. C* **2007**, *146*, 519–524. [CrossRef]
67. Kreutz, L.C.; Barcellos, L.J.G.; de Faria Valle, S.; de Oliveira Silva, T.; Anziliero, D.; dos Santos, E.D.; Pivato, M.; Zanatta, R. Altered hematological and immunological parameters in silver catfish fish (*Rhamdia quelen*) following short term exposure to sublethal concentration of glyphosate. *Fish Shellfish Immunol.* **2011**, *30*, 51–57. [CrossRef]
68. Persch, T.S.P.; Weimer, R.N.; Freitas, B.S.; Oliveira, G.T. Metabolic parameters and oxidative balance in juvenile (*Rhamdia quelen*) exposed to rice paddy herbicides: Roundup®, Primoleo®, and Facet®. *Chemosphere* **2017**, *174*, 98–109. [CrossRef]
69. Lushchak, O.V.; Kubrak, O.I.; Storey, J.M.; Storey, K.B.; Lushchak, V.I. Low toxic herbicide Roundup induces mild oxidative stress in goldfish tissues. *Chemosphere* **2009**, *76*, 932–937. [CrossRef] [PubMed]
70. Li, M.H.; Xu, L.D.; Liu, Y.; Chen, T.; Jiang, L.; Fu, Y.H.; Wang, J.S. Multi-tissue metabolic responses of goldfish (*Carassius auratus*) exposed to glyphosate-based herbicide. *Toxicol. Res.* **2016**, *5*, 1039–1052. [CrossRef]
71. Li, M.H.; Ruan, L.Y.; Zhou, J.W.; Fu, Y.H.; Jinag, L.; Zhao, H.; Wang, J.S. Metabolic profiling of goldfish (*Carassius auratis*) after long-term glyphosate-base herbicide exposure. *Aquatic. Toxicol.* **2017**, *188*, 159–169. [CrossRef] [PubMed]
72. Roy, N.M.; Carneiro, B.; Ochs, J. Glyphosate induces neurotoxicity in zebrafish. *Environ. Toxicol. Pharmacol.* **2016**, *42*, 45–54. [CrossRef] [PubMed]
73. Goulart, T.L.S.; Boyle, R.T.; Souza, M.M. Cytotoxicity of the association of pesticides Roundup Transorb® and Furadan 350 SC® on the zebrafish cell line, ZF-L. *Toxicol. Vitro.* **2015**, *29*, 1377–1384. [CrossRef]
74. Samanta, P.; Pal, S.; Mukherjee, A.K.; Glosh, A.R. Biochemical effects of glyphosate based herbicide, Excel Mera 71 on enzyme activities of acetylcholinesterase (AChE), lipid peroxidation (LPO), catalase (CAT), gluthation-S-transferase (GST) and protein content on teleostean fishes. *Ecotoxicol. Environ. Saf.* **2014**, *107*, 120–125. [CrossRef]
75. Le Du-Carrée, J.; Morin, T.; Danion, M. Impact of chronic exposure of rainbow trout, *Oncorhynchus mykiss*, to low doses of glyphosate or glyphosate-based herbicides. *Aquat. Toxicol.* **2021**, *230*, 105687. [CrossRef] [PubMed]
76. Zheng, T.; Jia, R.; Cao, L.; Du, J.; Gu, Z.; He, Q.; Xu, P.; Yin, G. Effects of chronic glyphosate exposure on antioxdative status, metabolism and immune response in tilapia (GIFT, *Oreochromis niloticus*). *Com. Biochem. Physiol. C* **2021**, *239*, 108878. [CrossRef] [PubMed]

77. Matozzo, V.; Munari, M.; Maseiro, L.; Finos, L.; Marin, M.G. Ecotoxicological hazard of a mixture of glyphosate and aminomethylphosphonic acid to the mussel *Mytilus galloprovincialis* (Lamarck 1819). *Sci. Rep.* **2019**, *9*, 14302. [CrossRef]
78. Sandrini, J.Z.; Rola, R.C.; Lopes, F.M.; Buffon, H.F.; Freitas, M.M.; Martins, C.M.; da Rosa, C.E. Effects of glyphosate on cholinesterase activity of the mussel *Perna perna* and the fish *Danio rerio* and *Jenynsia multidentata*: In vitro studies. *Aquat. Toxicol.* **2013**, *130–131*, 171–173. [CrossRef] [PubMed]
79. Matozzo, V.; Fabrello, J.; Masiero, L.; Ferraccioli, F.; Finos, L.; Pastore, P.; Di Gangi, I.M.; Bogialli, S. Ecotoxicological risk assessment for the herbicide glyphosate to non-target aquatic species: A case study with the mussel *Mytilus galloprovincialis*. *Environ. Pollut.* **2018**, *233*, 623–632. [CrossRef]
80. Iummato, M.M.; Di Fiori, E.; Sabatini, S.E.; Cacciatore, L.C.; Cochón, A.C.; del Carmen Ríos de Molina, M.; Juárez, Á.B. Evaluation of biochemical markers in the golden mussel *Limnoperna fortune* exposed to glyphosate acid in outdoor microcosms. *Ecotoxicol. Environ. Saf.* **2013**, *95*, 123–129. [CrossRef] [PubMed]
81. Di Fiori, E.; Pizarro, H.; dos Santos Afonso, M.; Cataldo, D. Impact of the invasive mussels *Limnoperna fortunei* on glyphosate concentration in water. *Ecotoxicol. Environ. Saf.* **2012**, *81*, 106–113. [CrossRef]
82. Séguin, A.; Mottier, A.; Perron, C.; Lebel, J.M.; Serpentini, A.; Costil, K. Sub-lethal effects of glyphosate-based commercial formulation and adjuvants on juvenile oysters (*Crassostrea gigas*) exposed for 35 days. *Mar. Pollut. Bull.* **2017**, *117*, 348–358. [CrossRef] [PubMed]
83. Contardo-Jara, V.; Klingelmann, E.; Wiegand, C. Bioaccumulation of glyphosate and its formulations Roundup Ultra in *Lumbriculus variegatus* and its effects on biotransformation and antioxidant enzymes. *Environ. Pollut.* **2009**, *157*, 57–63. [CrossRef]
84. Costa, M.J.; Monteiro, D.A.; Oliveira-Neto, A.L.; Rantin, T.F.; Kalinin, A.L. Oxidative stress biomarkers and heart function in bullfrog tadpoles exposed to Roundup Original. *Ecotoxicology* **2008**, *17*, 153–163. [CrossRef]
85. Barky, F.A.; Abdelsam, H.A.; Mahmoud, M.B.; Hamdi, S.A.H. Influence of Atrazine and Roundup pesticides on biochemical and molecular aspects of *Biomphalaria alexandrina* snails. *Pest. Biochem. Physiol.* **2012**, *104*, 9–18. [CrossRef]
86. Mohamed, A.H. Sublethal toxicity of Roundup to immunological and molecular aspects of *Biomphalaria alexandrina* to *Schistosoma mansoni* infection. *Ecotoxicol. Environ. Saf.* **2011**, *74*, 754–760. [CrossRef]
87. Levine, S.L.; von Mérey, G.; Minderhout, T.; Manson, P.; Sutton, P. Aminomethylphosphonic acid has low chronic toxicity to *Daphnia magna* and *Pimephales promelas*. *Environ. Toxicol. Chem.* **2015**, *34*, 1382–1389. [CrossRef] [PubMed]
88. Poppov, K.; Ronkkomaki, H.; Lajunens, L.H.J. Critical evaluation of stability constants of phosphonic acids. *Pure Appl. Chem.* **2001**, *73*, 1641–1677. [CrossRef]
89. Bonansea, R.I.; Filippi, I.; Wunderlin, D.A.; Marino, D.J.G.; Amé, M.V. The Fate of Glyphosate and AMPA in a Freshwater Endorheic Basin: An Ecotoxicological Risk Assessment. *Toxics* **2018**, *6*, 3. [CrossRef]
90. Battaglin, W.A.; Koplin, D.W.; Scribner, E.A.; Kuivila, K.M.; Sandstorm, M.W. Glyphosate, other herbicides, and transformation products in midwestern streams, 2002. *J. Am. Water Resour. Assoc.* **2005**, *41*, 323–332. [CrossRef]
91. Battaglin, W.A.; Rice, K.C.; Focazio, M.J.; Salmons, S.; Barry, R.X. The occurrence of glyphosate, atrazine, and other pesticides in vernal pool and adjacent streams in Washington, DC, Maryland, Iowa, and Wyoming, 2005–2006. *Environ. Monitor. Assess.* **2009**, *155*, 281–307. [CrossRef]
92. Directive 2000/60/EC of the European Parliament and of the Council. 23 October 2000 Establishing a Framework for Community Action in the Field of Water Policy. Available online: https://eur-lex.europa.eu/legal-content/CS/ALL/?uri=CELEX%3A32000L0060 (accessed on 22 December 2020).
93. Aparicio, V.C.; De Gerónimo, E.; Marino, D.; Primost, J.; Carriquiriborde, P.; Costa, J.L. Environmental fate of glyphosate and aminomethylphosphonic acid in surface waters and soil of agricultural basins. *Chemosphere* **2013**, *93*, 1866–1873. [CrossRef]
94. Di Poi, C.; Costil, K.; Bouchart, V.; Harm-Lemeille, M.P. Toxicity assessment of five emerging pollutants. Alone and in binary or ternary mixtures, towards three aquatic organisms. *Environ. Sci. Pollut.* **2018**, *25*, 6122–6134. [CrossRef]
95. Tajnaiová, L.; Vurm, R.; Kholomyeva, M.; Kobera, M.; Koci, V. Determination of the ecotoxicity of herbicides Roundup®Classis Pro and Garlon New in aquatic and terrestrial environments. *Plants* **2020**, *9*, 1203. [CrossRef]
96. EFSA (European Food and Safety Authority). Conclusion on the peer review of the pesticide risk assessment of the active substance glyphosate. *EFSA J.* **2015**, *13*, 4302. [CrossRef]
97. Reddy, K.N.; Rimando, A.M.; Duke, S.O. Aminomethylphosphonic acid, a metabolite of glyphosate, cases injury in glyphosate-treated, glyphosate-resistant soybean. *J. Agric. Food Chem.* **2004**, *52*, 5139–5143. [CrossRef] [PubMed]
98. Gomez, M.P.; Smedbol, E.; Chalifour, A.; Hénault-Ethier, L.; Labrecque, M.; Lepage, L.; Lucotte, M.; Juneau, P. Alteration of plant physiology by glyphosate and it by-product Aminomethylphosphonic acid: An overview. *J. Exp. Bot.* **2014**, *65*, 4691–4703. [CrossRef] [PubMed]
99. Serra, A.A.; Nuttens, A.; Larvor, V.; Renault, D.; Couée, I.; Sulmon, C.; Gouesbet, G. Low environmentally relevant levels of bioactive xenobiotics and associated degradation products cause cryptic perturbations of metabolism and molecular stress responses in Arabidopsis thaliana. *J. Exp. Bot.* **2013**, *64*, 2753–2766. [CrossRef]
100. Guilherme, S.; Santos, M.A.; Gaivao, I.; Pacheco, M. DNA and chromosomal damage induced in fish (*Anguilla anguilla* L.) by aminomethylphosphonic acid (AMPA)—The major environmental breakdown product of glyphosate. *Environ. Sci. Pollut. Res.* **2014**, *21*, 8730–8739. [CrossRef] [PubMed]

101. Matozzo, V.; Marin, M.G.; Maseiro, L.; Tremonti, M.; Biamonte, S.; Viale, S.; Finos, L.; Lovato, G.; Pastore, P.; Bogialli, S. Effects of aminomethylphosphonic acid, the main breakdown product of glyphosate, on cellular and biochemical parameters of the mussel *Mytilus galloprovincialis*. *Fish Shellfish Immunol.* **2018**, *83*, 321–329. [CrossRef]
102. Cheron, M.; Brischoux, F. Aminomethylphosphonic acid alters amphibian embryonic development at environmental concentrations. *Environ. Res.* **2020**, *190*, 109944. [CrossRef]
103. Commission Implementing Regulation (EU) 2017/2324. Renewing the Approval of the Active Substance Glyphosate in Accordance with Regulation (EC) No 1107/2009 of the European Parliament and of the Council Concerning the Placing of Plant Protection Products on the Market, and Amending the Annex to Commission Implementing Regulation (EU) No 540/2011. OJEU, L333/10. 2017. Available online: https://eur-lex.europa.eu/legal-content/EN/TXT/?uri=CELEX%3A32017R2324 (accessed on 19 September 2021).

 *applied sciences*

The Nasal Epithelium as a Route of Infection and Clinical Signs Changes, in Rainbow Trout (*Oncorhynchus mykiss*) Fingerlings Infected with *Aeromonas* spp.

Fabián Ricardo Gómez de Anda [1], Vicente Vega-Sánchez [1], Nydia Edith Reyes-Rodríguez [1], Víctor Manuel Martínez-Juárez [1], Juan Carlos Ángeles-Hernández [1], Ismael Acosta-Rodríguez [2], Rafael German Campos-Montiel [1,*] and Andrea Paloma Zepeda-Velázquez [1,*]

[1] Instituto de Ciencias Agropecuarias, Universidad Autónoma del Estado de Hidalgo, Av. Universidad km 1. Ex-Hda. de Aquetzalpa A.P. 32 CP., Tulancingo 43600, Hidalgo, Mexico; fabian_gomez9891@uaeh.edu.mx (F.R.G.d.A.); vicente_vega11156@uaeh.edu.mx (V.V.-S.); nydia_reyes@uaeh.edu.mx (N.E.R.-R.); victormj@uaeh.edu.mx (V.M.M.-J.); juan_angeles@uaeh.edu.mx (J.C.Á.-H.)

[2] Facultad de Ciencias Químicas, Universidad Autónoma de San Luis Potosí, Av. Dr. Manuel Nava, No. 6. Zona Universitaria. C.P., San Luis Potosí 78320, Mexico; iacosta@uaslp.mx

* Correspondence: rcampos@uaeh.edu.mx (R.G.C.-M.); andrea_zepeda@uaeh.edu.mx (A.P.Z.-V.); Tel.: +52-72-2247-8295 (R.G.C.-M. & A.P.Z.-V.)

Citation: Gómez de Anda, F.R.; Vega-Sánchez, V.; Reyes-Rodríguez, N.E.; Martínez-Juárez, V.M.; Ángeles-Hernández, J.C.; Acosta-Rodríguez, I.; Campos-Montiel, R.G.; Zepeda-Velázquez, A.P. The Nasal Epithelium as a Route of Infection and Clinical Signs Changes, in Rainbow Trout (*Oncorhynchus mykiss*) Fingerlings Infected with *Aeromonas* spp.. *Appl. Sci.* **2021**, *11*, 9159. https://doi.org/10.3390/app11199159

Academic Editors: Panagiotis Berillis and Božidar Rašković

Received: 25 August 2021
Accepted: 28 September 2021
Published: 1 October 2021

Abstract: The genus *Aeromonas* is a group of bacteria that is widely distributed in water bodies and belongs to the normal intestinal microbiota of aquatic and terrestrial animals. In the present work, rainbow trout fingerlings were experimentally infected by an immersion bath with different *Aeromonas* species. Subsequently, the behavior of the infected groups was observed and recorded. Infected fingerlings were evaluated by histopathology. The highest percentages of hyperpigmentation (18.88%) and inappetence (47.7%) were observed in fish infected with *A. salmonicida*, while abnormal swimming (83.33%) was recorded in fish infected with *A. bestiarum*. In histopathological findings, the highest percentages were observed in the olfactory epithelium (50.0%) for *A. lusitana* and *A. salmonicida* (41.1%)-infected fish. While, in the nervous system, the cerebral hemispheres (31.1%) in *A. media*-infected fish and the oblongata medulla (40.0%) in the *A. bestiarum*-infected fish presented the highest percentages. Meanwhile, *A. salmonicida* and *A. bestiarum* have the highest pathogenicity and virulence based on the histopathological findings in the olfactory epithelium and nervous system. Due to the proximity of the olfactory epithelium with the nervous tissue, it is possible that the infection generated by the *Aeromonas* species and the histopathological findings in the nervous tissue are reflected in different behavioral changes that suggest differences in the pathogenicity and virulence of the bacteria.

Keywords: *Aeromonas* spp.; rainbow trout; histopathology; bacteria; infection

1. Introduction

The genus *Aeromonas* is a group of bacteria of importance to aquaculture, which is frequently isolated in aquatic environments [1,2]. They can act as important pathogens in poikilotherm species because they are widely distributed in water bodies [3], and they are part of the normal gut microbiota of aquatic and terrestrial animals [4]. They have zoonotic potential, high pathogenic potential, and the ability to adapt to different environments [5]. Currently, 36 genetically identified *Aeromonas* species are known and isolated from aquatic environments [6]. This genus is divided into two groups: an immobile group, with *A. salmonicida* being the main species, and a mobile group that contains the rest of the *Aeromonas* species [5]. *A. bestiarum, A. caviae, A. hydrophila, A. piscicola, A. salmonicida,* and *A. veronii* are reported to be the species that are mostly isolated from pathological conditions in salmonids [7]. *Aeromonas* infection in different species of fish can occur at

69

different ages of those fish that are susceptible to the disease, the presentation of stress in the fish is an important factor for the development of the disease, however the participation of external factors such as changes in water temperature, large amounts of excreta in ponds, overcrowding, and others can favor the presentation of the disease [3,5].

The pathology associated with the infection of *A. Salmonicida*, *A. hydrophila*, *A. bestiarum*, and *A. veronii* in rainbow trout includes septic processes, the colonization of liver tissue and mortality [8], fish can develop exophthalmos, ocular ulceration, anorexia, the clinical signs included dermis lesion (ulcers, erosion, hemorrhage, darkened skin, loss of scales, furunculosis, and others), abdominal distension, fin rot, and in some cases, premaxilla bone exposition can be observed [1,8–10]. Infected young fish modify their behavior, including jumping out of the pond, the darkening of the skin, abnormal swimming, low swimming, lethargy and inappetence [1,5,11]. Sometimes, bloody discharge from the nostrils and death occurs [12]. In the fish into the nasal cavity, the olfactory organ or olfactory rosette, or just rosette is a chemosensory system, which are composed by pseudostratified epithelium with sensory (olfactory sensory neurons) and mucosal areas (goblet cells), on the lamina propria of the rosette are localized the terminal axons [13,14].

However, studies focused on nasal infection in *Aeromonas* have not been investigated. Bartkova et al. [15] carried out an infection in rainbow trout intraperitoneally with *A. salmonicida* and through real-time monitoring by bioluminescence, it was established that the colonization sites, were dorsal and pectoral fins; gills, stomach, intestine, anus; as well as the oral, nasal and eye cavity, these sites being the first to be exposed to the aquatic environment, as well as the skin. Currently, the immunization of fish against *Aeromonas* is carried out by intraperitoneal injection or by immersion bath [6]. Recently, the nasal route of *Yersinia ruckeri* was demonstrated as an entry point for bacteria to establish, infect and spread through the nervous system, which in turn causes erratic swimming and exophthalmia [16]. This suggests that pathogenic and virulent bacteria can cause nervous problems that affect behavior. On the other hand, the use of the nasal route as a vaccination route was investigated as an alternative for the immunization in teleost fish, on the case of rainbow trout, it was used as a biological model for the challenge with infectious hematopoietic necrosis virus [13].

The aim of the present research was to identify the presence of histopathological lesions observed in the rosette and nervous tissue, to identify a relation between the nasal tissue and nervous tissue with fish behavior, in rainbow trout fingerlings, experimentally infected with different *Aeromonas* species.

2. Materials and Methods

2.1. Ethics Statement

This experimental infection was approved by the Bioethics Commission and Animal Welfare Committee (Comisión de Bioética y Bienestar animal, CBBA) of the Facultad de Medicina Veterinaria y Zootecnia (FMVZ), Universidad Autónoma del Estado de México (UAEMex) 012-2012. A total of 1200 rainbow trout were obtained from a trout farm (El Zarco, Ocoyoacac, Mexico), previously reported by Zepeda-Velázquez and colleagues [8].

2.2. Animals and Husbandry

For the acclimation period, 1200 fish (3–5 cm and 2–3 g of weight) were maintained in 100-L (L) containers, as previously reported by Zepeda-Velázquez and colleagues [8]. A total of 120 fish, 30 for the control group and 90 for the three replicas of each experimental infection, were randomly distributed in a 6 L glass aquarium equipped with aeration (air pump ELITE 799), air stones positioned at opposite front ends of the aquarium and fresh water without chlorine. An immersion bath with 10 different *Aeromonas* species was performed at a bacterial concentration of 1.5×10^8 colony-forming units (CFU)/milliliter (mL) for 1 h in a 6 L glass aquarium with aeration; all groups were immersed independently [8]. For the maintenance and propagation of bacterial isolates, Farto and their colleagues' methodology was used [17]. After the immersion bath, rainbow trout fingerlings were put in a new 6 L glass aquarium with the same characteristics as mentioned above, and all glass aquarium were isolated by a plastic separation, to observe independent behavior between fishes in glass aquarium. The maintenance of the glass aquarium included 40% water replacement, feeding, and elimination of biological sediment, including feces and food rest. Fish were observed daily, and moribund fish and those with clinical signs were euthanized by an overdose of tricaine methane sulfonate (MS222; Finquel) [12].

2.3. Behavioral Observations and Clinical Signs

The evaluated behaviors are listed in Table 1. The experimental period lasted a total of 14 days [8]. For the first 5 h post-infection (pi) and all 14 days, all fish were observed for 10–15 min [18], and the number of fish that present behavioral modification were counted and register before maintenance. Moribund fish and those with severe clinical signs were immediately euthanized [8].

Table 1. *Aeromonas* species and histopathological lesions observed in nasal epithelium or rosette, telencephalon, and oblongata medulla changes (%), observed in rainbow trout fingerlings infected with *Aeromonas* spp.

Behavior or Clinical Sign	Definition	Reference(s)
Loss of appetite	Fish refuse to eat and/or lose interest in food (anorexia)	[12,17,19–21]
Abnormal swimming (Erratic swimming)	Loss of consistency, regularity and/or uniformity in movement; fish swimming in different directions	[17,22]
Hyperpigmentation and/or darkened skin	Fingerlings show an excess of melanin pigmentation in the skin, giving them a darker color	[1,9,12,23]

2.4. Histopathological Studies

Fish were fixed in 10% (volume/volume) neutral buffered formalin (pH 7.4), embedded in paraffin, and stained with hematoxylin and eosin [8,10].

2.5. Statistical Analysis

Data were analyzed with SPSS software (version 21, IBM Inc. Chicago, IL, USA). Differences in the percentages of behaviors, clinical signs and histopathological lesions in the rosette, telencephalon and medulla oblongata were analyzed using a Chi-square test (significance at $p < 0.05$) [18]. Differences between isolates of each behavior and clinical sign were determined using Tukey's multiple comparison tests, with a significance level of $p < 0.05$ [24].

Two independent correspondence analyses (CA) were implemented with the aim of establishing whether histopathological lesions and behavioral changes depend on *Aeromonas* species through a graphical display of their relationship. The CA were carried out in the package FactoMineR version 2.3 [25], of the statistical program R version 4.0.2. (R core team, 2016) [26] and its graphical representation was through a biplot drawn in the FactoExtra package [27]. The biplot approximates the distribution of a multivariate sample in a generally two-dimensional reduced-dimension space, where angles and proximity among variables provide a measure of association.

3. Results
3.1. Behaviors and Clinical Signs

The experimental infection process was separated from the 5-hour feed period for all groups. Fish infected with *A. salmonicida*, *A. veronii*, *A. bestiarum*, *A. hydrophila*, *A. caviae* and *A. sobria* showed at least one change in the evaluated behaviors or clinical sign. The presentation of darkening of the skin in fish infected with *A. salmonicida* (18.8%), *A. veronii* (15.5%) and, *A. hydrophila* (11.1%) was observed with statistical differences; meanwhile fingerlings infected with *A. allosaccharophila*, *A. bestiarum*, *A. caviae*, *A. lusitana*, *A. media*, *A. popoffii*, *A. sobria*, and the control group, did not present color changes ($p > 0.05$) (Table 1) (Figure 1a).

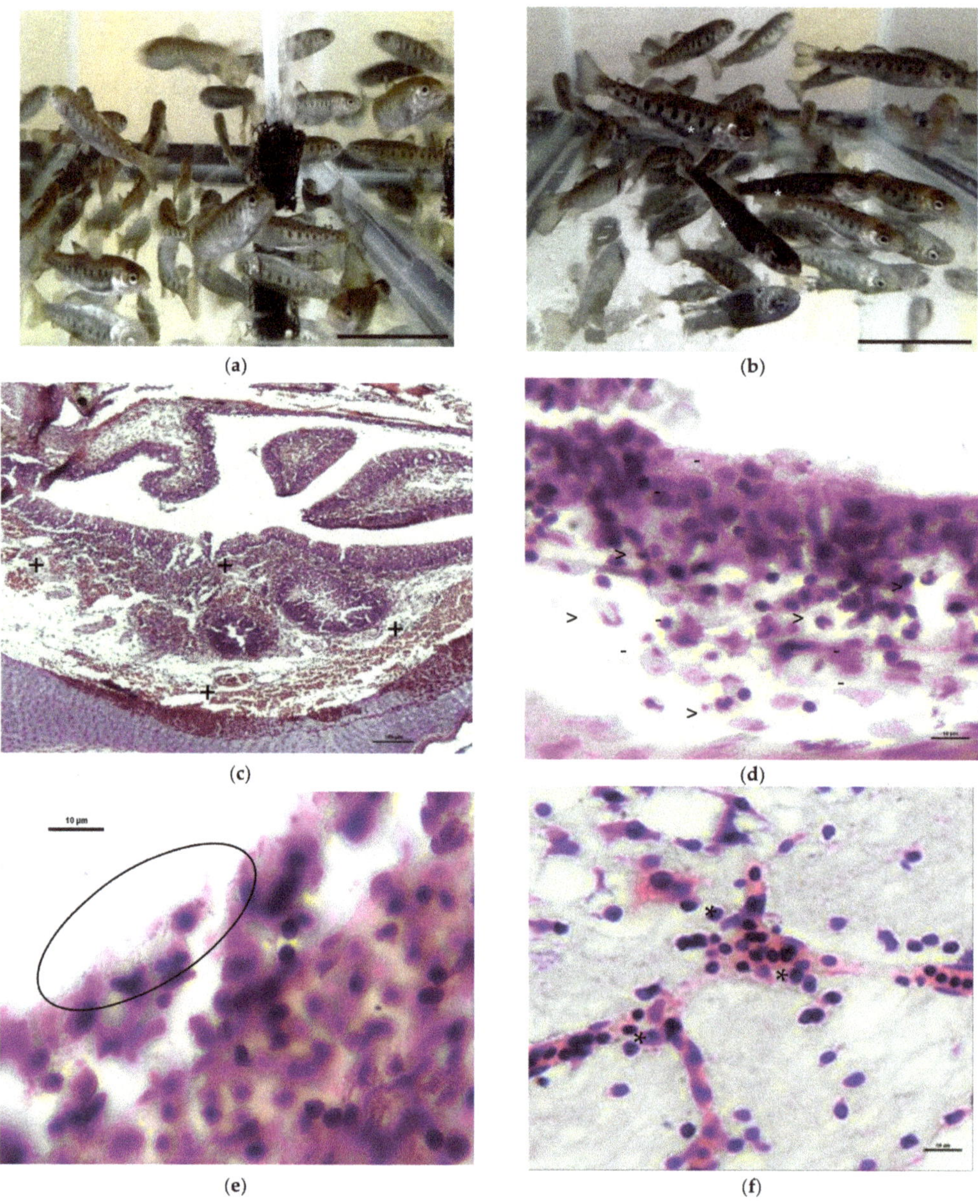

Figure 1. Rainbow trout fingerlings experimental infected by immersion with different *Aeromonas* species. (**a**) Fingerlings control group. Bar = 1 cm.; (**b**) Fingerlings present different darkening skin grades (asterisk), infected with *A. salmonicida*. Bar = 1 cm.; (**c**) nasal epithelium, moderate edematous and severe hemorrhagic rhinitis (plus sign) on fingerlings infected with *A. lusitana*; (**d**) nasal epithelium, suppurative rhinitis in fingerlings infected with *A. salmonicida*, neutrophil (greater-than sing) and macrophage (hyphen-minus) infiltration were observed; (**e**) nasal epithelium with bacterial shapes on the surface (circle), in fingerlings infected with *A. salmonicida*; and, (**f**) oblongata medulla, slight focal myelitis (asterisk) in fingerlings infected with *A. hydrophila*.

The lack of appetite or inappetence in fingerlings infected with *A. salmonicida* (47.7%), *A. hydrophila* and *A. sobria* (46.6%), *A. caviae* (38.8%) and, *A. bestiarum* (32.2%), was observed with a statistical difference ($p > 0.05$) for *A. allosaccharophila, A. lusitana, A. media, A. popoffii, A. veronii,* and the control group, who showed no changes in their feeding behavior or feeding regulation. Finally, fingerlings infected with *A. bestiarum* (83.3%), *A. hydrophila* (74.4%), *A. sobria* (67.7%), *A. salmonicida* (66.6%) and, *A. caviae* showed erratic behavior changes, a loss of consistency and a loss of swimming uniformity. No significant statistical differences ($p > 0.05$) were noted for the *A. allosaccharophila, A. lusitana, A. media, A. popoffii* and *A. veronii* groups or the control group (Table 2).

Table 2. Percentages of clinical signs or behavioral changes, present in fingerlings experimentally infected with different *Aeromonas* species.

Aeromonas Species	N	% of Fish with Clinical Sign or Behavioral Changes [1]		
		Darkened Skin/Hyperpigmentation	Lack of Appetite/Inappetence	Abnormal Swimming
A. allosaccharophila	90	0	1.11	25.55
A. bestiarum	90	6.66	32.22	83.33
A. caviae	90	0	38.88	60
A. hydrophila	90	11.11	46.66	74.44
A. lusitana	90	0	1.11	34.44
A. media	90	1.11	16.66	25.55
A. popoffii	90	0	3.33	31.11
A. salmonicida	90	18.88	47.7	66.66
A. sobria	90	1.11	46.66	67.77
A. veronii	90	15.55	28.88	38.88
Control group	90	0	0	8.88

[1] Groups in italics within the same column are significantly different ($p < 0.05$).

3.2. Histopathological Lesions in Nasal and Nervous Tissue

In the nasal epithelium, a significant number of fingerlings infected with *A. lusitana* (50%) and *A. salmonicida* (41.1%) showed different histopathological lesions. Congestion with different distribution patterns and high severity grades, focal hemorrhagic epithelium (Figure 1b), moderate multifocal epithelium and hyperplasia of the nasal epithelium and goblet cells, as well as a leukocyte infiltrate of mononuclear and polymorphonuclear cells, were observed (Figure 1c). On the other hand, the histopathological finds of *A. media* (35.5%), *A. allosaccharophila* (30%), *A. caviae* (28.8%), *A. hydrophila* (27.7%), *A. bestiarum* (26.6%), *A. sobria* (20%), *A. veronii* (11.1%), and *A. popoffii* (7.7%), present middle-to-less severity and distribution; edema, congestion, goblet cell hyperplasia and, leucocyte infiltration in different grades and, different distributions in the nasal epithelium were observed, with no significant histopathological lesions ($p > 0.05$). Histopathological lesions in control group were not observed (Table 3). Basophilic aggregates or bacterial aggregations were observed in all fishes experimentally infected with the different *Aeromonas* species. Basophilic aggregates or bacterial aggregations in the control group was not observed.

The main histopathological lesion in the telencephalon consisted of severe multifocal congestion for fish infected with *A. media* (31.1%), and a moderate degree for the groups infected by *A. salmonicida* (25.5%), and *A. caviae* (22.2%). The congestion was observed in focal and multifocal presentation, and also hemorrhagic lesions were present on focal and bifocal distribution, in moderate-to-slight presentation in *A. bestiarum* (21.1%), *A. hydrophila* and *A. lusitana* (18.8%). No significant histopathological changes were observed in *A. veronii* (10%), *A. sobria* (7.7%), *A. allosaccharophila*, and *A. popoffii* (3.3%), and control groups (Table 3).

Table 3. *Aeromonas* species and histopathological lesions observed in nasal epithelium or rosette, telencephalon, and oblongata medulla changes (%), observed in rainbow trout fingerlings infected with different *Aeromonas* species.

Aeromonas Species	N	% of Fish with Histopathological Lesions [1]		
		Rosette	Telencephalon	Oblongata Medulla
A. allosaccharophila	90	30	3.3	1.1
A. bestiarum	90	26.6	*21.1*	*40*
A. caviae	90	28.8	22.2	16.6
A. hydrophila	90	27.7	*18.8*	22.2
A. lusitana	90	*50*	*18.8*	22.2
A. media	90	35.5	*31.1*	22.2
A. popoffii	90	7.7	3.3	4.4
A. salmonicida	90	*41.1*	25.5	15.5
A. sobria	90	20	7.7	12.2
A. veronii	90	11.1	10	17.7
Control group	90	0	0	0

[1] Groups in italics within the same column are significantly different ($p < 0.05$).

On the other hand, the histopathological findings in the oblongata medulla were focal and multifocal congestion, and moderate-to-severe degrees of infection with *A. bestiarum* (40%). No significant histopathological lesions ($p > 0.05$) were observed for the groups infected with the species *A. hydrophila*, *A. lusitana*, and *A. media*, *A. veronii, A. caviae, A. salmonicida, A. sobria, A. popoffii,* and *A. allosaccharophila* (Table 3). The fingerling group infected with *A. salmonicida* presented a mild degree and focal congestion with perivascular mononuclear infiltrate were identified in one fingerling (Figure 1d), but basophilic aggregates were not observed. The percentages of bacterial isolation were obtained from the region in cerebellum and medulla oblongata. Fingerlings were infected with *A. bestiarum* (0.90%), *A. veronii* and *A. media* (0.80%), *A. lusitana* and *A. popoffii* (0.60%), *A. salmonicida* (0.30%), *A. hydrophila* (0.10%), but were not statistically representative; bacterial isolation of *A. allosaccharophila, A. caviae, A. sobria* and the control group were not obtained (data not shown).

3.3. Correspondence Analysis

To identify the associations between the *Aeromonas* species and histopathological lesions and behavioral changes two correspondence analysis (CA) were implemented. CA is an exploratory analysis that reduces the complexity of the initial database while retaining most information and providing the graphical output's relationships, favoring their interpretation. Table 4 depicts the results of X^2 analysis, which reveals a strong dependency ($p < 0.001$) between *Aeromonas* species with histopathological lesions and behavioral changes.

Table 4. Results of X^2 analysis of histopathological lesions and behavioral changes observed in rainbow trout fingerlings infected with several *Aeromonas* species.

Item	Histopathological Lesions	Behavioral Changes
X^2 (observed value)	68.20	123.64
X^2 (critical value)	31.41	31.41
DF	20	20
p-value	0.0001	0.0001
α	0.05	0.05

To the histopathological lesions CA, the first and second dimensions account for 79.94 and 20.05 percent of inertia, respectively; both dimensions represent for 100 % of cumulative inertia, indicating that the totality of variability was explained by the information contained within the first two axis (Table 2). The CA focused on behavioral changes associates with

Aeromonas species reveals that the two dimensions provides an adequate description of data. The first dimension accounts for 71.17 % and, for the second dimension, 28.82 % of inertia, together accounting for 100 % of the variability of analyzed data (Table 5).

Table 5. Dimension inertia of correspondence analysis of histopathological lesions and behavioral changes observed in rainbow trout fingerlings infected with several *Aeromonas* species.

Item	Histopathological Lesions	
Eigen value	0.079	0.020
Inertia (%)	79.94	20.05
Cumulative (%)	79.94	100.00
Item	**Behavioral changes**	
	Dimension 1	Dimension 2
Eigen value	0.095	0.038
Inertia (%)	71.17	28.82
Cumulative (%)	71.17	100.00

Figure 2a shows CA maps of histological lesions and the quality of representation in the row and columns of the first dimension. The most of *Aeromonas* species are well represented, with the exception of *A. popoffii*, *A. sobria*, and *A. hydrophila*. With respect to the site of lesions, the telencephalon category showed a low quality of representation; therefore, the first dimension explains most of the deviation from the expected values of categories oblongata medulla and nasal epithelium. This means that the variability of first component can be explained by differences between oblongata medulla and nasal epithelium. According to the level of contribution of column categories of CA map of behavioral changes, the first dimension can be explained by a negative correlation among *A. veronii* and *A. salmonicida* (allocated in negative pole) with *A. lusitana* and *A. popoffii* (allocated in positive pole). On other hand, *A. hydrophila* and *A. bestiarum* showed a poor representation, had a lower contribution, and was allocated near to the centroid. The contribution of row to the definition of the first dimension is shown in Figure 2b. The category hyperpigmentation shows the highest contribution to define the first dimension. However, inertia of databases can be better explained by the opposition of hyperpigmentation and inappetence of the second dimension. Finally, swimming abnormal behavior showed the lowest contribution, little representation on the AC map, and was close to centroid.

The higher proportion of inertia explained by the first two dimensions in both CAs allows the proper interpretation of biplots shown in the Figure 2a and b. In the biplot of histopathological lesion CA, we see that *A. popoffii* and *A. hydrophila* are close to the origin point (Figure 3a). Therefore, we can conclude that these *Aeromonas* species cannot be different based on data from our study. Oblongata medulla and telencephalon are away from the origin; hence, these are good discriminators between *Aeromonas* species. *A. bestiarum* and *A. veronii* species are close to oblongata medulla lesions and the angle formed for these categories is very small, which means that they are associated. According to the CA biplot, lesions on telencephalon are associated with *A. media* and *A. caviae* species. Based on the angle formed by olfactory epithelium and *A. allosaccharophila*, we can conclude that they are closely associated. Lower strength of association is depicted between olfactory epithelium lesions and control and *A. lusitana* species. *A. salmonicida* and *A. sobria* species did not show a particular site of lesions based on analyzed data. And finally, the Figure 3b showed the biplot of CA focused on behavioral changes, which displays that changes of behavior are clearly separated. Hyperpigmentation and inappetence are located far to the centroid and can be a good discriminator between *Aeromonas* species. This means that most *Aeromonas* species can be grouped according to the type of affection on behavior. *A. veronii* and *A. salmonicida* are located on the left-hand side of the graph and according to their angles are strongly associated with hyperpigmentation changes. Abnormal swimming is located on the right-side of biplot and their manifestation in infected animals is considerably associated with *A. lusitana*, *A. allosaccharophila*, and *A.*

popoffii species. Inappetence changes showed an adequate representation to be far to the centroid; however, the exhibiting of these changes was not associated with a specific *Aeromonas* species, with a slight relationship with *A. media*, *A. caviae*, and *A. sobria* species.

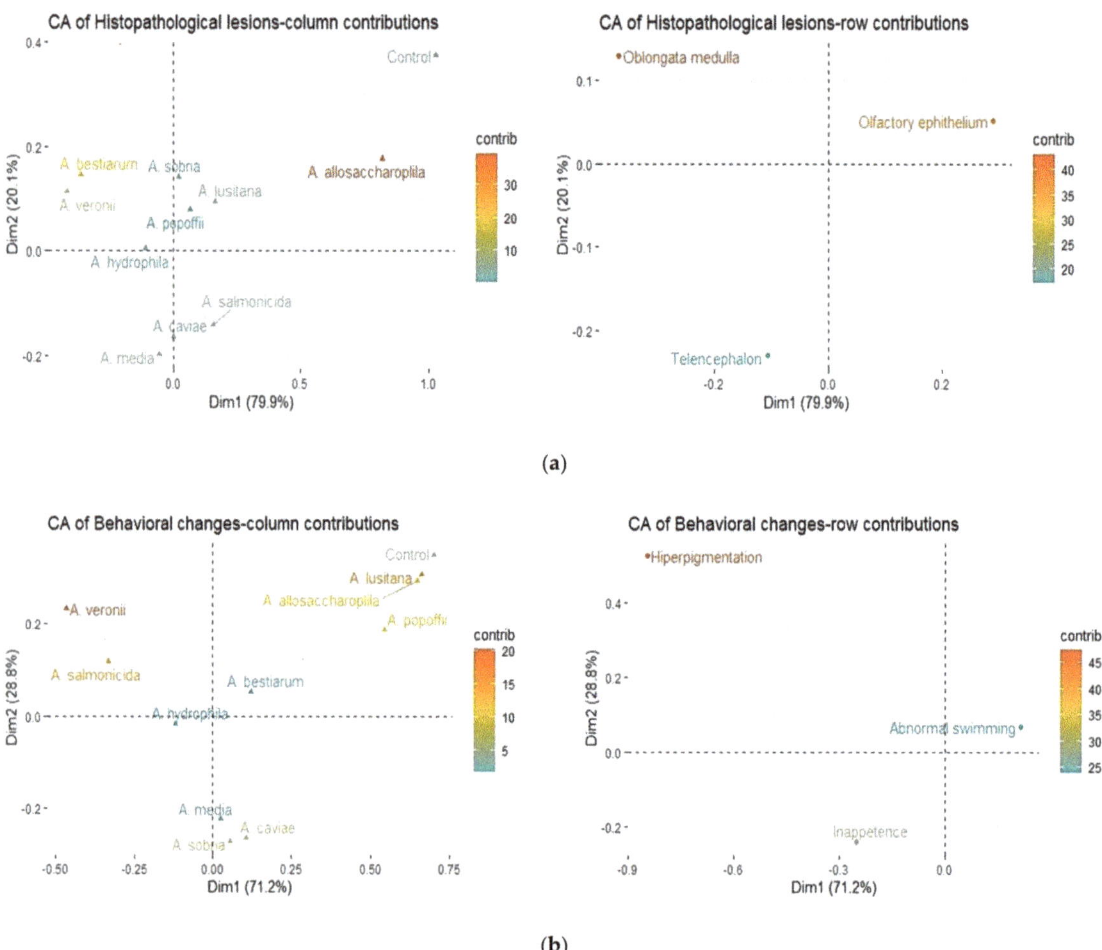

Figure 2. Correspondence analysis map. (**a**) Histopathological lesions and quality of representation to row and columns of dimen-sion 1; (**b**) behavioral changes and quality of representation to row and columns of dimension 1. The row and column categories are color coded according with the contribution to dimension 1.

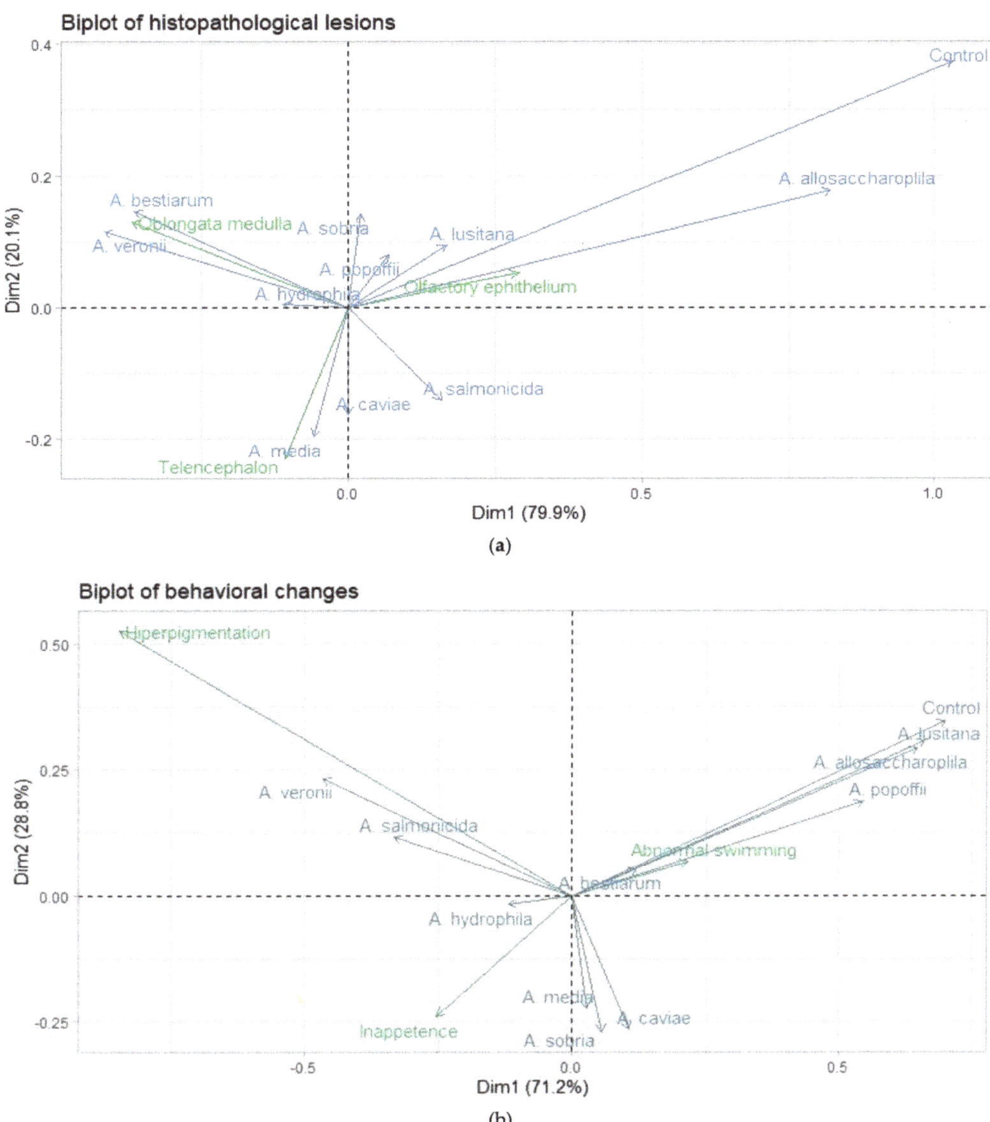

Figure 3. Bioplot analysis map. (**a**) Histopathological tissue localization and the histopathological lesions produce by different *Aeromonas* species in rainbow trout fingerlings of representation to row and columns of dimension 1; (**b**) behavioral changes observed in fingerlings experimentally infected with different *Aeromonas* species and quality of representation to row and columns of dimension 1.

4. Discussion

The genus *Aeromonas* is a group of bacteria that are widely distributed in different ecosystems, such as different types of food, animals (dogs, cats, horses, fish, mollusks, among others), infections in humans (gastritis, skin infections, bacteremia/septicemia processes, among others) and aquatic [6]. It is well known that the development of disease, and the clinical manifestations in a living organism, depend mainly on the susceptibility and immunity of the host, the virulence of the pathogenic agent and the environmental conditions, which result in an alteration of the homeostasis of the host and, therefore, the state of disease [22].

The nasal route of fish, also known as the rosette, is composed of pseudostratified epithelium, which contains mucous portions, sensory and vascular tissue; and it´s one of the entry routes that different pathogens can use to infect and colonize the host and that has the ability to establish an innate immune response [13].

In the experimental infection of fingerlings with different *Aeromonas* species, the presence of histological lesions in the rosette were observed, mainly of the vascular type, identifying hemorrhages with different degrees of severity and distribution, as well as goblet cell hyperplasia in different degrees of severity and polymorphonuclear leucocyte infiltrates. This acute inflammatory response was observed in fish infected with the species of *A. lusitana*, *A. allosaccharophila*, *A. popoffii*, and *A. sobria*, these species being the ones that caused severe histological lesions. On the other hand, the fish infected with *A. bestiarum*, *A. caviae*, *A. hydrophila*, *A. media*, and *A. salmonicida*, the lesions were slight; the control group did not present histopathological lesions.

Salinas et al., [14] observed that the administration of a nasal vaccine against infectious hematopoietic necrosis and enteric red mouth disease, inducing histological changes in nasal tissue, identifying infiltrates of leukocytes in different severity degrees and the presence of vascular changes, related to the inflammatory process, without the presentation of hemorrhages.

The histological findings present in this work were consistent with Lapatra et al. [13], who carried out the nasal infection of the infectious hematopoietic necrosis and enteric red mouth disease, infecting each nostril separately, to implement a vaccination strategy, goblet cell hyperplasia, leukocyte infiltrate and vascular changes were observed in the nasal tissue, which consisted in the formation of new tissue.

It was reported that the bacterium needs a second site of infection, the presentation of meningitis and encephalitis in salmonid species infected with *Renibacterium salmoninarum* is localized to sites with greater blood supply, including the heart, reticuloendothelial organ, and encephalon, as observed in asymptomatic fish [28]. The presentation of hemorrhages and the inflammatory response in the rosette, against the experimental infection by immersion with different *Aeromonas* species, demonstrated that it i not only the vascular change derived from the inflammatory response, but also of the capacity of each *Aeromonas* species to possess its own pathogenicity and virulence capacity.

The olfactory system of all vertebrates is made up of sensory neurons; in fish, the rosette is attached to the olfactory bulb by terminal axons, which project from the olfactory tract to different parts of the brain [29]. In the present work, the olfactory bulb was not analyzed histologically due to the size of the trout fingerlings. Ye et al. [30] identified that the olfactory bulb has an important role in the detection of pathogens, as well as in the regulation of the immune response in addition to identifying the genes that are responsible for the secretion of inter-leucine (IL) 1β and IL-8, present during the inflammation process. Song et al. [31] identified that part of the inflammatory response due to *A. hydrophila* infection in the intestine is closely related to the expression of IL8 and IL1β levels. This suggests that infection with *Aeromonas* species can infect the nasal route, and the nasal route has an important role in the early stages of infection and triggers the immune response, however more studies are needed to accept this hypothesis.

The production of vaccines in the aquaculture industry is focused on the stimulation of humoral immunity, by stimulating the mucosal immune response present in the mucosal tissue distributed in the body, while the intranasal application of vaccines in fish has generated, step by step, a fertile field as an alternative route of immunization in aquaculture [14]. In the case of *Aeromonas* spp., active and inactive vaccines were administered by immersion bath or intraperitoneally [32]. Therefore, due to the clinical importance of some *Aeromonas* species, they could be candidates for use in the generation of bacterins. As in the case of infectious hematopoietic necrosis and enteric red mouth disease [13]; however, more studies are needed to support this hypothesis, so this work will serve as a basis for future studies.

The presence of sensory neurons in the rosette, and their connection between the olfactory bulb and the brain, suggest they can induce alterations in feeding behavior, reproductive behavior and alarm reactions [29]. The histopathological findings in telencephalon consisted of moderate multifocal congestion for fish infected with *A. media*, and a moderate-to-severe degree for the groups infected by *A. salmonicida*, *A. bestiarum*, *A. caviae*, *A. hydrophila*, and *A. lusitana*, which presented moderate focal congestion; and, in the oblongata medulla were focal and multifocal congestion, and moderate to severe degrees of infection with *A. bestiarum* and *A. veronii*. No significant histopathological lesions ($p > 0.05$) were observed for the groups infected with the species *A. allosaccharophila*, *A. caviae*, *A. hydrophila*, *A. lusitana*, *A. media*, *A. popoffii*, *A. sobria*, and the control group.

The telencephalon in teleost participates in appetite regulation factors and the regulation of hormone inducing genes, such as corticotrophin releasing hormone 1a, which causes an anorectic effect; and adrenergic receptors, which participate in sex-specific swimming behavior [30]. On the other hand, the medulla oblongata participates in progressive hearing loss, in balance disorders and in the feeding role of fish [33]. Strom et al. [16] demonstrated that experimental infection with *Yersinia ruckeri* O1 biotype 2 causes lesions in the nasal epithelium and nervous system of fish infected via an immersion bath; this shows that bacterial pathogenesis begins with direct contact of the pathogen and the nasal epithelium close to the nervous system, causing problems in coordination and abnormal behavior. In the control group, the presence of congestion in the nasal epithelium was observed in two small fingerlings, because the hierarchy behavior of trout at feeding or the social status in the aquarium cause injuries in some subordinated rainbow trout; this induces stress by the biogenesis of cortisol [34].

The color change in teleost can be caused by multiple factors, stress being one of the main ones; continuous stress favors an increase in the secretion of the hormone cortisol, adrenocorticotropic hormone and alpha melanophore-stimulating hormone (Msh), which increases the number of chromophores and favors the presentation of skin darkening in stressed fish [35]. In the present study, hyperpigmentation or darkened skin was observed in fish infected with the species of *A. salmonicida* (18.8%) (Figure 1a), *A. veronii* (15.5%), and *A. hydrophila* (11.1%). In a study carried out by Aydin and Ciltas [19], the experimental infection of *A. hydrophila* in rainbow trout, which presented darkening skin, anorexia, abnormal movements, as well as other authors, who observed similar signology (Table 1). However, in the case of *A. salmonicida*, hyperpigmentation or darkened skin is not frequently reported as a clinical sign. In the case of *A. veronii*, it was reported that dying fish have pale dermis [36].

Other factors that induce skin color change in fish include biotic and abiotic external factors as nutrition, social interactions, luminosity, and UV incidence; individual genetic factors, cellular responses, nervous and hormonal responses [35]. Darkening or pale skin of fish infected with different species of *Aeromonas*, should not be considered as clinical signs of the disease.

In clinical case reports of fish infected with *Aeromonas* spp., it was reported that fish may present anorexia and abnormal swimming, sometimes described as swimming near the water surface [37]. Wang et al. [1] reported that fish naturally infected with *A. veronii* showed changes in their way of swimming, described as slow movements of diseased fish. The results of this work, reported that fingerlings infected with *A. bestiarum*, *A. caviae*, *A. hydrophila*, *A. salmonicida*, and *A. sobria*, are the species that cause erratic swimming (Tables 1 and 2). Histological findings and behavioral changes were identified in fish infected by *A. salmonicida* and *A. bestiarum*, the most pathogenic and virulent species. Although the presence of basophilic aggregates was not observed in nervous tissue, the bacterial isolation of the species *A. bestiarum*, *A. veronii*, *A. media*, *A. lusitana*, *A. popoffii*, *A. salmonicida*, and *A. hydrophila* was obtained (data not shown). This agrees with Bartkova et al., [15] who reported the isolation of the kidney, spleen and brain from fish infected by immersion, and the distribution of *A. salmonicida* in dorsal, pectoral, and caudal area; anal fin, anal opening, gills, oral and nasal mucosa; and in eyes, without the presence of basophilic aggregates in the tissue.

The absence of bacterial structures in the histological sections of nerve tissue suggests the participation of virulence factors present in the *Aeromonas* genus, among them: the production of lipopolysaccharides (LPS), which favor the differences in virulence between the different species and the nonspecific inflammatory response due to the differences between the O-antigen LPS; the S layer that is related to the pathogenicity functions, the hemolysis that cause the formation of pores in the target cell; and proteases that also favor bacterial pathogenicity and the promotion of bacterial invasion, through direct damage due to the activation of proteolytic toxins [6].

5. Conclusions

Based on the identification of histopathological lesions of rainbow trout fingerlings infected with different *Aeromonas* species, it was observed that the species with the greatest capacity to cause the presentation of suppurative rhinitis are *A. allosaccharophila*, *A. lusitana*, *A. popoffii* and, *A. sobria*. While the species of *A. bestiarum*, *A. veronii*, and *A. hydrophila*, possess the ability to generate vascular changes in telencephalon; and the species of *A. media*, *A. caviae*, and *A. salmonicida*, cause vascular changes in medulla oblongata, in both cases, these nervous tissues are linked to the behavioral changes and clinical signs identified in experimentally infected fish. This suggests the pathogenic and virulent ability of the bacteria to establish itself in the nasal epithelium and affect the nervous system.

Author Contributions: Conceptualization, A.P.Z.-V. and R.G.C.-M.; methodology, V.V.-S.; software, J.C.Á.-H.; validation, N.E.R.-R., F.R.G.d.A., V.M.M.-J., I.A.-R. and V.V.-S.; formal analysis, J.C.Á.-H.; investigation, V.M.M.-J. and I.A.-R.; data curation, J.C.Á.-H. and A.P.Z.-V.; writing—original draft preparation, A.P.Z.-V.; writing—review and editing, A.P.Z.-V. and R.G.C.-M.; visualization, N.E.R.-R. and V.V.-S.; supervision, A.P.Z.-V. All authors have read and agreed to the published version of the manuscript.

Funding: This research received no external funding.

Institutional Review Board Statement: The study was conducted according to the guidelines of the Declaration of Helsinki and approved by the Bioethics Commission and Animal Welfare Committee (Comisión de Bioética y Bienestar animal, CBBA) of the Universidad Autónoma del Estado de México (UAEMex), Facultad de Medicina Veterinaria y Zootecnia (FMVZ) (012-2012).

Informed Consent Statement: Not applicable.

Data Availability Statement: Not applicable.

Acknowledgments: This investigation study was realized in CIESA installations (Centro de Investigación y Estudios Avanzados en Salud Animal) of the Universidad Autónoma del Estado de México. We appreciate the support of Edgardo Soriano-Vargas, in this research work.

Conflicts of Interest: The authors declare no conflict of interest.

References

1. Wang, B.; Mao, C.; Feng, J.; Li, Y.; Hu, J.; Jiang, B.; Gu, Q.; Su, Y. A First report of *Aeromonas veronii* infection of the sea bass, *Lateolabrax maculatus* in China. *Front. Veter-Sci.* **2021**, *7*, 600587. [CrossRef]
2. Srivastava, A.; Mistri, A.; Mittal, S.; Kumar, A.M. Alterations in the epidermis of the carp, *Labeo rohita* (Cyprinidae: Cypriniformes), infected by the bacteria, *Aeromonas hydrophila*: A scanning electron microscopic, histopathological and immunohistochemical investigation. *J. Fish Dis.* **2020**, *43*, 941–953. [CrossRef]
3. Janda, J.M.; Abbott, S.L. The genus *Aeromonas*: Taxonomy, pathogenicity, and infection. *Clin. Microbiol. Rev.* **2010**, *23*, 35–73. [CrossRef] [PubMed]
4. Huber, I.; Spanggaard, B.; Appel, K.F.; Rossen, L.; Nielsen, T.; Gram, L. Phylogenetic analysis in situ identification of the in-testinal microbial community of rainbow trout (*Oncorhynchus mykiss*, Walbaum). *J. Appl. Microbiol.* **2004**, *96*, 117–132. [CrossRef] [PubMed]
5. Austin, B.; Austin, D.A. *Bacterial Fish Pathogens, Diseases of Farmed and Wild Fish*, 4th ed.; Praxis Publishing Ltd.: Chichester, UK, 2007; pp. 123–153.

6. Fernández-Bravo, A.; Figueras, M.J. An update on the genus *Aeromonas*: Taxonomy, epidemiology, and pathogenicity. *Microorganisms* **2020**, *8*, 129. [CrossRef] [PubMed]
7. Dallaire-Dufresne, S.; Tanaka, K.H.; Trudel, M.V.; Lafaille, A.; Charette, S.J. Virulence, genomic features, and plasticity of *Aeromonas salmonicida* subsp. salmonicida, the causative agent of fish furunculosis. *Veter-Microbiol.* **2014**, *169*, 1–7. [CrossRef] [PubMed]
8. Zepeda-Velázquez, A.P.; Sánchez, V.V.; Ortega-Santana, C.; Rubio-Godoy, M.; de Oca-Mira, D.M.; Soriano-Vargas, E. Pathogenicity of Mexican isolates of *Aeromonas* sp. in immersion experimentally-infected rainbow trout (*Oncorhynchus mykiss*, Walbaum 1792). *Acta Trop.* **2017**, *169*, 122–124. [CrossRef] [PubMed]
9. Fuentes, R.J.M.; Pérez, H.J.A. Aislamiento de *Aeromonas hydrophila* en truchas Arcoíris (*Oncorhynchus mykiss*). *Vet. Méx.* **1998**, *29*, 117–119.
10. Zepeda-Velázquez, A.P.; Vega-Sánchez, V.; Salgado-Miranda, C.; Soriano-Vargas, E. Histopathological findings in farmed rainbow trout (*Oncorhynchus mykiss*) naturally infected with 3 different *Aeromonas* species. *Can. J. Vet. Res.* **2015**, *79*, 250–254.
11. Marinho-Neto, F.A.; Claudiano, G.S.; Aguinaga, J.Y.; Quiroz, V.A.C.; Kobashigawa, K.K.; Cruz, N.R.N.; Moraes, F.R.; Moraes, J.R.E. Morphological, microbiological and ultrastructural aspects of sepsis by *Aeromonas hydrophila* in *Piaractus mesopotamicus*. *PLoS ONE* **2019**, *14*, e0222626. [CrossRef] [PubMed]
12. Noga, E.J. *Fish Disease, Diagnosis, and Treatment*, 2nd ed.; Wiley-Blackwell: Iowa, IA, USA, 2010.
13. LaPatra, S.; Kao, S.; Erhardt, E.B.; Salinas, I. Evaluation of dual nasal delivery of infectious hematopoietic necrosis virus and enteric red mouth vaccines in rainbow trout (*Oncorhynchus mykiss*). *Vaccine* **2015**, *33*, 771–776. [CrossRef] [PubMed]
14. Salinas, I.; LaPatra, S.; Erhardt, E. Nasal vaccination of young rainbow trout (*Oncorhynchus mykiss*) against infectious hematopoietic necrosis and enteric red mouth disease. *Dev. Comp. Immunol.* **2015**, *53*, 105–111. [CrossRef]
15. Bartkova, S.; Kokotovic, B.; Dalsgaard, I. Infection routes of *Aeromonas salmonicida* in rainbow trout monitored in vivo by real-time bioluminescence imaging. *J. Fish Dis.* **2017**, *40*, 73–82. [CrossRef]
16. Strøm, H.K.; Ohtani, M.; Nowak, B.; Boutrup, T.S.; Jones, B.; Raida, M.K.; Bojesen, A.M. Experimental infection by *Yersinia ruckeri* O1 biotype 2 induces brain lesions and neurological signs in rainbow trout (*Oncorhynchus mykiss*). *J. Fish Dis.* **2017**, *41*, 529–537. [CrossRef]
17. Farto, R.; Milton, D.L.; Bermúdez, M.B.; Nieto, T.P. Colonization of turbot tissues by virulent and avirulent *Aeromonas salm-onicida* subsp *salmonicida* strains during infection. *Dis. Aquat. Organ.* **2011**, *95*, 167–173. [CrossRef]
18. Nordgreen, J.; Tahamtani, F.M.; Janczak, A.M.; Horsberg, T.E. Behavioural effects of the commonly used fish anaesthetic tricaine methanesulfonate (MS-222) on zebrafish (*Danio rerio*) and its relevance for the acetic acid pain test. *PLoS ONE* **2014**, *9*, e92116. [CrossRef]
19. Aydin, S.; Çiltaş, A. Systemic infections of *Aeromonas hydrophila* in rainbow trout (*Oncorhynchus mykiss* Walbaum): Gross pathology, bacteriology, clinical pathology, histopathology, and chemotherapy. *J. Anim. Vet. Adv.* **2004**, *3*, 810–819.
20. Aydoğan, A.; Metin, N. Investigation of pathological findings infected with *Aeromonas salmonicida* in rainbow trout (*Oncorhynchus mykiss* Walbaum, 1972). *Kafkas Univ. Vet. Fak Derg* **2010**, *16*, S325–S328.
21. Inglis, V.; Roberts, R.J.; Bromage, N.R. *Bacterial Diseases of Fish*, 1st ed.; Iowa State University Press: Iowa City, IA, USA, 1993; pp. 80–121.
22. Lago, E.P.; Nieto, T.P.; Farto, R. Virulence factors of *Aeromonas salmonicida* subsp. salmonicida strains associated with infections in turbot Psetta maxima. *Dis. Aquat. Org.* **2012**, *99*, 145–151. [CrossRef] [PubMed]
23. Roberts, R.J. *Fish Pathology*, 3rd ed.; Wiley-Blackwell: Iowa, IA, USA, 2001.
24. Miller, K.M.; Günther, P.O.; Li, S.; Kaukinen, H.K.; Ming, J.T. Molecular indices of viral disease development in wild migrating salmon. *Conserv Physiol.* **2017**, *5*, cox036. [CrossRef] [PubMed]
25. Lê, S.; Josse, J.; Husson, F. FactoMineR: An R package for multivariate analysis. *J. Stat. Softw.* **2008**, *25*. [CrossRef]
26. R Core Team. *R: A Language and Environment for Statistical Computing*. R Foundation for Statistical Computing. Vienna, Austria, 2013.
27. Kassambara, A.; Mundt, F. Factoextra: Extract and visualize the results of multivariate data analyses. *R Package Version* **2017**, *1*, 337–354.
28. Speare, D. Differences in patterns of meningoencephalitis due to bacterial kidney disease in farmed Atlantic and chinook salmon. *Res. Veter- Sci.* **1997**, *62*, 79–80. [CrossRef]
29. Hamdani, E.H.; Døving, K.B. The functional organization of the fish olfactory system. *Prog. Neurobiol.* **2007**, *82*, 80–86. [CrossRef] [PubMed]
30. Ye, C.; Xu, S.; Hu, Q.; Hu, M.; Zhou, L.; Qin, X.; Jia, J.; Hu, G. Structure and function analysis of various brain subregions and pituitary in grass carp (*Ctenopharyngodon idellus*). *Comp. Biochem. Physiol. Part D Genom. Proteom.* **2020**, *33*, 100653. [CrossRef] [PubMed]
31. Song, X.; Zhao, J.; Bo, Y.; Liu, Z.; Wu, K.; Gong, C. *Aeromonas hydrophila* induces intestinal inflammation in grass carp (*Ctenopharyngodon idella*): An experimental model. *Aquaculture* **2014**, *434*, 171–178. [CrossRef]
32. Bøgwald, J.; Dalmo, R.A. Review on Immersion Vaccines for Fish: An Update. *Microorganisms* **2019**, *7*, 627. [CrossRef] [PubMed]
33. Yang, Y.; Wang, X.; Liu, Y.; Fu, Q.; Tian, C.; Wu, C.; Shi, H.; Yuan, Z.; Tan, S.; Liu, S.; et al. Transcriptome analysis reveals enrichment of genes associated with auditory system in swim bladder of channel catfish. *Comp. Biochem. Physiol. Part D Genom. Proteom.* **2018**, *27*, 30–39.

34. Kostyniuk, D.J.; Culbert, B.M.; Mennigen, J.A.; Gilmour, K.M. Social status affects lipid metabolism in rainbow trout, *Oncorhynchus mykiss*. *Am. J. Physiol. Integr. Comp. Physiol.* **2018**, *315*, R241–R255. [CrossRef] [PubMed]
35. Vissio, P.G.; Darias, M.J.; Di Yorio, M.P.; Sirkin, D.I.P.; Delgadin, T.H. Fish skin pigmentation in aquaculture: The influence of rearing conditions and its neuroendocrine regulation. *Gen. Comp. Endocrinol.* **2021**, *301*, 113662. [CrossRef] [PubMed]
36. Dong, H.T.; Techatanakitarnan, C.; Jindakittikul, P.; Thaiprayoon, A.; Taengphu, S.; Charoensapsri, W.; Khunrae, P.; Rattanarojpong, T.; Senapin, S. *Aeromonas jandaei* and *Aeromonas veronii* caused disease and mortality in Nile tilapia, *Oreochromis niloticus* (L.). *J. Fish Dis.* **2017**, *40*, 1395–1403. [CrossRef] [PubMed]
37. Sreedharan, K.; Philip, R.; Sarojani, I.; Singh, B. Characterization, and virulence potential of phenotypically diverse *Aeromonas veronii* isolates recovered from moribund freshwater ornamental fishes of Kerala, India. *Antonie Van Leeuwenhoek* **2013**, *103*, 53–67. [CrossRef]

Article

Lipoxygenase Enzymes, Oligosaccharides (Raffinose and Stachyose) and 11sA4 and A5 Globulins of Glycinin Present in Soybean Meal Are Not Drivers of Enteritis in Juvenile Atlantic Salmon (*Salmo salar*)

Artur N. Rombenso [1,*], David Blyth [1], Andrew T. James [2], Teisha Nikolaou [3] and Cedric J. Simon [3]

[1] Agriculture & Food, Livestock & Aquaculture Program, Bribie Island Research Centre, CSIRO, Woorim, QLD 4507, Australia; David.Blyth@csiro.au

[2] Agriculture & Food, Crops Program, Queensland Bioscience Precinct, CSIRO, St. Lucia, QLD 4067, Australia; Andrew.James@csiro.au

[3] Agriculture & Food, Livestock & Aquaculture Program, Queensland Bioscience Precinct, CSIRO, St. Lucia, QLD 4067, Australia; teisha.lea@hotmail.com (T.N.); Cedric.Simon@csiro.au (C.J.S.)

* Correspondence: artur.rombenso@csiro.au

Citation: Rombenso, A.N.; Blyth, D.; James, A.T.; Nikolaou, T.; Simon, C.J. Lipoxygenase Enzymes, Oligosaccharides (Raffinose and Stachyose) and 11sA4 and A5 Globulins of Glycinin Present in Soybean Meal Are Not Drivers of Enteritis in Juvenile Atlantic Salmon (*Salmo salar*). *Appl. Sci.* **2021**, *11*, 9327. https://doi.org/10.3390/app11199327

Academic Editors: Panagiotis Berillis and Božidar Rašković

Received: 10 September 2021
Accepted: 2 October 2021
Published: 8 October 2021

Publisher's Note: MDPI stays neutral with regard to jurisdictional claims in published maps and institutional affiliations.

Abstract: Soybean meal has been largely investigated and commercially used in fish nutrition. However, its inclusion levels have been carefully considered due to the presence of antinutritional factors, which depending on a series of factors might induce gut inflammation damaging the mucosal integrity and causing enteritis. Several strategies including genetic engineering have been applied attempting to reduce or eliminate some of the antinutritional factors. Accordingly, we assessed the intestinal health of juvenile Atlantic salmon fed high levels of speciality soybean genotypes with reduced-to-no content amounts of lipoxygenases, altered glycinin profile and reduced levels of oligosaccharides. No major signs of enteritis, only indication of enteritis progression, was noticed in the soybean meal-based diets illustrated by mild changes in distal intestine morphology. Whereas fish, fed fishmeal control feeds, displayed normal distal intestine integrity. Speciality soybean types did not improve intestinal health of juvenile Atlantic salmon suggesting these antinutrients are not drivers of the intestinal inflammatory process in this species. No additional benefits in terms of production performance or blood biochemistry were noticed in the speciality soybean types compared to the traditional soybean.

Keywords: antinutritional factors; soybean; gut health

1. Introduction

Soybean meal (SBM) has been largely investigated and commercially used in fish nutrition. Although SBM offers several advantages to the aquafeed industry including worldwide availability, competitive pricing, consistent nutritional quality, and an acceptable amino acid profile it also displays some constraints, mainly associated with antinutritional factors [1,2]. There is a long list of antinutritional factors including saponins, lectins, phytic acid, oligosaccharides, isoflavones, and allergens, among others [1]. These antinutrients are known to impair feed intake, palatability, growth performance, digestive enzymes and in some instances induce gut inflammation damaging the mucosal integrity and causing enteritis [1,3–8]. The degree of physiological impairments induced by dietary SBM is linked to the SBM inclusion level, blends of raw materials, duration of feeding SBM-based feeds, and the species sensitivity to antinutritional factors [3–6,9,10]. Intestinal damage induced by SBM has been reported mostly in distal intestine and liver tissues across several fish species including Atlantic salmon *Salmo salar* [3,4,9,10], Totoaba *Totoaba macdonaldi* [5,6], Seriola spp *Seriola lalandi*, *Seriola dorsalis* [7,11–13], common Carp *Cyprinus carpio* [14,15], and Largemouth Bass *Micropterus salmoides* [8]. Distal intestine histology shows changes in the length of the mucosal fold, reduction in the number of supranuclear vacuoles of the enterocytes and thickness of the lamina propria, among other pathohistological modifications [3,5,6,10,12,16,17].

Among the fish species, salmon appears to be one of the most sensitive to the antinutritional factors presented in SBM developing a condition known as soybean meal-induced enteropathy, which exhibits similar changes as those described above [3,4,9,17]. Some salmon species such as pink salmon *Oncorhynchus gorbuscha* appears to be more resistant to antinutritional factors present in dietary SBM than chinook *O. tshawytscha* and Atlantic salmon *S. salar* [18]. In the early 2000's, Buttle et al. [19], suggested the binding mechanism of soybean agglutinin (lectin) to Atlantic salmon intestinal epithelium as a primary contributor to pathological changes in this tissue. Saponins are other top candidates of key antinutritional factors present in soybean meal. A dose-response study reported increasing inflammatory process in Atlantic salmon distal intestine with greater dietary soybean saponins [20].

Several strategies have been applied attempting to reduce or eliminate some of the antinutritional factors in SBM including extrusion, fermentation, pre-processing techniques, and genetic engineering. For example, extrusion with shorter barrel retention times and higher temperatures improved utilization of SBM-rich diets (52% SBM) in salmonids [21]. Another commercial strategy is to increase the protein fraction of SBM through concentration (SPC) or isolation (SPI). A relatively small effort has been done in the genetic space focusing on selecting non-GM SBM for specific genotypes in aquafeeds. Recently, the removal of trypsin inhibitor, lectin and allergen P34/Gly m Bd 30 k from a soybean cultivar failed to alleviate inflammatory processes in Atlantic salmon [9]. Collectively, these studies suggest the challenge in identifying specific antinutritional factors present in soybean responsible for enteritis induction and highlight the complexity of interactions with compounds present in other plant ingredients largely used in aquafeed formulations. As a result, the salmon aquafeed industry has adopted ingredient inclusion limits and more processed soy protein products such as soy protein concentrate and isolate. However, from a cost-effective perspective, finding approaches to minimize or eliminate the soybean-induced enteritis in fish nutrition is worthwhile.

As part of the Australian Soybean improvement program, CSIRO has developed speciality SBM genotypes with reduced-to-no content amounts of lipoxygenases [22], altered glycinin profile [23], and reduced levels of oligosaccharides for human consumption. Accordingly, we assessed the intestinal health of juvenile Atlantic salmon fed high levels of these speciality SBM genotypes.

2. Materials and Methods

2.1. Formulations and Feed Manufacture

Dietary treatments are presented in Table 1. A fish meal-based diet (45%) was used as a control treatment, whereas the experimental diets contained 29% of fishmeal and 30% of SBM from three distinct genotypes of similar genetic background and matched as closely as possible for protein content: standard soybean meal (STD SBM); a soybean genotype homozygous for the gy4 allele conditioning null 11sA4 and 11sA5 globulins of Glycinin, homozygous for the $l \times 1$, $l \times 2$ and $l \times 3$ alleles conditioning absence of seed lipoxygenases and homozygous for the rs2 allele conditioning near absence of seed raffinose and stachyose (TLP SBM); and a soybean genotype homozygous for the gy4 allele conditioning null 11sA4 and 11sA5 globulins of Glycinin (11sA4 null SBM).

All macro ingredients were milled to <750 μm, and well mixed with the remaining dry ingredients, and then extruded through a Baker-Perkins MPV24 twin-screw extruder. Each feed was manufactured using a 1.5 mm Ø die (~2.5 mm Ø pellets) using standard CSIRO Extrusion protocols (Table 2). The pellets were dried at 60 °C for 12 h, after which they were vacuum infused with their specific allocation of oil. All feeds were kept in frozen storage (−20 °C) throughout the feeding trial. All uneaten feed was removed from the collection tank of all treatments 1 h following feeding, and the collected waste feed was then dried to allow calculation of apparent feed intake and feed conversion ratio.

Table 1. Dietary formulation and proximate composition.

Ingredients g kg^{-1}	FM Control	STD SBM	TLP SBM	11sA4 Null SBM
Fishmeal [a]	450.0	290.0	290.0	290.0
Wheat flour	223.0	110.0	93.0	106.0
Wheat gluten	120.0	89.0	106.0	93.0
Blood meal [a]	60.0	60.0	60.0	60.0
Fish oil [a]	70.0	70.0	70.0	70.0
Poultry oil [a]	70.0	70.0	70.0	70.0
Stay-C 35% [b]	1.0	1.0	1.0	1.0
Vitamin mineral premix [c]	6.0	6.0	6.0	6.0
Standard soybean meal (STD SBM) [d]	0.0	300.0	0.0	0.0
Soybean meal triple lipoxygenase plus (TLP SBM) [d]	0.0	0.0	300.0	0.0
Soybean meal 11sA4 null (11sA4 null SBM) [d]	0.0	0.0	0.0	300.0
Methionine [e]	0.0	3.0	3.0	3.0
Taurine [f]	0.0	1.0	1.0	1.0
Proximate composition (g kg^{-1})				
Dry matter	959	960	954	954
Protein	506	492	494	505
Lipid	166	171	179	170
Ash	68	63	62	62
Gross energy (kJ g^{-1})	23.7	24.2	24.3	24.3
Amino acid (g kg^{-1})				
ASP	39	43	45	43
SER	20	22	23	21
GLU	88	87	93	89
GLY	26	24	26	23
HIS	16	15	15	15
ARG	22	25	25	26
THR	19	19	20	19
ALA	25	24	24	24
PRO	29	28	29	28
CYS	5	5	5	5
TYR	14	19	20	19
VAL	23	23	24	23
MET	11	11	11	10
LYS	26	28	28	28
ILE	17	18	19	18
LEU	38	38	39	38
PHE	22	23	25	23
TAU	3	3	3	3

[a] Ridley, Aquafeeds, Queensland, Australia. [b] DSM, Heerlen, Netherlands. [c] Rabar Pty Ltd., Queensland, Australia. [d] CSIRO, Australia. [e] Redox, Queensland, Australia. [f] Bulk Nutrients, Tasmania, Australia.

Table 2. Dietary extrusion parameters.

Parameter	FM Control	STD SBM	TLP SBM	11sA4 Null SBM
RPM	220	220	220	220
Feed (g min^{-1})	72	62	59	53
H2O (g min^{-1})	16	22	20	20
Torque (%)	8	6	6	6
SME (kJ kg^{-1})	99	69	73	80
Zone 1 (Cone)	120	120	120	120
Zone 2	95	95	95	95
Zone 3	85	85	85	85
Zone 4 (Inlet)	70	70	70	70
Die Diameter (mm)	1.5	1.5	1.5	1.5

RPM = revolutions per minute. SME = specific mechanical energy.

2.2. Experimental System and Feeding Trial

Sixteen tanks with 300 L at Bribie Island Research Centre (BIRC, Queensland, Australia) were used in this experiment with ~3 L/min flow of continuously aerated, recirculating freshwater at 15 °C ± 0.16 (mean ± SEM). Photoperiod was set at 12L:12D. Twenty-five salmon of 37.3 g ± 0.42 (mean ± SD) were randomly allocated to each tank. Fish were fed to apparent satiation twice daily for 56 days. This research was approved by the CSIRO Queensland Animal Ethics Committee—AEC Number: 2018-44.

2.3. Analytical Methods

Chemical analyses were carried out to confirm proximate composition of dietary treatments, and main ingredients (Tables 1 and 3) [24]. Samples were dried at 105 °C for 12 h to determine gravimetrically dry matter and followed by ashing at 550 °C for 12 h. Total nitrogen was measured by combustion (CHNS auto-analyzer, Leco Corp., St. Joseph, MI, USA) and crude protein was calculated by nitrogen conversion (%N × 6.25). Total lipid was gravimetrically determined via Folch extraction [25]. Finally, gross energy was measured via an adiabatic bomb calorimeter (Parr 6200, Par Instrument Company, Moline, IL, USA).

Table 3. Key ingredients proximate and amino acid composition (g kg^{-1}).

	Fishmeal	STD SBM	TLP SBM	11sA4 null SBM
Proximate composition (g kg^{-1})				
Dry matter	911	949	918	926
Protein	719	495	521	462
Lipid	117	119	116	118
Ash	126	54	52	58
Amino acid (g kg^{-1})				
ASP	67.7	46.0	48.5	48.7
SER	29.8	19.6	20.8	21.6
GLU	97.4	69.5	73.4	69.1
GLY	45.2	16.0	16.3	16.7
HIS	16.2	9.1	10.2	9.3
ARG	39.7	24.6	26.0	25.0
THR	30.8	14.3	15.2	16.3
ALA	40.7	16.3	17.5	17.6
PRO	31.6	19.3	20.5	19.7
CYS	6.5	5.5	5.2	5.3
TYR	24.4	11.7	13.9	13.2
VAL	34.4	16.7	17.5	16.7
MET	22.8	2.8	3.0	3.1
LYS	58.9	21.1	23.7	23.4
ILE	29.0	15.8	16.7	17.5
LEU	52.6	27.6	30.4	29.8
PHE	30.1	18.2	18.8	20.4
TAU	7.8	ND	ND	ND

ND = not detected.

Total amino acid (TAA) quantification was performed by mass detection following high performance reverse-phase liquid chromatography with pre-column derivatization with 6-aminoquinolyl-N-hydroxysuccinimidyl (AQC). Analyses were undertaken on a Shimadzu Nexera X2 series UHPLC (Shimadzu Corporation, Kyoto, Japan), coupled with a Shimadzu 8030 Mass Spectrometer using a modification of the Waters AccQ-tag system (Waters Corporation, Milford, MA). Bovine serum albumin (BSA) (ICN), milk powder (NIST SRM 1549a) and a well characterized aquafeed were used as reference materials. Samples or reference materials were hydrolyzed using phenolic 6N HCl at 112 °C according to the protocol for complex feed samples outlined by Waters Corp. (1996).

2.4. Production Parameters

The following production parameters were calculated for the 28 and 56 days of the feeding trial; feed conversion ratio (FCR), hepatosomatic index (HSI), and condition factor (K). No differences in production performance parameters were observed among the dietary treatments throughout the feeding trial.

$$FCR = \frac{FI}{WG} \tag{1}$$

$$HSI = \frac{W_{Liver}}{W_{Fish}} * 100 \tag{2}$$

$$K = 100 * \frac{W_{Fish}}{L_{Fork}^{3}} \tag{3}$$

where FCR = feed conversion ratio, FI = apparent feed intake (g), WG = weight gain (g), HSI = hepatosomatic index, W_{Liver} = wet weight of liver (g), W_{Fish} = fish whole weight (g), K = condition factor, L_{Fork} = fish fork length (mm)

2.5. Histology

Upon completion of the feeding trial, three fish per tank were randomly selected and euthanized via AQIS overdose (~70 ppm). The distal intestine, i.e., last 2 cm section of the intestine, of each fish was sampled and stored in Davidson's solution for 24 h, and then transferred to 70% ethanol until further analysis. For each fish sample, the distal intestine was divided into three sections and gradually dehydrated in ethanol, clarified in benzene and embedded in paraffin. As a result, a complete intestinal annular ring from each fish (nine per treatment) was cut into three sections ($n = 36$) and mounted onto individual glass slides for histological assessment.

Transversal sections of 3 µm were cut using a rotary microtome (Leica RM2245), stained with hematoxylin and eosin (H&E). The slides were blind examined after randomization, using the Zeiss Axiocam light microscope. The pictures were taken using the camera function on the Zeiss Axiocam microscope and then processed and analyzed using Zeiss Zen Light (Version 3.1) image analysis software.

To assess the degree of intestinal damage, a semiquantitative scoring system was used. In this scoring system, three parameters were quantified independently based on [10]: (1) the appearance and length of mucosal folds (MF); (2) the degree of widening of the lamina propria (LP); (3) the abundance of goblet cells (GC). For each of these parameters a score was given on a scale of 1 to 5. An increasing score value represents a greater degree of intestinal damage.

2.6. Statistical Analysis

All data were analyzed for normality and equality by Levene's tests, respectively, and then subjected to one-way ANOVA (analysis of variance, NCSS 12.0). When significant effects were identified, the post-hoc Tukey's HSD pairwise comparison test was used to determine difference among means with a significance level of 0.05.

3. Results

Juvenile Atlantic salmon fed the fishmeal control dietary treatment displayed normal distal intestine integrity (Figure 1A). There were no major signs of enteritis, only an indication of enteritis progression was noticed in the SBM-based diets illustrated by mild changes in distal intestine morphology, including reduced number and length of mucosal folds, enlargement of the apical zone of mucosal folds, thickening of lamina propria, and changes in abundance of goblet cells (Figure 1B–D). The removal of lipoxygenases, 11sA4 and A5 globulins of glycinin, and oligosaccharides from SBM failed to prevent morphological changes linked to the inflammatory process. Histology scoring demonstrated statistically higher scores of goblet cells and lamina propria in the SBM treatments than the fishmeal control

(Table 4). Scoring of mucosal folds was higher in the speciality SBM groups compared to the control group.

Blood biochemistry was largely unaffected by the dietary treatments. Out of the sixteen parameters analyzed, only albumin and protein were statistically higher in TLP SBM than STD SBM (Table 5).

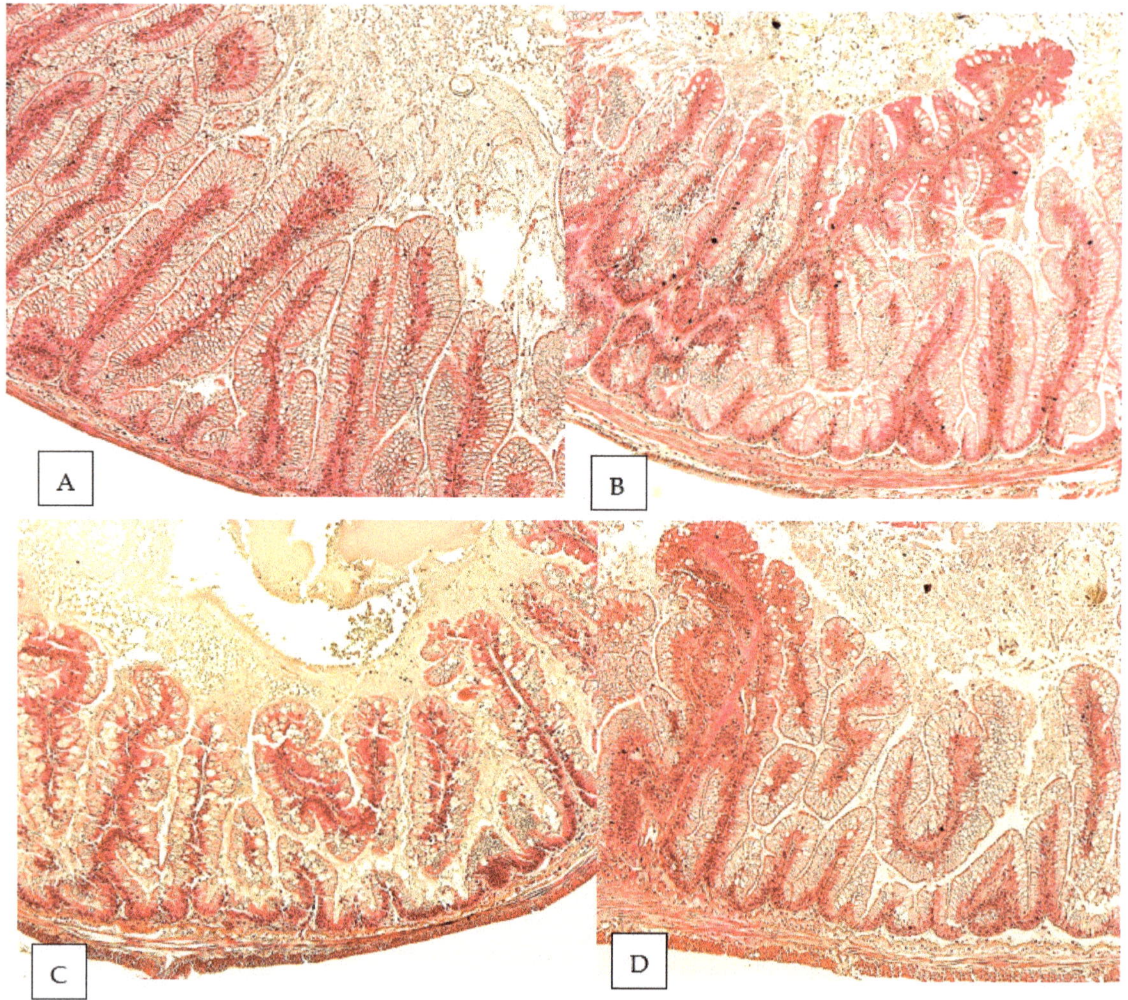

Figure 1. Light microscopic images illustrating morphological changes in the distal intestine associated with inflammatory process in Atlantic salmon fed different soybean meal types for 56 days ((**A**)—control fishmeal, (**B**)—STD SBM—standard soybean meal, (**C**)—TLP SBM – triple null soybean meal absent of seed lipoxygenases and homozygous for the rs2 allele conditioning and near absent of seed raffinose and stachyose, and (**D**)—11sA4 null SBM—soybean meal conditioning null 11sA4 and 11sA5 globulins of Glycinin).

Survival was high across the dietary treatments (97–100%; Table 6). No differences in production performance parameters, including CV, final weight, FCR, K, and HSI were noticed between the fishmeal control group and the SBM-based groups at day 28 and day of feeding trial.

Table 4. Semiquantitative scoring system (mean ± SD) of three parameters based on [10]: (1) the appearance and length of mucosal folds (MF); (2) the degree of widening of the lamina propria (LP); (3) the abundance of goblet cells (GC). For each of these parameters a score was given on a scale of 1 to 5. An increasing score value represents a greater degree of intestinal damage.

Scoring	FM Control	STD SBM	TLP SBM	11sA4 Null SBM	*p*-Value
Goblet cells	1.9 ± 0.3 b	3.2 ± 0.4 a	3.4 ± 0.5 a	3.7 ± 0.5 a	<0.001
Lamina propria	1.8 ± 0.4 b	3.2 ± 0.4 a	3.5 ± 0.5 a	3.6 ± 0.5 a	<0.001
Mucosal fold	1.9 ± 0.5 b	2.7 ± 0.7 ab	2.9 ± 0.6 a	3.3 ± 0.7 a	0.001

Table 5. Sixteen blood chemistry parameters of juvenile Atlantic salmon fed a fishmeal control diet (FM control) and different soybean meal types-based diets (STD SBM—standard soybean meal, C—TLP SBM—triple null soybean meal absent of seed lipoxygenases and homozygous for the rs2 allele conditioning and near absent of seed raffinose and stachyose, and D—11sA4 null SBM—soybean meal conditioning null 11sA4 and 11sA5 globulins of Glycinin) for 56 days.

Blood Chemistry Parameters	FM Control	STD SBM	TLP SBM	11sA4 Null SBM	*p*-Value
Albumin (g L^{-1})	20.3 ± 0.6 ab	20.0 ± 1.0 b	22.3 ± 1.1 a	20.7 ± 0.6 ab	0.044
Alkaline phosphatase (U L^{-1})	21.3 ± 6.1	18.7 ± 4.7	24.3 ± 2.1	15.7 ± 5.7	0.242
Anion gap (mmol L^{-1})	59.3 ± 3.5	56.0 1.0	53.0 ± 2.6	58.0 ± 8.7	0.463
AST (U L^{-1})	301.3 ± 53.6	301.7 ± 28.7	359.5 ± 2.1	315.7 ± 31.5	0.361
Bicarbonate (mmol L^{-1})	6.3 ± 0.6	5.3 ± 1.1	5.7 ± 0.6	6.7 ± 2.1	0.577
Chloride (mmol L^{-1})	123.3 ± 8.7	120.3 ± 1.5	125.3 ± 5.1	120.0 ± 1.0	0.561
Cholesterol (mmol L^{-1})	13.2 ± 2.9	9.8 ± 1.1	12.4 ± 0.9	11.9 ± 1.4	0.236
Creatine kinase (U L^{-1})	7286 ±1425	7992 ± 1198	7446 ± 1719	8089 ± 1674	0.890
Globulin (g L^{-1})	18.7 ± 2.1	18.7 ± 1.1	21.0 ± 1.0	20.3 ± 0.6	0.142
Glucose (mmol L^{-1})	5.2 ± 0.8	4.2 ± 0.3	4.2 ± 0.2	4.0 ± 0.4	0.055
Phosphate (mmol L^{-1})	3.1 ± 0.3	2.7 ± 0.3	2.6 ± 0.3	2.5 ± 0.3	0.103
Potassium (μm mmol^{-1})	4.2 ± 0.6	4.1 ±0.6	4.1 ± 0.5	4.0 ± 0.4	0.963
Protein (g L^{-1})	39.0 ± 2.6 ab	38.7 ± 1.5 b	43.3 ± 1.5 a	41.0 ± 1.0 ab	0.042
Ratio	1.1 ± 0.1	1.1 ± 0.1	1.0 ± 0.1	1.0 ± 0.0	0.344
Sodium (mmol L^{-1})	184.7 ± 11.6	177.7 ± 0.6	179.7 ± 3.0	180.3 ± 5.1	0.683
Triglyceride (mmol L^{-1})	2.9 ± 0.4	3.5 ± 0.5	3.1 ± 0.8	3.5 ± 0.3	0.377

Table 6. Production performance at 28 and 56 days of juvenile Atlantic salmon fed a fishmeal control diet (FM control) and different soybean meal types-based diets (STD SBM—standard soybean meal, C—TLP SBM—triple null soybean meal absent of seed lipoxygenases and homozygous for the rs2 allele conditioning and near absent of seed raffinose and stachyose, and D—11sA4 null SBM—soybean meal conditioning null 11sA4 and 11sA5 globulins of Glycinin).

Production Performance	FM Control	STD SBM	TLP SBM	11sA4 Null SBM
0 day				
Initial weight (g)	37.3 ± 0.0	37.4 ± 0.0	37.3 ± 0.1	37.3 ± 0.0
Initial CV (%)	12.1 ± 0.5	15.7 ± 3.4	11.1 ± 0.8	11.5 ± 0.5
28 days				
Survival (%)	99 ± 0.0	97 ± 0.0	99 ± 0.0	100
CV (%)	12.3 ± 1.3	13.0 ± 0.5	12.1 ± 0.7	13.1 ± 0.9
Mid weight (g)	49.6 ± 0.6	52.1 ± 0.7	49.8 ± 0.7	51.4 ± 0.6
FCR	1.0 ± 0.0	1.0 ± 0.0	1.1 ± 0.1	1.0 ± 0.0
K	1.3 ± 0.07	1.4 ± 0.09	1.5 ± 0.09	1.4 ± 0.01
HSI	1.8 ± 0.16	1.7 ± 0.17	1.6 ± 0.19	1.9 ± 0.40
56 days				
Survival (%)	98.5 ± 1.5	97.0 ± 1.5	98.5 ± 1.5	100
CV (%)	15.2 ± 1.5	15.5 ± 0.8	14.6 ± 1.0	15.2 ± 0.7
Final weight (g)	59.4 ± 1.5	66.1 ± 1.2	65.1 ± 2.1	64.4 ± 0.7
FCR	1.0 ± 0.0	0.9 ± 0.0	1.0 ± 0.0	0.9 ± 0.0
K	1.5 ± 0.17	1.4 ± 0.02	1.4 ± 0.04	1.3 ± 0.03
HSI	1.6 ± 0.19	1.4 ± 0.11	1.4 ± 0.07	1.6 ± 0.25

* CV = coefficient of variance. FCR = feed conversion ratio. K = condition factor. HSI = hepatosomatic index.

4. Discussion

Soybean meal-induced enteritis was not observed in juvenile Atlantic salmon, despite the high dietary SBM inclusion level of 30%. Nevertheless, intestinal health impairment was characterized by mild changes in distal intestine morphology (i.e., changes in length, shape, and number of mucosal folds, changes in thickness of lamina propria and changes in number of goblet cells), indicating enteritis progression. Speciality soybean types lacking 11sA4 and A5 globulins of Glycinin, lipoxygenases and oligosaccharides did not further reduce the intestinal inflammatory process compared to STD SBM, although inflammation was mild overall. Similarly, a recent thorough study removing three proteinaceous antinutrients, namely trypsin inhibitor, lectin and the allergen P34/Gly m Bd 30 k, from soybean meal did not mitigate enteritis in Atlantic salmon [9]. The authors suggested extrusion technology inactivated these compounds during feed manufacturing. Although soybean agglutinin, a type of lectin, has been reported to bind to Atlantic salmon intestinal epithelium contributing to pathological changes in gut health, the study did not describe the feed manufacturing procedure [19]. It is unlikely that extrusion technology inactivated the compounds investigated in the present study due to their chemical compositions and exposure to lower mechanical energy during extrusion conditions compared to those reported by [9] (specific mechanical energy 69–80 vs. 1872–3060 kJ/kg, respectively).

Soybean-induced enteritis in Atlantic salmon has been widely reported. However, the main drivers of enteritis and their positive and negative interactions with other compounds present in feed formulations remain unclear. A summary of seventeen studies around soybean-induced enteritis in Atlantic salmon is provided in Tables 7–9. Due to the complexity of the matter, there is a wide breadth of experimental designs with different stocking densities, fish size, duration, water temperature, salinity, feeding ration and experimental systems. For example, most literature is focused on smaller fish ranging from 41–442 g and some on larger fish 500–927 g. Similarly, soybean type and pre-processing and dietary composition varies throughout the literature. Soybeans are from different parts of the globe and are under several pre-processing conditions such as dehulled, toasted, defatted, and solvent-extracted, included at various levels 8–34% in feeds containing crude protein of 35–47%, total lipids 23–31%, and gross energy 15–24 MJ/kg. Regardless of this variation in the experimental design, most histology analyses are focused on the distal intestine describing morphological changes of standardized parameters including mucosal folds, supranuclear vacuoles, lamina propria, eosinophilic granulocytes, sub-epithelial mucosa, and connective tissue, and continue to be one of the most reliable tools to detect this histopathological condition. Other parameters such as reduced feed intake and growth commonly used as indicators of enteritis are not as reliable with contradicting findings throughout the literature. Most studies highlight the inflammatory process in the intestine of Atlantic salmon fed soybean-based feeds leading to enteritis; however, there are exceptions where only minor to mild intestinal damage were reported, including the present findings. Collectively, these studies illustrate the challenge and high complexity in tackling this issue in fish nutrition. Understandably, the aquafeed industry does not prioritize this research topic and adopt a more conservative approach of using moderate inclusion levels of plant ingredients avoiding any potential intestinal health impairments caused by antinutritional factors. It is likely extremely difficult to identify the key antinutritional factors in the major plant ingredients and their interactions with compounds from other components of the formulations having the effect. Interestingly, the intestinal health research of fish fed soybean-based feeds has gone beyond the traditional highly carnivorous species reaching a wide range of species from different feed, salinity and water temperature preferences, including for example common carp [14,15], grass carp [26], kingfish [7,12], totoaba [5,6], and largemouth bass [8].

Table 7. Summary of the experimental design of soybean meal-induced enteritis studies with Atlantic salmon (*Salmo salar*).

	Experimental Design										
IBW (g)	SGR (% BW)	FCR	Tank Volume (L)	Stocking Density (Fish/Tank)	Duration (Days)	Salinity	Temp. (°C)	Feeding	System	Refeed to Control	Ref.
927			2000	25	21	FW	12–13		Semi-RAS		[3]
280	0.94–1.05	0.81–0.94	450	54	60	SW	8.4				[4]
41		0.7	250	55	56	FW	14		RAS		[9]
300			400	50	20	SW	8 and 12	120%	RAS		[10]
550			27,000		120	3.2	5	120%	Net pens		[16]
550			27,000	80	42	SW	10.8 and 8.2	120%	Net pens	Yes	[17]
535 | 140 | 166 *			1900 | 650 | 850	50	21	SW	9	1% BW	Flow-through		[18]
54			1000	20	7	FW	15				[19]
442			1000	22		SW	12–9		RAS		[20]
900			125 000	300	300	SW	8.2		Net pens		[27]
213 | 202			400 | 100	30 | 20	62 | 44	SW	8.3 | 9	120%	RAS		[28]
396			400	25	28	SW	12	110%	RAS		[29]
500–600			1000	25	21	SW	9		RAS		[30]
500–600			1000	25–30	21	SW	9		RAS		[31]
80			400	70	53	SW	8.6	120%	Flow-through		[32]
60	1.36–1.47	0.71–0.74	500	75	93	SW	10.9	120%	Flow-through		[33]
207	0.7–0.9	0.92–1.17	600	50	84	FW	7	115%	Flow-through		[34]

IBW = initial body weight in g; SGR = specific growth rates in percentage of body weight; FCR = feed conversion ratio; Temp. = temperature; Ref. = references; RAS = recirculating aquaculture systems; SW = seawater; FW = freshwater; * numbers represent the following species Atlantic salmon, Chinook salmon and pink salmon.

Table 8. Summary of soybean types and dietary treatments of soybean meal-induced enteritis studies with Atlantic salmon (*Salmo salar*).

SBM Information			Dietary Treatments							
Location	Type	CP (%)	Pellet Type	CP (%)	TL (%)	GE (MJ /kg)	SBM Type/Inclusion (%)	SBM Type/Inclusion (%)	SBO Inclusion (%)	Ref.
Norway	With hulls, toasted and extracted		Extruded	35	28		SBM 30			[3]
Norway	Solvent-extracted, toasted		Extruded	40–44	22–24	22–23	SBM—8, 12, 15, 19, 27			[4]
USA	Triple null and standard		Extruded	40	26	24	SBM 25	Triple null SBM 27		[9]
Netherlands	Extracted		Extruded	45	30		SBM 20			[10]
				43	20	15	FFSBM 30	SPC 28	0–10	[16]
							SBM 33		0–8.5	[17]
			Pelleted	37	23		SBM 20			[18]
	Soybean agglutinin						Soybean agglutinin 3.5			[19]
	Soya saponins			42–44	29–30	24	Soy saponins—0, 2, 4, 6, 10			[20]
	Dehulled solvent extracted SBM		Extruded	40	22		SBM 17 and 34			[27]
Norway	Defatted		Extruded				DSBM 20 I molasses			[28]
NA, EU, SA		44–49	Extruded	42	25	23	20			[29]
	Extracted			43	28	24	SBM 20			[30]
	Extracted			43	28	24	SBM 20			[31]
Norway	Defatted		Extruded	47	26	23	SBM 25 + soy saponins 0.17	Lupins + soy saponin concentrate 0.17 + soy saponins 0.11		[32]
Switzerland			Extruded	46	25	24	SBM 20	Pre-processed SBM 20		[33]
	Dehulled defatted SBM		Pelleted	40	31	24	SBM 30			[34]

Ref. = references; NA = North America, EU = Europe, SA = South America; FFSBM = full-fat soybean meal; SBM = soybean meal; DSBM—defatted soybean meal; SPC = soy protein concentrate; SBO = soybean oil.

Table 9. Summary of the main findings and parameters investigated of soybean meal-induced enteritis studies with Atlantic salmon (*Salmo salar*).

Growth Impairment	Feed Intake	Level of DI Inflammation	Enzymes	Time Sampling (Days)	Tissues Analyzed	Parameters	Ref.
SBM yes SPC no		SBM +++			PI and DI	MSA/LSC, ESA/LSC, LPSA/LSC, ESA/LPSA, GC/E, LM	[16]
Yes		+++		2, 7, 14 and 21	DI	MF	[17]
		++	5′ N, Mg-ATPase, ALP, ACP, NSE, LAP, AAP	21	MI and DI		[3]
Low SBM no High SBM yes	No changes	+, ++ and +++	ALP, LAP, maltase, isomaltase, lactase and sucrase		MI and DI	MF, SNV, LP, leycocyte	[4]
		+, ++ and +++			DI	MF, SNV, LP, CT	[28]
		++ and +++			DI	MF, GC, LP SNV, EG, SM	[29]
		++ and +++			DI	MF, GC, LP SNV, EG, SM	[10]
		+, ++, and +++	Pancreatic (trypsin, chymotrypsin, elastase, and lipase), chyme (LAP), brush border membrane (LAP and maltase)	0, 1, 2, 3, 5, 7, 10, 14, 17, and 21	PI, MI and DI		[30]
		+, ++, and +++		1, 2, 3, 5 and 7	DI		[31]
Low saponins no Mid-high saponins yes	Low saponins no Mid-high saponins yes	+, ++, and +++	Trypsin activity, bile acids, brush border membrane enzyme activity (LAP)		PI and DI	MF, LP, enterocyte vacuolization, GC, nucleos position within the enterocytes	[20]
		+++			DI	MF, SNV, LP, CT	[32]
		+, ++, and +++		7, 14 and 21	DI	Inflammation score, SM and microbiome	[18]
No	No changes				DI	MF, GC, LP, SNV, EG, SM	[33]
No	No changes	+ and ++	Brush border LAP, trypsin activity		DI	MF, SNV, SM, LP, microbiota, gene expression	[9]
		+ and ++			PI, MI and DI	MF, SNV, LP, CT,	[19]
SBM 17 no SBM 34 yes						Body composition and blood biochemistry	[27]
Yes	No changes	++ and +++			DI	MF, SNV, LP, CT,	[34]

Ref. = references; Level of enteritis: + mild, ++ moderate, and +++ high; 5′ N = 5′-nucleotidase; Mg-ATPase = Mg^{2+} dependent adenosine triphosphatase; ALP = alkaline phosphatase; ACP = acid phosphatase; NSE = non-specific esterase; LAP = leucine aminopeptidase, AAP = alanine aminopeptidase; PI = proximal intestine; MI = mid-intestine; DI = distal intestine; MF = mucosal fold; MSA = mucosal surface area; LSC = length mucosal stratum compactum; ESA = epithelial surface area; LPSA = lamina propria surface area; GC = goblet cells; E = 100 um epithelium; LM = length microvilli; SNV = supranuclear vacuoles; LP = lamina propria; EG = eosinophilic granulocytes; CT = connective tissue; SM = sub-epithelial mucosa.

This is not the first study where animal growth was not impaired by dietary SBM displaying equivalent performance as the fishmeal SBM-free feeds. As Krogdahl et al. [27] has demonstrated, no detrimental growth effects in feeding 20% SBM to juvenile Atlantic salmon are present; however, more aggressive inclusion levels of 40% reduced growth are evident as compared to the fishmeal control feeds. Indeed, this pattern has been described with other fish species, including California yellowtail *Seriola dorsalis* [7]. Removal of certain antinutritional factors from SBM also did not affect Atlantic salmon growth response compared to the standard SBM [9]. Conservative inclusion levels of standard SBM at the expense of fishmeal appears to be suitable as long as no intestinal health impairment is noticed.

High levels of dietary SBM resulted in mild intestinal inflammation indicating enteritis progression. Speciality soybean types lacking lipoxygenases, altered glycinin profile and oligosaccharides did not improve intestinal health of juvenile Atlantic salmon suggesting these antinutrients are not drivers of the intestinal inflammatory process in this species. No additional benefits in terms of production performance or blood biochemistry were noticed in the speciality soybean types compared to the traditional soybean. The present findings contribute to and summarize the growing literature in the antinutrient space, providing more insights to future research.

Author Contributions: Conceptualization, A.N.R., A.T.J. and C.J.S.; Data curation, A.N.R., D.B. and T.N.; Formal analysis, A.N.R.; Funding acquisition, A.T.J. and C.J.S.; Methodology, D.B. and T.N.; Writing—original draft, A.N.R.; Writing—review & editing, D.B., A.T.J. and C.J.S. All authors have read and agreed to the published version of the manuscript.

Funding: This research was funded by Commonwealth Scientific and Industrial Research Organisation grant number SIP331.

Institutional Review Board Statement: Not applicable.

Informed Consent Statement: This research was approved by the CSIRO Queensland Animal Ethics Committee—AEC Number: 2018-44.

Data Availability Statement: Data available upon request.

Acknowledgments: The authors thank Russell McCulloch and Roger Chong for assistance with the histology analysis and interpretation. We also thank Barney Hines, Nicholas Bourne and our late colleague Aijun Yang for the chemical analyses.

Conflicts of Interest: The authors declare no conflict of interest.

References

1. Gatlin, D.M.; Barrows, F.T.; Brown, P.; Dabrowski, K.; Gaylord, T.G.; Hardy, R.W.; Herman, E.; Hu, G.; Krogdahl, A.; Nelson, R.; et al. Expanding the utilization of sustainable plant products in aquafeeds: A review. *Aquac. Res.* **2007**, *38*, 551–579. [CrossRef]
2. NRC (National Research Council). *Nutrient Requirements of Fish and Shrimp*; Natlional Academic Press: Cambridge, MA, USA, 2011. [CrossRef]
3. Bakke-McKellep, A.M.; Press, C.M.; Baeverfjord, G.; Krogdahl, A.; Landsverk, T. Changes in immune and enzyme histochemical phenotypes of cells in the intestinal mucosa of Atlantic salmon, *Salmo salar* L., with soybean meal-induced enteritis. *J. Fish Dis.* **2000**, *23*, 115–127. [CrossRef]
4. Krogdahl, A.; Bakke-McKellep, A.M.; Baeverfjord, G. Effects of graded levels of standard soybean meal on intestinal structure, mucosal enzyme activities, and pancreatic responses in Atlantic salmon (*Salmo salar* L.). *Aquac. Nutr.* **2003**, *9*, 361–371. [CrossRef]
5. Fuentes-Quesada, J.P.; Viana, M.T.; Rombenso, A.N.; Guerrero-Renteria, Y.; Nomura-Solis, M.; Gomez-Calle, V.; Lazo, J.P.; Mata-Sotres, J.A. Enteritis induction by soybean meal in *Totoaba macdonaldi* diets: Effects on growth performance, digestive capacity, immune response and distal intestine integrity. *Aquaculture* **2018**, *495*, 78–89. [CrossRef]
6. Fuentes-Quesada, J.P.; Cornejo-Granados, F.; Mata-Sotres, J.A.; Ochoa-Romo, J.P.; Rombenso, A.N.; Gurerrero-Renteria, Y.; Lazo, J.P.; Pohlenz, C.; Ochoa-Leyva, A.; Viana, M.T. Prebiotic agavin in juvenile tototaba, *Totoaba macdonaldi* diets, to relieve soybean meal-induced enteritis: Growth performance, gut histology and microbiota. *Aquac. Nutr.* **2020**. [CrossRef]
7. Viana, M.T.; Rombenso, A.N.; Del Rio-Zaragoza, O.B.; Nomura, M.; Diaz-Arguello, R.; Mata-Sotres, J.A. Intestinal impairment of the California yellowtail, *Seriola dorsalis*, using soybean meal in the diet. *Aquaculture* **2019**, *513*, 734443. [CrossRef]

8. Rossi, W., Jr.; Allen, K.M.; Habte-Tsion, H.-M.; Meesala, K.-M. Supplementation of glycine, prebiotic, and nucleotides in soybean meal-based diets for largemouth bass (*Micropterus salmoides*): Effects on production performance, whole-body nutrient composition and retention, and intestinal histopathology. *Aquaculture* **2021**, *532*, 736031. [CrossRef]

9. Krogdahl, A.; Kortner, T.M.; Jaramillo-Torres, A.; Gamil, A.A.A.; Chikwati, E.; Li, Y.; Schmidt, M.; Herman, E.; Hymowitz, T.; Teimouri, S.; et al. Removal of three proteinaceous antinutrients from soybean does not mitigate soybean-induced enteritis in Atlantic salmon (*Salmo salar*, L.). *Aquaculture* **2020**, *514*, 734495. [CrossRef]

10. Uran, P.A.; Schrama, J.W.; Rombout, J.H.W.M.; Obach, A.; Jensen, L.; Koppe, W.; Verreth, J.A.J. Soybean meal-induced enteritis in Atlantic salmon (*Salmo salar* L.) at different temperatures. *Aquac. Nutr.* **2008**, *14*, 324–330. [CrossRef]

11. Mata-Sotres, J.A.; Viana, M.T.; Tinajero, A.; Del Rio-Zaragoza, O.; Skrzynska, A. Estrés nutricional en peces y los nuevos desafíos de la nutrición. In *Revista Bio Ciencias–Coloquio de Nutrigenomica y Biotecnología Acuicola 2020*; Universidad Autónoma de Nayarit: Tepic, Mexico, 2021; p. e1132. [CrossRef]

12. Stone, D.A.J.; Bellgrove, E.J.; Forder, R.E.A.; Howarth, G.S.; Bansemer, M.A. Inducing subacute enteritis in yellowtail kingfish *Seriola lalandi*: The effect of dietary inclusion of soybean meal and grape seed extract on hindgut morphology and inflammation. *N. Am. J. Aquac.* **2018**, *80*, 59–68. [CrossRef]

13. Legrand, T.P.R.A.; Wynne, J.W.; Weyrich, L.S.; Oxley, A.P.A. Investigation both mucosal immunity and microbiota in response to gut enteritis in yellowtail kingfish. *Microorganisms* **2020**, *8*, 1267. [CrossRef]

14. Uran, P.A.; Goncalvez, A.A.; Taverne-Thiele, J.J.; Schrama, J.W.; Verreth, J.A.J.; Rombout, J.H.W.M. Soybean meal induces intestinal inflammation in common carp (*Cyprinus carpio* L.). *Fish Shellfish Immunol.* **2008**, *25*, 751–760. [CrossRef] [PubMed]

15. Van Der Marel, M.; Propsting, M.J.; Battermann, F.; Jung-Schroers, V.; Hubner, A.; Rombout, J.H.W.M.; Steinhagen, D. Differences between intestinal segments and soybean meal induced changes in intestinal mucus composition of common carp *Cyprinus carpio* L. *Aquac. Nutr.* **2014**, *20*, 12–24. [CrossRef]

16. Van Den Ingh, T.S.G.A.M.; Krogdahl, A.; Olli, J.J.; Hendriks, H.G.C.J.M.; Koninkx, J.G.J.F. Effects of soybean-containing diets on the proximal and distal intestine in Atlantic salmon (*Salmo salar*) L. a morphological study. *Aquaculture* **1991**, *94*, 297–305. [CrossRef]

17. Baeverfjord, G.; Krogdahl, A. Development and regression of soybean meal induced enteritis in Atlantic salmon, *Salmo salar* L.; distal intestine: A comparison with the intestines of fasted fish. *J. Fish Dis.* **1996**, *19*, 375–387. [CrossRef]

18. Booman, M.; Forster, I.; Vederas, J.C.; Groman, D.B.; Jones, S.R.M. Soybean meal-induced enteritis in Atlantic salmon (*Salmo salar*) and Chinook salmon (*Oncorhynchus tshawystscha*) but not in pink salmon (*O. gorbuscha*). *Aquaculture* **2018**, *483*, 238–243. [CrossRef]

19. Buttle, L.G.; Burrells, A.C.; Good, J.E.; Williams, P.D.; Southgate, P.J.; Burrells, C. The binding of soybean agglutinin (SBA) to the intestinal epithelium of Atlantic salmon, *Salmo salar* and Rainbow trout, *Oncorhynchus mykiss*, fed high levels of soybean meal. *Vet. Immunol. Immunopathol.* **2001**, *80*, 237–244. [CrossRef]

20. Krogdahl, A.; Gajardo, K.; Kortner, T.M.; Penn, M.; Gu, M.; Berge, G.M.; Bakke, A.M. Soya saponins induce enteritis in Atlantic salmon (*Salmo salar* L.). *J. Agric. Food Chem.* **2015**, *63*, 3887–3902. [CrossRef] [PubMed]

21. Barrows, F.T.; Stone, D.A.J.; Hardy, R.W. The effects of extrusion conditions on the nutritional value of soybean meal for rainbow trout (*Oncorhynchus mykiss*). *Aquaculture* **2007**, *265*, 244–252. [CrossRef]

22. Yang, A.; Smyth, H.; Chalija, M.; James, A. Sensory quality of soymilk and tofu from soybeans lacking lipoxygenases. *Food Sci. Nutr.* **2016**, *4*, 207–215. [CrossRef]

23. James, A.T.; Yang, A. Interactions of protein content and globulin subunit composition of soybean proteins in relation to tofu gel properties. *Food Chem.* **2016**, *194*, 284–289. [CrossRef]

24. AOAC. *Official Methods of Analysis*; Association of Analytical Chemists: Arlington, VA, USA, 2005.

25. Folch, J.; Less, M.; Sloane-Stanley, G.H. A simple method for the isolation and purification of total lipid from animal tissues. *J. Biol. Chem.* **1957**, *276*, 497–507. [CrossRef]

26. Feng, L.; Ni, P.-J.; Jiang, W.-D.; Wu, P.; Liu, Y.; Jiang, J.; Kuang, S.-Y.; Tang, L.; Tang, W.-N.; Zhang, Y.-A.; et al. Decreased enteritis resistance ability by dietary low or excess levels of lipids through impairing the intestinal physical and immune barriers function of young grass carp (*Ctenopharyngodon Idella*). *Fish Shellfish Immunol.* **2017**, *67*, 493–512. [CrossRef] [PubMed]

27. Olli, J.J.; Krogdahl, A.; Vabano, A. Dehulled solvent-extracted soybean meal as a protein source in diets for Atlantic salmon, *Salmo salar* L. *Aquac. Res.* **1995**, *26*, 167–174. [CrossRef]

28. Knudsen, D.; Uran, P.; Arnous, A.; Koppe, W.; Frokler, H. Saponing-containing subfractions of soybean molasses induce enteritis in the distal intestine of Atlantic salmon. *J. Agric. Food Chem.* **2007**, *55*, 2261–2267. [CrossRef] [PubMed]

29. Uran, P.A.; Schrama, J.W.; Jaafari, S.; Baardsen, G.; Rombout, J.H.W.M.; Koppe, W.; Verreth, J.A.J. Variation in commercial sources of soybean meal influences the severity of enteritis in Atlantic salmon (*Salmo salar* L.). *Aquac. Nutr.* **2009**, *15*, 492–499. [CrossRef]

30. Chikwati, E.M.; Sahlmann, C.; Holm, H.; Penn, M.H.; Krogdahl, A.; Bakke, A.M. Alterations in digestive enzyme activities during the development of diet-induced enteritis in Atlantic salmon, *Salmo salar* L. *Aquaculture* **2013**, *402–403*, 28–37. [CrossRef]

31. Sahlmann, C.; Sutherland, B.J.G.; Kortner, T.M.; Koop, B.F.; Krogdahl, A.; Bakke, A.M. Early response of gene expression in the distal intestine of Atlantic salmon (*Salmo salar* L.) during the development of soybean meal induced enteritis. *Fish Shellfish Immunol.* **2013**, *34*, 599–609. [CrossRef]

32. Silva, P.F.; McGurk, C.; Knudsen, D.L.; Adams, A.; Thompson, K.D.; Bron, J.E. Histological evaluation of soya bean-induced enteritis in Atlantic salmon (*Salmo salar* L.): Quantitative image analysis vs. semi-quantitative visual scoring. *Aquaculture* **2015**, *445*, 42–56. [CrossRef]

33. Jacobsen, H.J.; Samuelsen, T.A.; Girons, A.; Kousoulaki, K. Different enzyme incorporation strategies in Atlantic salmon diet containing soybean meal: Effects on feed quality, fish performance, nutrient digestibility and distal intestinal morphology. *Aquaculture* **2018**, *491*, 302–309. [CrossRef]
34. Refstie, S.; Korsoen, O.J.; Storebakken, T.; Baeverfjord, G.; Lein, I.; Roem, A.J. Differing nutritional responses to dietary soybean meal in rainbow trout (*Oncorhynchus mykiss*) and Atlantic salmon *(Salmo salar)*. *Aquaculture* **2000**, *190*, 49–63. [CrossRef]

Article

Gill Histopathology as a Biomarker for Discriminating Seasonal Variations in Water Quality

Zoran Marinović [1,*], **Branko Miljanović** [2], **Béla Urbányi** [1] and **Jelena Lujić** [3]

[1] Department of Aquaculture, Institute of Aquaculture and Environmental Safety, Hungarian University of Agriculture and Life Sciences, 2100 Gödöllő, Hungary; Urbanyi.Bela@uni-mate.hu

[2] Department of Biology and Ecology, Faculty of Sciences, University of Novi Sad, 21000 Novi Sad, Serbia; branko.miljanovic@dbe.uns.ac.rs

[3] Center for Reproductive Genomics, Department of Biomedical Sciences, Cornell University, Ithaca, NY 14850, USA; jelenalujic@cornell.edu

* Correspondence: zor.marinovic@gmail.com

Featured Application: The study highlights the utility of using histopathological alterations in gills of an indigenous fish species (common bream) in differentiating seasonal variations in water quality. The methodology displayed in this study can be used to obtain more compelling and comprehensive environmental monitoring programs.

Abstract: Histopathological alterations in various fish organs have a pronounced value in aquatic toxicology and are widely used in environmental monitoring. The aim of this study was to evaluate whether histopathological alterations in fish gills can discriminate seasonal variations in environmental conditions within the same aquatic ecosystem, and if so, which alterations contributed the most to seasonal differentiation. Microscopic examination of common bream *Abramis brama* gills displayed various alterations in gill structure, including epithelial hypertrophy, hyperplasia, mucous and chloride cell alterations, epithelial lifting, necrosis, hyperemia and aneurism. These alterations were subsequently quantified by a semi-quantitative analysis in order to detect differences in the intensity of the mentioned alterations. Epithelial hypertrophy, hyperplasia, epithelial lifting and necrosis varied significantly between seasons with only necrosis being significantly higher in the first season. Discriminant canonical analysis displayed that epithelial hyperplasia, mucous cell alterations, epithelial lifting and necrosis contributed the most to discrimination between seasons. Overall, this study demonstrates that histopathological biomarkers in fish gills can be used in discriminating seasonal variations in water quality within the same aquatic ecosystem.

Keywords: environmental monitoring; histopathological biomarkers; histopathological alterations; fish gills

Citation: Marinović, Z.; Miljanović, B.; Urbányi, B.; Lujić, J. Gill Histopathology as a Biomarker for Discriminating Seasonal Variations in Water Quality. *Appl. Sci.* **2021**, *11*, 9504. https://doi.org/10.3390/app11209504

Academic Editors: Panagiotis Berillis and Božidar Rašković

Received: 8 September 2021
Accepted: 4 October 2021
Published: 13 October 2021

Publisher's Note: MDPI stays neutral with regard to jurisdictional claims in published maps and institutional affiliations.

1. Introduction

Increased anthropogenic activity in terms of rising industrial, mining and agricultural activity over the last few decades has caused increased water pollution. The most common methods for monitoring water quality in aquatic ecosystems are measurements of pollutant concentrations in water and sediment [1]. Since these methods are not cost-efficient and they do not provide information on biological reactions of aquatic organisms, a vast array of biomarkers has been developed in order to qualitatively and quantitatively assess biological reactions of aquatic organisms to water pollution [2].

Histopathological alterations in various fish organs have a pronounced value in aquatic toxicology [3] and are widely used in the monitoring of aquatic pollution [4–6]. Among them, gills are particularly useful in environmental monitoring given that they have a large surface area, thin epithelium, are in direct contact with water and are generally considered to be among most affected organs by waterborne compounds [7,8]. So far, many studies have assessed gill structural changes in response to waterborne pollutants and environmental pollution, and have recognized them as valuable biomarkers of water pollution [3,5,9–11].

Histopathological alterations are determined as reliable biomarkers in differentiating levels of pollution between different aquatic ecosystems [4,5,12,13]. However, there are very few studies which account for seasonal differences in gill reactions and which analyze the differences in gill reaction to seasonal variations in water quality. The aim of the present study was to evaluate whether histopathological alterations in bream gills can discriminate seasonal variations in water quality within the same aquatic ecosystem. The present study was a part of a much larger monitoring study conducted on the Tamiš River; therefore, it did not aim to make a strong correlation between specific water quality parameters or specific pollutants and gill alterations, but rather to assess whether gill histopathological analysis can be utilized for the monitoring of seasonal differences in water quality. As a study organism we used freshwater bream, *Abramis brama*, an indigenous freshwater fish species which is commonly found in the Danube River basin [14]. Key features that make this species a good bioindicator are its bottom-dwelling nature, as it is exposed to pollution from both water and sediment during feeding on benthic animals [14], as well as its stationary nature, as it exhibits high site fidelity [15,16], and thus alterations observed would most likely be a depiction of local environmental conditions. Therefore, an additional aim was to evaluate if freshwater bream could be a good study organism for pollution monitoring. In achieving the mentioned aims, we can further expand our knowledge on biological responses of aquatic organisms to variations in environmental conditions and can apply this knowledge to environmental monitoring.

2. Materials and Methods

2.1. Sampling Site and Water Quality Data

The Tamiš River is located in the eastern part of the Vojvodina province, Republic of Serbia. It is the longest river in the Banat region and its basin covers a total of 10,352 km^2 (Figure 1). Anthropogenic activity has had a significant impact on the hydrologic regime and water as well as sediment quality of the Tamiš River [17]. The main sources of pollution are irrigation channels which are not maintained properly, fish and livestock farms, sewage effluents and effluents from the mineral industry of the city of Sečanj. Water quality data were obtained from the Hydrological Yearbook of Republic of Serbia [18] and through personal communication with Republic Hydrometeorological Service of Serbia (RHSS) for 2010. We used water quality parameters for the Jaša Tomić water sampling site (located around 10 km upstream from the Sečanj fish sampling site and around 15 km upstream from the Banatski Despotovac fish sampling site; Figure 1).

2.2. Sample Collection and Processing

A total of 33 freshwater bream, *Abramis brama,* individuals (total length: 31.4 ± 0.7 cm; weight: 333.8 ± 24.7 g) were caught along the Tamiš River at the Banatski Despotovac (45°16′58.6″ N, 20°37′51.7″ E) and Sečanj (45°21′28.6″ N, 20°46′22.2″ E) sites (Figure 1). They were caught during October (hereafter: season 1) and April of the next year (hereafter: season 2) with gill nets of various mesh sizes and standard electrofishing devices. Fish were sacrificed with a quick blow to the head and the mid-part of the second gill arch from the left side of each individual was sampled and fixed in 10% formalin. Samples were later decalcified, dehydrated in graded ethanol series, cleared in xylene and embedded into paraffin blocks [19]. Five-micron-thick sections were cut, and three sections were placed on glass slides and stained with the standard hematoxylin and eosin (H&E) staining technique. Sections were examined by a Primostar light microscope (Carl Zeiss, Heidenheim, Germany) and photographed by an AxioCam MRc 5 digital camera (Carl Zeiss).

2.3. Histopathological Analysis

Ten randomly selected gill filaments per individual were first subjected to qualitative evaluation and thereafter to semi-quantitative analysis. Semi-quantitative analysis included calculation of alteration indices for the most pronounced alterations. The eight most pronounced alterations chosen for this analysis were epithelial hypertrophy, epithelial hyperplasia, mucous cell alterations (including both hypertrophy and hyperplasia), chloride cell alterations (also including both hypertrophy and hyperplasia), epithelial lifting, necrosis, hyperemia and aneurism. Alteration indices were calculated according to Bernath et al. [20]. Briefly, the importance factor for each alteration presents the pathological relevance (how the alteration affects the organ function) of the given alteration, and it ranges from 1 (minimal importance—the alteration is reversible when exposure to the irritant stops) to 3 (marked importance—the alteration is irreversible). To each alteration a score value ranging from 0 to 6 was assigned. Namely, a score value was assigned for each of the 10 filaments analyzed based on the extent of that particular alteration—0 meaning that the change was not detected; 6 indicating that the alteration was present throughout the whole filament. Thereafter, a score value for an individual fish was calculated as a mean of score values assigned to the 10 filaments. An alteration index (AI) was calculated by multiplying the importance factor and score value for the given alteration according to the formula $AI = IF \times SV$, where IF is the importance factor (0–3) and SV is the score value (0–6). Additionally, reaction patter indices (I_{RP}) were calculated by summing all alteration indices for the given pattern according to the formula $I_{RP} = \sum_{RP} AI$, where AI are alteration indices of alterations belonging to that particular reaction pattern. A gill index was calculated by summing all eight alteration indices ($I_G = \sum_{G} AI$).

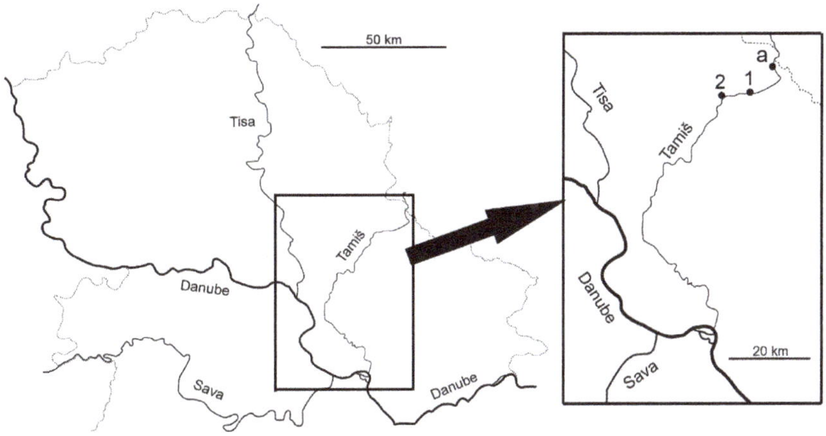

Figure 1. Map of the Vojvodina province (Republic of Serbia) and Tamiš River. a—Jaša Tomić water quality sampling station; 1—Sečanj sampling site; and 2—Banatski Despotovac sampling site. Cited and modified from Lujić et al. [10] with permission from the publisher.

2.4. Statistical Analysis

All measurements were expressed as mean $\pm$ S.E. Normality of variance was checked using the Shapiro–Wilks normality test. One-way analysis of variance (ANOVA), followed by Tukey's HSD post hoc test was used to determine whether differences in alteration indices between sites and seasons were significant ($p < 0.05$). Discriminant canonical analysis (DCA) was conducted to determine the contribution of each histopathological alteration to seasonal differentiation. Given that the water quality was assessed only at the Jaša Tomić water quality sampling site, and not at every fish sampling site separately, both the B. Despotovac and Sečanj sites from season 2, alongside the Sečanj site from season 1, were submitted as grouping factors in order to test the possible influence of sites on variation in histological alterations. All statistical analyses were conducted in STATISTICA v12.0 software (Statsoft Inc., Tulsa, OK, USA).

3. Results and Discussion

3.1. Water Quality

Water quality analysis displayed certain differences in several parameters between seasons (Table 1). Seasonal differences were mainly encompassed by levels of electroconductivity, ammonia, chlorides, Cu, Fe and Pb. According to WHO guidelines [21], only the levels of Fe from the Tamiš River were above the guideline limit in both seasons, while Pb values were just below the limit during season one. On the other hand, according to maximum admissible values proposed by Svobodova et al. [22], levels of Fe, Cu and chlorine from the Tamiš River were elevated in both seasons and could potentially cause tissue damage in cyprinid fish. Chemical analysis of water can provide valuable data for interpretation of observed histopathological alterations by pointing out possible pollutants which could induce observed alterations [10].

Table 1. Water quality parameters for the Jaša Tomić site in different seasons [18].

Parameter	Season 1	Season 2	WHO [21]
Water temperature (°C)	8.2	14.8	-
Water transparency (mm)	150	-	-
Suspended matters (mg/L)	39	35	-
Dissolved oxygen (mL/L)	10.4	9.5	-
Alkalinity (mmol/L)	1.7	1.6	-
pH	7.6	7.5	-
Electroconductivity (µS/cm)	235	579	-
Ammonia (mg/L)	0.2	0.05	No guidelines
Nitrates (mg/L)	0.77	0.56	-
Nitrites (mg/L)	0.01	0.013	-
Orthophosphates (mg/L)	0.028	0.032	-
Chlorides (mg/L)	12	7.3	No guidelines
Sulphates (mg/L)	23	25	-
Fe (mg/L)	1.24	1.78	1
Mn (mg/L)	0.07	0.05	0.5
Zn (mg/L)	0.04	0.04	5
Cu (mg/L)	0.02	0.007	2
Total Cr (mg/L)	0.002	0.002	0.05
Pb (mg/L)	0.009	<0.001	0.01
Cd (µg/L)	0.2	0.2	3
Hg (µg/L)	0.1	<0.1	1
Ni (mg/L)	0.006	0.004	0.02
As (mg/L)	0.001	<0.001	0.01

3.2. Histopathological Observations

Microscopic examination of bream gills demonstrated alterations in the gill structure (Figure 2). Epithelial hyperplasia (Figure 2A), hypertrophy (Figure 2B), epithelial lifting (Figure 2C) and necrosis (Figure 2D) were the most pronounced alterations detected in the present study. In some instances, hemorrhage (Figure 2A) or lymphocyte infiltration indicative of mild inflammation were also observed (Figure 2B). These alterations are commonly observed in histopathological studies. Mallatt [8] has hypothesized that epithelial hypertrophy and necrosis are more frequently the result of heavy metal exposure, which could be correlated with the higher Fe and Cu concentrations observed in the water of the Tamiš River. However, ascribing the cause of certain gill histopathological alterations to specific environmental conditions or pollutants is not possible, as gills mostly react in a generalized way; therefore, many different factors can cause the same structural alterations [8,23,24].

Furthermore, even though parasitic examination was not in the scope of the study, a very small number of *Trichodina* sp. were found in a few fish from both seasons.

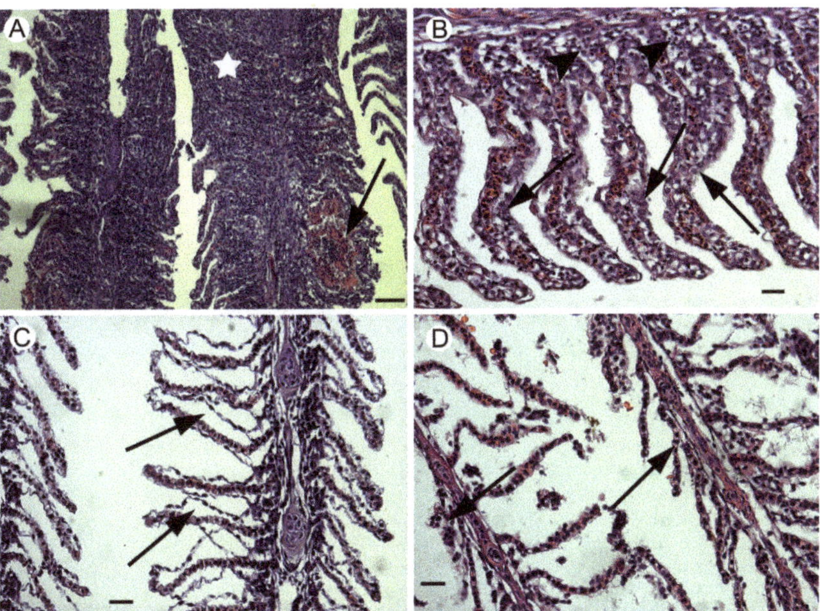

Figure 2. Histopathological alterations observed in gills of *Abramis brama* during the two investigated seasons in the Tamiš River. (**A**) Hyperplasia of epithelium leading to fusions of lamellae (star) and the focal rupture of secondary lamellae leading to hemorrhage (arrow). (**B**) Epithelial hypertrophy (arrows) as well as leukocyte infiltration indicative of inflammation. (**C**) Epithelial lifting (arrows). (**D**) Rupture of the primary and secondary epithelium and necrosis (arrows). H&E. Scale bars: A = 50 μm; B, C and D = 20 μm.

Structural alterations mainly arise from two principal reasons: either as a defense mechanism, or as a result of direct damage cause by environmental factors/xenobiotics. Epithelial hyperplasia, hypertrophy and lifting are considered to be defense mechanisms of gills as the former decreases the respiratory surface, while the latter two increase the diffusion distance, thus hindering xenobiotic uptake [7,8]. These alterations are reversible, meaning that the gill tissue is expected to recover once exposure to irritants ends [20]. On the other hand, alterations such as necrosis, aneurism or hemorrhage are considered to be a result of xenobiotic action, and are generally considered to be irreversible. Furthermore, many parasites can also induce histopathological alterations such as hyperplasia, telangiectasia, inflammation or even necrosis [25,26]. Therefore, conducting parasitic investigations prior to histopathological investigation is imperative for successfully delineating possible effects of the environment vs. the effects of parasites. Additionally, special attention needs to be given to possible artifacts and misdiagnosis, as certain alterations, such as epithelial lifting, can be the result of handling artifacts and can therefore be misinterpreted as a lesion [27].

3.3. Seasonal Variation in Histopathological Alterations

Epithelial hypertrophy, hyperplasia, lifting and necrosis varied significantly between seasons (Table 2). Necrosis was significantly higher in season one, while other three alterations were higher in season two. Gill indices and reaction pattern indices on the other hand did not display seasonal variability (Figure 3A). Discriminant canonical analysis provided two canonical functions to explain differentiation between seasons (Table 3). Differentiation between seasons was mainly contributed by epithelial hyperplasia, mucous cell alterations, epithelial lifting and necrosis. The first canonical function explained most of the variance (75.6%) and was contributed by epithelial hyperplasia, mucous cell alterations, epithelial lifting and necrosis, while the second canonical function accounted for only 24.4% of variance and was mostly associated with epithelial hypertrophy. An obvious separation of samples from season one was observed along the first canonical function ($p < 0.01$) (Figure 3B), while there was no significant separation of centroids along the second canonical function ($p = 0.19$). Considering no alterations showed significant differences between B. Despotovac and Sečanj (season 2) sites and there was a slight, but not significant separation of these two sites along the second canonical function, it can be concluded that the influence of sites on variation in gill alterations was relatively negligible.

Table 2. Histopathological alterations and alteration indices for gills of fish sampled in the Tamiš River in different seasons.

Alteration (Pathological Relevance)	Season (Site)			All Effects ANOVA		
	Season 1	Season 2		F	d.f.	p
	Sečanj (N = 5)	B. Despotovac (N = 8)	Sečanj (N = 20)			
Epithelial hypertrophy (1)	1.28 ± 0.20 [a]	1.52 ± 0.20 [a,b]	2.30 ± 0.20 [b]	5.08	2	0.01
Epithelial hyperplasia (2)	0.48 ± 0.23 [a]	3.40 ± 0.64 [b]	3.01 ± 0.35 [b]	6.54	2	<0.01
Mucous cell alterations (2)	2.00 ± 1.56 [a]	1.34 ± 0.65 [a]	1.05 ± 0.33 [a]	0.48	2	0.62
Chloride cell alterations (2)	0.00 ± 0.00 [a]	0.05 ± 0.05 [a]	0.03 ± 0.02 [a]	0.37	2	0.70
Epithelial lifting (1)	0.92 ± 0.21 [a]	2.02 ± 0.54 [a,b]	2.65 ± 0.31 [b]	3.47	2	0.04
Necrosis (3)	3.84 ± 1.05 [a]	0.09 ± 0.09 [b]	1.47 ± 0.61 [a,b]	3.99	2	0.03
Hyperemia (1)	1.88 ± 0.38 [a]	1.57 ± 0.31 [a]	2.40 ± 0.31 [a]	1.41	2	0.26
Aneurism (1)	0.00 ± 0.00 [a]	0.17 ± 0.09 [a]	0.11 ± 0.03 [a]	1.51	2	0.24

Importance factors are indicated in parenthesis. Within rows, for a given parameter, different letters in the superscript indicate significant statistical difference (i.e., the presence of the same letter in two columns within the same row indicates no significant differences, while different letters in two columns within the same row indicate a significant difference of $p < 0.05$).

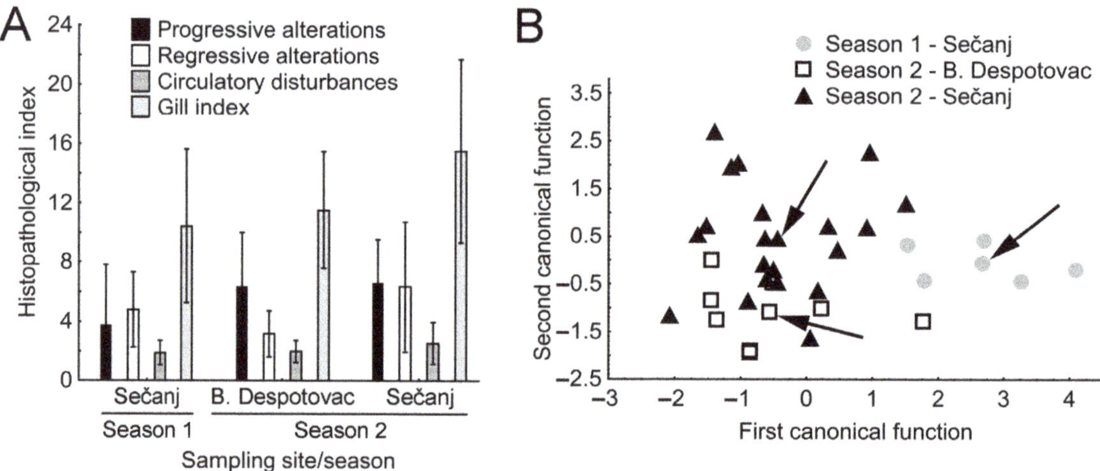

Figure 3. Semi-quantitative and statistical analyses of gill alterations in fish sampled in the Tamiš River. (**A**) Gill indices and reaction pattern (progressive, regressive and circulatory) indices for gills sampled in different seasons. (**B**) Scatterplot of canonical scores for histopathological alterations of fish captured in different seasons. Arrows indicate centroids.

Epithelial lifting, hyperplasia and necrosis were alterations that displayed the most significant seasonal variation in several studies [6,11,28]. When analyzing seasonal variation in common bream, Kostić et al. [6] and Kostić-Vuković et al. [11] also did not report any significant differences in gill or reaction pattern indices, while Barišić et al. [28], analyzing Vardar chub, and Santos et al. [23], analyzing barbell *Luciobarbus bocagei* and Douro nase *Pseudochondrostoma duriense,* observed seasonal differences in progressive, regressive and circulatory reaction patterns. This can point to the differences in species sensitivity to the changing environment. It is important to note that dominant alterations observed in season 1 were irreversible, while dominant alterations observed in season 2 were mostly reversible, indicating defensive mechanisms of the tissue, especially since Fe values were higher in season 2.

Table 3. Discriminant canonical analysis (DCA) for gill histopathological alterations in fish captured in Tamiš River during different seasons.

Alterations	Canonical Function 1	Canonical Function 2
Epithelial hypertrophy	−0.086	0.861
Epithelial hyperplasia	−0.861	−0.427
Mucous cell alterations	0.568	−0.154
Chloride cell alterations	−0.021	−0.184
Epithelial lifting	−0.557	0.262
Necrosis	0.547	0.349
Hyperemia	−0.017	0.255
Aneurism	−0.026	−0.023
Eigenvalue	1.407	0.455
Cum. prop.	0.756	1.000
Canonical R	0.765	0.559
Wilks' lambda	0.256	0.687
d.f.	16	7
p	0.007	0.192

Importance factors are indicated in parenthesis. Within rows, for a given parameter, different letters in the superscript indicate significant statistical difference (i.e., the presence of the same letter in two columns within the same row indicates no significant differences, while different letters in two columns within the same row indicate a significant difference of $p < 0.05$).

The time of sampling during season 1 was in October, when fish are preparing for the winter hibernation (they usually swim to the riverbed and retain a lower level of metabolism until spring). However, in season 2 (April), fish gills displayed different alterations, among them epithelial lifting and hyperplasia, which are considered to be defensive mechanisms. We postulate that alterations with high pathological importance had occurred before the winter hibernation, with fish having been exposed to pollutants during the whole year, but that they recovered during winter hibernation and after it (in spring when fish are very active, and during spawning season) they are ready to better cope with waterborne pollutants and react in a defensive manner. A similar pattern of more severe gill alterations occurring during autumn compared to the spring was observed in another study on common bream [11], but also in roach [24] and Vardar chub [28]. Moreover, studies regarding metal accumulation in fish tissues corroborate this claim, since fish accumulate most of the metals during summer and the least amounts during winter and spring [11,29–31]. These differences in accumulation rate are most likely due to temperature differences, since in summer fish have a higher metabolic rate, hence accumulating most of the metals. Since accumulation levels are low during winter due to their low metabolic rate, this enables the tissue to recover, and at spring they are in a better position to handle pollution pressure.

Many studies report that different pollutants may cause similar gill alterations, suggesting that gills react non-specifically and that gill alterations present a reflection of a generalized stress response [8,23,24]. However, the use of quantitative or semi-quantitative analysis has great advantage over simple qualitative/descriptive analysis as it can quantify the intensity of the given alterations, and therefore be more informative. Quantification can enable comparison of alteration intensity between localities and/or seasons, and therefore indirectly indicate differences in water quality. This indirect assessment of water quality/pollution is very important as not all xenobiotics (such as heavy metals, PAHs, cyanotoxins, estrogen modulators and other) can be quantified during monitoring studies. Therefore, having biomarkers which can effectively discriminate between water quality/pollution between different sites or between seasons on the same site is imperative. Additionally, in this study we have focused on the histopathology of the gills; however, future studies should also include other biomarkers, such as oxidative stress, immunological, biotransformation or reproductive parameters for a more holistic approach. Special attention should be given to parasitic investigations prior to conducting histopathological analysis as many parasites can also induce histopathological alterations, but also to good laboratory and sampling practices as some common histopathological alterations could be the result of artefacts, and therefore can be misdiagnosed.

4. Conclusions

The present study demonstrates that histopathological biomarkers of bream gill alterations can be used in discriminating seasonal variations in water quality within the same aquatic ecosystem and its effects on organisms that live in it. These results contribute to the knowledge of biological responses to variations in environmental conditions and implicate the importance of histopathological biomarkers in environmental monitoring. On the other hand, it is proved that bream could be used in further environmental pollution studies. Given the value of these biomarkers, we further suggest that they should be included as standard methods in environmental monitoring, alongside chemical and physical analysis.

Author Contributions: Conceptualization, J.L. and B.M.; methodology, J.L. and Z.M.; investigation, J.L. and Z.M.; writing—original draft preparation, Z.M.; writing—review and editing, J.L., B.M. and B.U.; supervision, B.U. and J.L. All authors have read and agreed to the published version of the manuscript.

Funding: The research was financially supported by the funds of the Ministry of Education and Science of the Republic of Serbia (project number: 143058), funds of European Union and the city of Pančevo (project number: 06SER02/03/007-8), the Ministry of Innovation and Technology of Hungary within the framework of the Thematic Excellence Programme 2020, Institutional Excellence Subprogramme (TKP2020-IKA-12) and by the EFOP-3.6.3-VEKOP-16-2017-00008 project co-financed by the European Union and the European Social Fund.

Institutional Review Board Statement: Not applicable.

Informed Consent Statement: Not applicable.

Data Availability Statement: All relevant data are within the manuscript.

Acknowledgments: The authors want to express their gratitude to colleagues from the Department of Biology and Ecology, University of Novi Sad, Serbia, that helped during this research, especially the project coordinators Radmila Kovačević and Ivana Teodorović, as well as the hydrobiological team for help during sampling.

Conflicts of Interest: The authors declare no conflict of interest.

References

1. Vasanthi, L.A.; Revathi, P.; Mini, J.; Munuswamy, N. Integrated use of histological and ultrastructural biomarkers in Mugil cephalus for assessing heavy metal pollution in Ennore estuary, Chennai. *Chemosphere* **2013**, *91*, 1156–1164. [CrossRef]
2. van der Oost, R.; Beyer, J.; Vermeulen, N.P.E. Fish bioaccumulation and biomarkers in environmental risk assessment: A review. *Environ. Toxicol. Pharmacol.* **2003**, *13*, 57–149. [CrossRef]
3. Pereira, S.; Pinto, A.L.; Cortes, R.; Fontaínhas-Fernandes, A.; Coimbra, A.M.; Monteiro, S.M. Gill histopathological and oxidative stress evaluation in native fish captured in Portuguese northwestern rivers. *Ecotoxicol. Environ. Saf.* **2013**, *90*, 157–166. [CrossRef]
4. Nascimento, A.A.; Araújo, F.G.; Gomes, I.D.; Mendes, R.M.M.; Sales, A. Fish gills alterations as potential biomarkers of environmental quality in a eutrophized tropical river in South-Eastern Brazil. *Anat. Histol. Embryol.* **2012**, *41*, 209–216. [CrossRef]
5. Rašković, B.; Poleksić, V.; Višnjić-Jeftić, Ž.; Skorić, S.; Gačić, Z.; Djikanović, V.; Jarić, I.; Lenhardt, M. Use of histopathology and elemental accumulation in different organs of two benthophagous fish species as indicators of river pollution. *Environ. Toxicol.* **2015**, *30*, 1153–1161. [CrossRef]

6. Kostić, J.; Kolarević, S.; Kračun-Kolarević, M.; Aborgiba, M.; Gačić, Z.; Paunović, M.; Višnjić-Jeftić, Ž.; Rašković, B.; Poleksić, V.; Lenhardt, M.; et al. The impact of multiple stressors on the biomarkers response in gills and liver of freshwater breams during different seasons. *Sci. Total Environ.* **2017**, *601–602*, 1670–1681. [CrossRef] [PubMed]
7. Roberts, R.J.; Rodger, H.D. The patophysiology and systemic pathology of teleosts. In *Fish Pathology*; Roberts, R.J., Ed.; Wiley and Sons, Ltd.: Hoboken, NJ, USA, 2012; pp. 62–143.
8. Mallatt, J. Fish gill structural changes induced by toxicants and other irritants: A statistical review. *Can. J. Fish. Aquat. Sci.* **1985**, *42*, 630–648. [CrossRef]
9. Lenhardt, M.; Poleksić, V.; Vuković-Gačić, B.; Rašković, B.; Sunjog, K.; Kolarević, S.; Jarić, I.; Gačić, Z. Integrated use of different fish related parameters to assess the status of water bodies. *Slov. Vet. Res.* **2015**, *52*, 5–13.
10. Lujić, J.; Matavulj, M.; Poleksić, V.; Rašković, B.; Marinović, Z.; Kostić, D.; Miljanović, B. Gill reaction to pollutants from the Tamiš River in three freshwater fish species, *Esox lucius* L. 1758, *Sander lucioperca* (L. 1758) and *Silurus glanis* L. 1758: A comparative study. *Anat. Histol. Embryol.* **2015**, *44*, 128–137. [CrossRef]
11. Kostić-Vuković, J.; Kolarević, S.; Kračun-Kolarević, M.; Višnjić-Jeftić, Ž.; Rašković, B.; Poleksić, V.; Gačić, Z.; Lenhardt, M.; Vuković-Gačić, B. Temporal variation of biomarkers in common bream *Abramis brama* (L., 1758) exposed to untreated municipal wastewater in the Danube River in Belgrade, Serbia. *Environ. Monit. Assess.* **2021**, *193*, 465. [CrossRef] [PubMed]
12. Abdel-Moneim, A.M.; Al-Kahtani, M.A.; Elmenshawy, O.M. Histopathological biomarkers in gills and liver of *Oreochromis niloticus* from polluted wetland environments, Saudi Arabia. *Chemosphere* **2012**, *88*, 1028–1035. [CrossRef]
13. Van Dyk, J.C.; Cochrane, M.J.; Wagenaar, G.M. Liver histopathology of the sharptooth catfish *Clarias gariepinus* as a biomarker of aquatic pollution. *Chemosphere* **2012**, *87*, 301–311. [CrossRef]
14. Simonovic, P. *Ribe Srbije [Fishes of Serbia]*; NNK International, Zavod za Zaštitu Prirode Srbije: Belgrade, Serbia, 2001.
15. Lyons, J.; Lucas, M.C. The combined use of acoustic tracking and echosounding to investigate the movement and distribution of common bream (*Abramis brama*) in the River Trent, England. In *Aquatic Telemetry*; Thorstad, E.B., Fleming, I.A., Næsje, T.F., Eds.; Springer: Dordrecht, The Netherlands, 2002; pp. 265–273.
16. Gardner, C.J.; Deeming, D.C.; Wellby, I.; Soulsbury, C.D.; Eady, P.E. Effects of surgically implanted tags and translocation on the movements of common bream *Abramis brama* (L.). *Fish. Res.* **2015**, *167*, 252–259. [CrossRef]
17. Marković, S.; Tomić, P.; Bogranović, Z.; Svirčev, Z.; Lazić, L.; Jovanović, M.; Pavić, D. Pokazatelji i uzroci smanjenja kvaliteta vode u jugoslovenskom delu toka reke Tamiš. [Indicators and causes of decrease in water quality in the Yugoslavian part of the river Tamiš]. In *Naš Tamiš [Our Tamiš]*; Marković, S., Svirčev, Z., Eds.; University of Novi Sad: Novi Sad, Serbia, 1998; pp. 11–18.
18. [RHSS] Republic Hydrometeorological Service of Serbia. *Hydrological Yearbook, 3. Water Quality*; Republic Hydrometeorological Service of Serbia: Belgrade, Serbia, 2010.
19. Suvarna, S.K.; Layton, C.; Bancroft, J.D. *Bancroft's Theory and Practice of Histological Techniques*, 8th ed.; Elsevier: Amsterdam, The Netherlands, 2019; ISBN 9780702068645.
20. Bernet, D.; Schmidt, H.; Meier, W.; Burkhardt-Holm, P.; Wahli, T. Histopathology in fish: Proposal for a protocol to assess aquactic pollution. *J. Fish Dis.* **1999**, *22*, 25–34. [CrossRef]
21. [WHO] World Health Organization. *Guidelines for Drinking-Water Quality*, 4th ed.; World Health Organization: Geneva, Switzerland, 2017.
22. Svobodova, Z.; Lloyd, R.; Machova, J.; Vykusova, B. *Water Quality and Fish Health*; Food and Agriculture Organization of the United Nations: Rome, Italy, 1993.
23. Santos, R.M.B.; Monteiro, S.M.; Cortes, R.M.V.; Pacheco, F.A.L.; Sanches Fernandes, L.F. Seasonal effect of land use management on gill histopathology of Barbel and Douro Nase in a Portuguese watershed. *Sci. Total Environ.* **2021**, *764*, 142869. [CrossRef] [PubMed]
24. Haaparanta, A.; Valtonen, E.T.; Hoffmann, R.W. Gill anomalies of perch and roach from four lakes differing in water quality. *J. Fish Biol.* **1997**, *50*, 575–591. [CrossRef]
25. Santos, M.A.; Jerônimo, G.T.; Cardoso, L.; Tancredo, K.R.; Medeiros, P.B.; Ferrarezi, J.V.; Gonçalves, E.L.T.; da Costa Assis, G.; Martins, M.L. Parasitic fauna and histopathology of farmed freshwater ornamental fish in Brazil. *Aquaculture* **2017**, *470*, 103–109. [CrossRef]
26. Hossain, M.K.; Hossain, M.D.; Rahman, M.H. Histopathology of some diseased fishes. *J. Life Earth Sci.* **2007**, *2*, 47–50. [CrossRef]
27. Wolf, J.C.; Baumgartner, W.A.; Blazer, V.S.; Camus, A.C.; Engelhardt, J.A.; Fournie, J.W.; Frasca, S.; Groman, D.B.; Kent, M.L.; Khoo, L.H.; et al. Nonlesions, misdiagnoses, missed diagnoses, and other interpretive challenges in fish histopathology studies: A guide for investigators, authors, reviewers, and readers. *Toxicol. Pathol.* **2015**, *43*, 297–325. [CrossRef]
28. Barišić, J.; Dragun, Z.; Ramani, S.; Filipović Marijić, V.; Krasnići, N.; Čož-Rakovac, R.; Kostov, V.; Rebok, K.; Jordanova, M. Evaluation of histopathological alterations in the gills of Vardar chub (*Squalius vardarensis* Karaman) as an indicator of river pollution. *Ecotoxicol. Environ. Saf.* **2015**, *118*, 158–166. [CrossRef]
29. Kargin, F. Metal concentrations in tissues of the freshwater fish *Capoeta barroisi* from the Seyhan River (Turkey). *Bull. Environ. Contam. Toxicol.* **1998**, *60*, 822–828. [CrossRef] [PubMed]
30. Duman, F.; Kar, M. Temporal variation of metals in water, sediment and tissues of the European chup (*Squalius cephalus* L.). *Bull. Environ. Contam. Toxicol.* **2012**, *89*, 428–433. [CrossRef] [PubMed]
31. Ibrahim, A.T.A.; Omar, H.M. Seasonal variation of heavy metals accumulation in muscles of the African catfish *Clarias gariepinus* and in River Nile water and sediments at Assiut Governorate, Egypt. *Egypt. J. Biol. Earth Sci.* **2013**, *3*, B236–B248.

applied sciences

Article

Cardiac and Cerebellar Histomorphology and Inositol 1,4,5-Trisphosphate (IP$_3$R) Perturbations in Adult *Xenopus laevis* Following Atrazine Exposure

Jaclyn Asouzu Johnson [1,*], Pilani Nkomozepi [2], Prosper Opute [3] and Ejikeme Felix Mbajiorgu [1]

[1] School of Anatomical Sciences, University of the Witwatersrand, Johannesburg 2193, South Africa; ejikeme.mbajiorgu@wits.ac.za
[2] Department of Human Anatomy and Physiology, University of Johannesburg, Johannesburg 2006, South Africa; pilanin@uj.ac.za
[3] Department of Animal and Environmental Biology, University of Benin, Benin City 1154, Nigeria; prosper.opute@uniben.edu
* Correspondence: jaclyn.johnson@wits.ac.za; Tel.: +277-34-694-543

Abstract: Despite several reports on the endocrine-disrupting ability of atrazine in amphibian models, few studies have investigated atrazine toxicity in the heart and cerebellum. This study investigated the effect of atrazine on the unique Ca^{2+} channel-dependent receptor (Inositol 1,4,5-trisphosphate; IP$_3$R) in the heart and the cerebellum of adult male *Xenopus laevis* and documented the associated histomorphology changes implicated in cardiac and cerebellar function. Sixty adult male African clawed frogs (*Xenopus laevis*) were exposed to atrazine (0 µg/L (control), 0.01 µg/L, 200 µg/L, and 500 µg/L) for 90 days. Thereafter, heart and cerebellar sections were processed with routine histological stains (heart) or Cresyl violet (brain), and IP$_3$R histochemical localization was carried out on both organs. The histomorphology measurements revealed a significant decrease in the mean percentage area fraction of atrial (0.01 µg/L and 200 µg/L) and ventricular myocytes (200 µg/L) with an increased area fraction of interstitial space, while a significant decrease in Purkinje cells was observed in all atrazine groups ($p < 0.008$, 0.001, and 0.0001). Cardiac IP$_3$R was successfully localized, and its mean expression was significantly increased (atrium) or decreased (cerebellum) in all atrazine-exposed groups, suggesting that atrazine may adversely impair cerebellar plasticity and optimal functioning of the heart due to possible disturbances of calcium release, and may also induce several associated cardiac and neural pathophysiologies in all atrazine concentrations, especially at 500 µg/L.

Keywords: atrazine; Purkinje; cerebellum; myocytes; toxicology; IP$_3$Rs

Citation: Asouzu Johnson, J.; Nkomozepi, P.; Opute, P.; Mbajiorgu, E.F. Cardiac and Cerebellar Histomorphology and Inositol 1,4,5-Trisphosphate (IP$_3$R) Perturbations in Adult *Xenopus laevis* Following Atrazine Exposure. *Appl. Sci.* **2021**, *11*, 10006. https://doi.org/10.3390/app112110006

Academic Editors: Panagiotis Berillis and Božidar Rašković

Received: 22 September 2021
Accepted: 20 October 2021
Published: 26 October 2021

1. Introduction

Atrazine is designed specifically to impair the physiology of weeds [1]; however, the physical and chemical properties of atrazine, such as its half-life, slow degradation [2], and solubility in different mediums, contribute to its persistence in the environment and, consequently, its contamination of ground, surface, and drinking water [3–5]. This suggests a constant presence of atrazine and its degradation products in the environment and, therefore, in drinking water and aquatic and terrestrial environments [3], with inevitable pathological consequences.

The effects of atrazine exposure have been widely reported such as impaired development in fish [6] and perturbed immunology and growth in *Xenopus laevis* [7,8]. However, most information on atrazine-induced effects in adult animal models have focused mostly on its endocrine-disrupting abilities, particularly in *Xenopus laevis* gonadal morphology [9–11]. Furthermore, atrazine and its metabolites can bio-accumulate in aquatic microbiota with adverse aquatic ecosystem implications [12] and can induce several organ

107

toxicities in non-target organisms. In organs such as the cerebellum, mostly physiological and biochemical effects of atrazine have been reported relative to perturbations of cerebellar pathophysiology and cerebral oxidative stress in quails and rats [13–15], as well as behavioral and neurotransmitter deficits in mice [16]. However, complementary reports on structural toxicity (histopathology) is lacking in the cerebellum, while atrazine cardiac toxicity in frogs has only been previously reported by us [17]. Furthermore, atrazine is widely known for its ability to disrupt estrogen and androgen receptors [18], but little is known about atrazine-induced effects on the histopathology of voltage-gated receptors (IP$_3$Rs) in the heart (peripheral) and cerebellum (central) and the differential effects in both organs relative to their crucial functions.

The frog heart and brain contain an abundance of excitable cells with voltage-gated channels on their membranes. Sodium, potassium, and calcium are physiologically important ions, and their homeostatic concentrations in human serum levels and frog heart tissue are vital for normal neuronal and cardiac function [19,20]. In cardiomyocytes, each calcium ion released is accompanied by sodium transportation via sodium–potassium pumps [21,22]. Additionally, calcium release is directly required for the initiation of cardiac excitation–contraction coupling and the activation of neuronal dendritic development, neuronal survival, and synaptic plasticity [23,24]. Furthermore, cardiac cells and cerebellar neurons depend on IP$_3$Rs for modulating calcium levels in response to external stimuli [25,26]. Physiologically, myocytes and neurons are dependent on the efflux of calcium into the cytosol for excitation–contraction coupling and modulation of synaptic transmission, respectively, particularly cerebellar Purkinje cells, where IP$_3$Rs are predominantly expressed [27]. Pathologically, the IP$_3$Rs levels and genes in the heart and cerebellum have been associated with hypertrophy and spinocerebellar ataxia, respectively [28,29]. However, histopathological evidence of atrazine IP$_3$R toxicity in the *Xenopus* frog cerebellum and heart is lacking.

Hence, this study is directed at investigating the possible adverse impact of atrazine on the IP$_3$Rs and its expression in the heart (peripherally located) and cerebellum (centrally located) of the adult *Xenopus* frog at different concentrations of environmentally relevant levels of 0.01 µg/L, 200 µg/L [7,30], and 500 µg/L, which has been reported in ambient surface water of sugarcane growing areas [31]. Since IP$_3$Rs play critical roles in cardiac and cerebellar functions in frog species, the possible toxicity of atrazine on IP$_3$Rs and its expression in the heart and cerebellum of adult *Xenopus* frogs will provide invaluable additional information on the atrazine pathophysiology of these organs.

2. Materials and Methods

2.1. Animal Husbandry

Forty-eight (48) male African *Xenopus laevis* between 9 and 10 months of age were purchased from local breeders (Knysner, Cape Town, South Africa) and used in the study. Six frogs per group were housed in metal cisterns (225 × 24 × 21 cm) containing 60 L of water at the University of Witwatersrand and allowed to acclimate to their environment for 2 weeks with a 12 h light and 12 h dark cycle before the experiment began. A variable room heating system was used to maintain the water temperature at 22 ± 2 °C and the water was recycled (100%) thrice weekly to achieve an adequate hygienic environment. The animals were permitted unlimited access to nutritious commercial fish pellets (Koi food; Daro Pet Products, Johannesburg, South Africa), which were augmented with beef liver pieces weekly. Ethical approval for this work was obtained from Gauteng Province Nature Conservation (CPF6 0115 (2015) and CPF6 0120, 2016) and the University of the Witwatersrand Animals Ethics Research Committee (2014/32/D), and the experiment was conducted in line with the approved standard procedures.

2.2. Treatment

The treatment procedure has previously been reported elsewhere [17]. Atrazine powder (99.5%; Accu Standard Incorporated, New Haven, CT, USA) was used to prepare

30 mg/L of standard solution in distilled water and then diluted to make 60 L of atrazine according to the varying experimental concentrations in four different 200 L metal cisterns/aquariums. The animals (male frogs of post-anuran metamorphosis age) were then exposed to 0 (control), 0. 01 µg/L, 200 µg/L, and 500 µg/L of atrazine solution for 90 days. Since the *Xenopus* frog has a semi-permeable skin, the exposure of *Xenopus laevis* frogs to atrazine (90 days) in tanks accurately simulates environmental exposure to atrazine in ponds and ground water. The atrazine exposure treatments were duplicated between July–September 2015 and February–April 2016.

To guarantee a consistent atrazine exposure level throughout the experiment, a solid phase micro-extraction (SPME) coupled to gas chromatography-mass spectrometry (GC-MS, GC 6890, MS 5975, Agilent Technologies, CA, USA) was utilized in the weekly measurement of atrazine levels in the reservoirs. Therefore, 50 mL of water was collected (before and after each water recycle) from each reservoir and passed through a 200 mg Bond Elut Plexa (Chemetrix, Johannesburg) solid-phase extraction cartridge. The results indicated an absence of atrazine in the control, but in the treated groups, the atrazine concentrations (% of atrazine above or below the desired concentration) were within the range of $\pm6\%$ in the 0.01 µg/L group, $\pm3.1\%$ in the 200 µg/L group, and $\pm1.6\%$ in the 500 µg/L group, showing insignificant changes in atrazine levels in the respective treatment reservoirs.

At the end of the treatment period, the animals were sacrificed after deep anesthesia (0.2 benzocaine) in bell jar. Whole brains and hearts were harvested from the animals and fixed in 4% paraformaldehyde. Whole fixed hearts were randomly selected, processed, and embedded in paraffin; thereafter, longitudinal 5 µm sections were cut onto silane-coated slides. The brains were processed overnight in 30% sucrose, mounted onto a frozen cryostat with optimum cutting compound (OCT), and then coronal sections were made at 50 microns into 0. 1 M phosphate buffer (PB) at pH 7.4 in 24-well plates. Starting from the caudal part of the optic lobe and cranial to the choroid plexus and fourth ventricle, five brain sections were collected per animal per group.

2.3. Routine Histological Saining of Cardiac and Cerebellar Sections

Brain sections were mounted on gelatin-coated slides and then placed in defatting solution (50% chloroform and 50% alcohol) overnight. Thereafter, the sections were stained with Cresyl violet for the analysis of cerebellar cytoarchitecture. The heart sections were routinely stained using the H&E protocol.

2.4. IP$_3$R Immunohistochemistry

Cardiac sections were processed using the immunohistochemistry protocol for paraffin sections. Antigen retrieval was performed in citrate buffer (pH 6), washed in phosphate-buffered saline (PBS), blocked in 1% hydrogen peroxide (in methanol), and rinsed in phosphate-buffered saline (PBS). Normal goat serum (5% in PBS) was used for incubation at room temperature, after which sections were incubated at 4 °C in primary antibody (rabbit polyclonal antibody raised against rat IP3R; Abcam, Invitrogen, CA, USA, AB5804), diluted at a ratio of 1:500 overnight. Sections were further rinsed and incubated in secondary antibody (goat anti-rabbit antibody IgGs) at room temperature, diluted at a ratio of 1:1000, rinsed and incubated in ABC, and then rinsed again and incubated in DAB working solution. The sections were further counterstained in hematoxylin, and then dehydrated, cleared, and cover slipped. A similar protocol was adopted to immunolocalize IP$_3$R in brain sections using free floating immunohistochemistry for brain sections [25] with incubation of primary antibody at 4 °C for 48 h with 0.1 M phosphate buffer as a diluent for all solutions. The immunolocalized sections were viewed with a light microscope. Negative control sections were incubated with PBS instead of the primary antibody.

2.5. Quantitative Analysis

High-resolution microscopic images ($\times 40$ and $\times 100$) of H&E- and Cresyl violet-stained cardiac and cerebellar sections, in addition to IP_3R immunolabeled cardiac and cerebellar sections, were digitally captured using a Leica ICC50 HD video camera linked to a Leica DM 500 microscope (Leica Biosystems, Pacheco, CA, USA). A total of 20 camera fields were analyzed per group. In each photomicrograph of H&E-stained sections, the total percentage area fraction (Af) occupied by myocytes (Afm) and interstitial spaces (Afi) was determined using the grid method and the cell counter plugin of ImageJ. Afm and Afi were calculated using the following equation:

$$\text{Afm or Afi} = (\text{App} \times \sum p)/(\text{A slide}) \times 100 \tag{1}$$

where App = area per point, $\sum p$ = sum of points, and A slide = set scale area.

Counts (density) of granule and Purkinje cells in the Cresyl violet-stained sections and counts of IP_3R expression in the cardiac and cerebellum immunolabeled sections were carried out using the cell counter plugin of ImageJ. Furthermore, myocytes were semiquantitatively scored for thin or wavy fibers; accordingly, 0 was assigned for no observations of thin or wavy fibers, 1 for small areas of wavy fibers, 2 for multiple areas of wavy fibers, and 3 for large areas of wavy fibers.

2.6. Statistical Analysis

The cardiac histomorphology, cerebellar cell density, and total number of cells that expressed IP_3R in the heart and cerebellum were recorded as the mean $\pm$ SEM. IBM SPSS version 27 was used for analyzing the data and Microsoft Excel was used for plotting graphs. The data were tested for normality using the Shapiro–Wilks's test, and parametric data were analyzed by one-way ANOVA tests, while a post-hoc Bonferroni's test was conducted to determine the difference between the groups. Results with $p < 0.05$ were regarded as significant.

3. Results

3.1. Gross Morphology of the Hearts and Weight of the Brains

The morphological examinations of adult frog hearts showed a normal heart shape and configuration, a normal size of the cardiac muscle, and a firm lumen for the control group. The heart lumen in the 200 µg/L group was enlarged (largest among the treated groups), followed by the 0.01 µg/L group, and both groups presented serrated fibers extending from the thin ventricular wall compared to the control (Figure 1). The hearts in the 500 µg/L group appeared slightly larger than those of the 0.01 µg/L, but with an apparent smaller lumen and thicker ventricle wall compared to the control. The weight of the hearts has been previously reported [17] to be significantly increased in the 200 µg/L group compared to the control and other atrazine exposed groups (Figure 1B). The brain weight was reduced in all concentrations of atrazine exposure, with the lowest decrease in the 0.01 µg/L group. However, no significant difference was observed between groups (Figure 1C).

3.2. Cardiac Myocyte Histomorphology

A significant increase in the area fraction of the interstitial space and a reduction in the area fraction of myocytes were observed in the atrium of the 0.01 µg/L and 200 µg/L groups compared to the control, while the area fraction of myocytes significantly increased in the 500 µg/L group compared to the 0.01 µg/L group. Furthermore, the semiquantitative waviness of the myocytes in the atrium was significantly increased in all atrazine-exposed groups compared to the control, as well as in the 0.01 µg/L group compared to the 200 µg/L group (Figure 2A–F).

The area of the ventricular myocytes significantly reduced ($p = 0.0001$) in the 200 µg/L group compared to the control and the other groups, while the area of the interstitial space in the ventricle significantly ($p = 0.0001$, 0.034, and 0.0001 respectively) increased. The semiquantitative scoring of ventricle myocytes revealed a significant (p = 0.029 and 0.0001, respectively) increase in waviness in all the atrazine exposed groups compare to the control and also with increased waviness in 200 μg/L group compared to 0.01 μg/L group (Figure 3A–F).

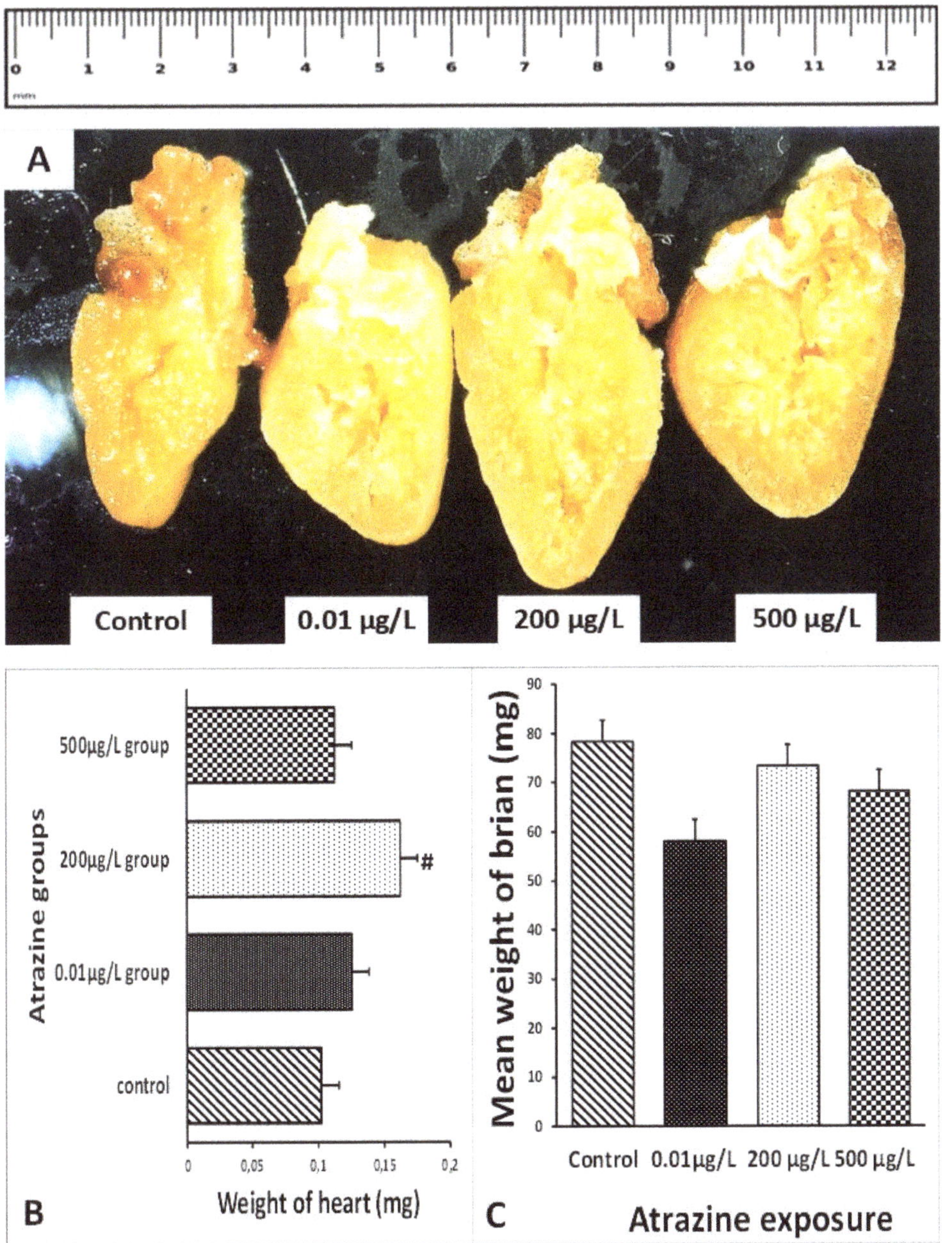

Figure 1. Photographs of the dose-dependent effect of atrazine on the gross morphology of adult frog hearts, and mean weights of the heart and brain of the treatment groups compared to the control. (**A**) Apparently enlarged hearts in all atrazine groups and enlarged lumen in the 0.01 μg/L and 200 μg/L groups. (**B**) "#" significant increase in heart weight in the 200 μg/L group compared to the other groups [25]. (**C**) No significant difference in brain weight.

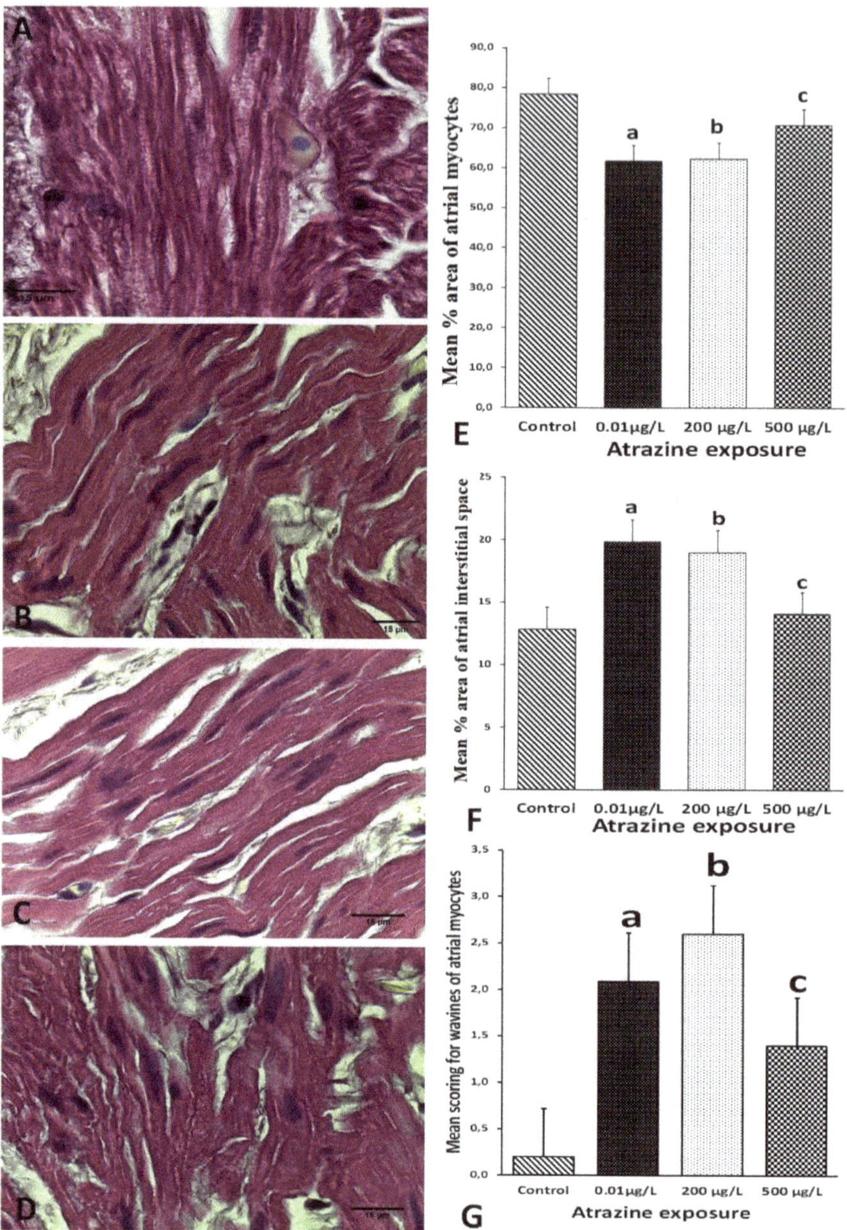

Figure 2. Photomicrograph of myocytes in the atrium of adult frogs, showing the effects of atrazine in the exposure groups. (**A**) Control, (**B**) 0.01 µg/L group, (**C**) 200 µg/L group, and (**D**) 500 µg/L group, ×1000. Scale bar = 15 µm. (**E**) "a" and "b" are significantly decreased compared to the control, while "c" is significantly increased compared to "a" but similar to the control in the area of AM ($p = 0.0001$). (**F**) "a" and "b" are significantly increased compared to the control, while "c" is not significantly different from the control in the area of AIS ($p = 0.003$ and 0.014, respectively; Bonferroni post-hoc test). (**G**) "a," "b," and "c" are significantly increased compared to the control ($p = 0.0001$ and 0.023, respectively), as is "b" compared to "c" ($p = 0.004$; Dunn's pairwise test, adjusted using Bonferroni correction) in the semiquantitative scoring of atrial waviness. AIS: % area fraction of atrial interstitial space; AM: % area fraction of atrial myocyte.

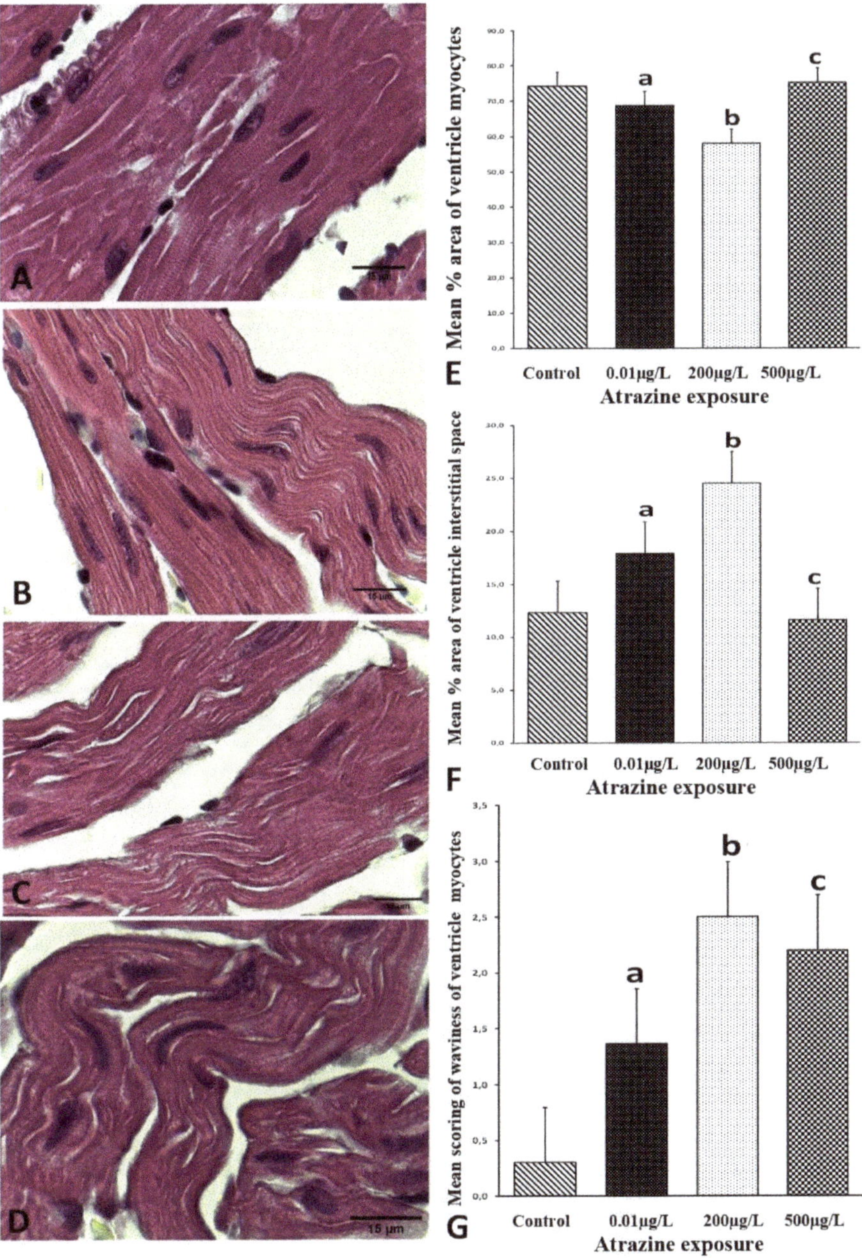

Figure 3. Photomicrograph of myocytes in the ventricle of adult frogs, showing the effects of atrazine in the exposure groups. (**A**) Control, (**B**) 0.01 μg/L group, (**C**) 200 μg/L group, and (**D**) 500 μg/L group, ×1000. Scale bar = 15 μm. (**E**) "b" is significantly reduced compared to "a," "c," and the control, while "c" is significantly increased compared to "a" but similar to control in the area of VM ($p = 0.0001$). (**F**) "b" is significantly increased compared to "a," "c," and the control in the area of VIS ($p = 0.0001$, 0.034, and 0.0001 respectively; Dunn's pairwise test, adjusted using Bonferroni correction). (**G**) "a," "b," and "c" are significantly increased compared to the control ($p = 0.029$ and 0.0001, respectively) and in "b" compared to "a" ($p = 0.017$, Bonferroni post-hoc test) in the semiquantitative scoring of atrial waviness.

3.3. Expression of IP₃Rs in Cardiac Tissue

In the ventricles and atrium, IP₃Rs were clearly expressed mostly in the perinuclear membrane of the cardiac myocytes of all groups (Figures 4 and 5), but no expression was observed in the negative controls (primary antibody omitted; image not included).

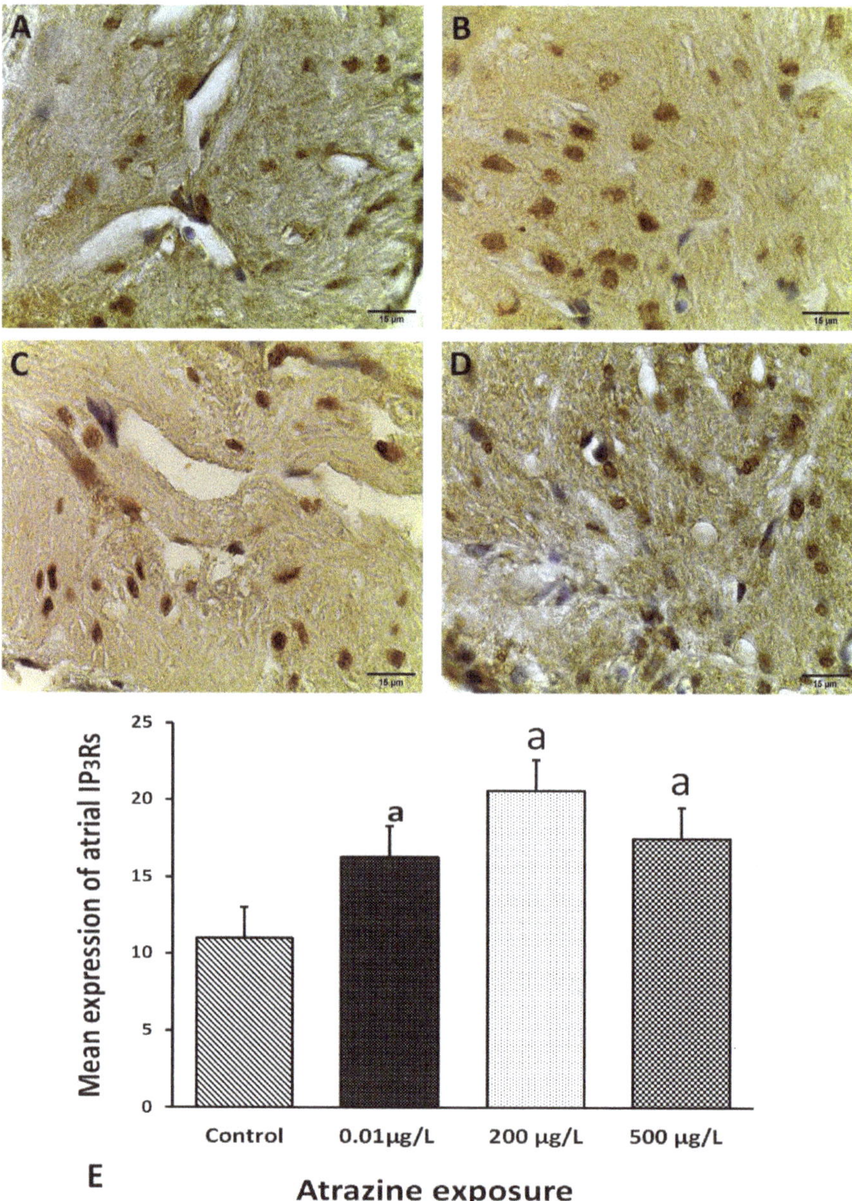

Figure 4. Photomicrograph of IP₃R expression in the atrium of adult frogs, showing the effects of atrazine in the different groups. (**A**) Control, (**B**) 0.01 μg L^{-1} group, (**C**) 200 μg L^{-1} group, and (**D**) 500 μg L^{-1} group, ×1000. Scale bar = 15 μm. (**E**) Significant difference between the groups (Kruskal–Wallis, p = 0.001). "a," is significantly increased compared to the control (p = 0.03, 0.005, and 0.0001; Dunn's pairwise test, adjusted using Bonferroni correction), ×1000. Scale bar = 15 μm.

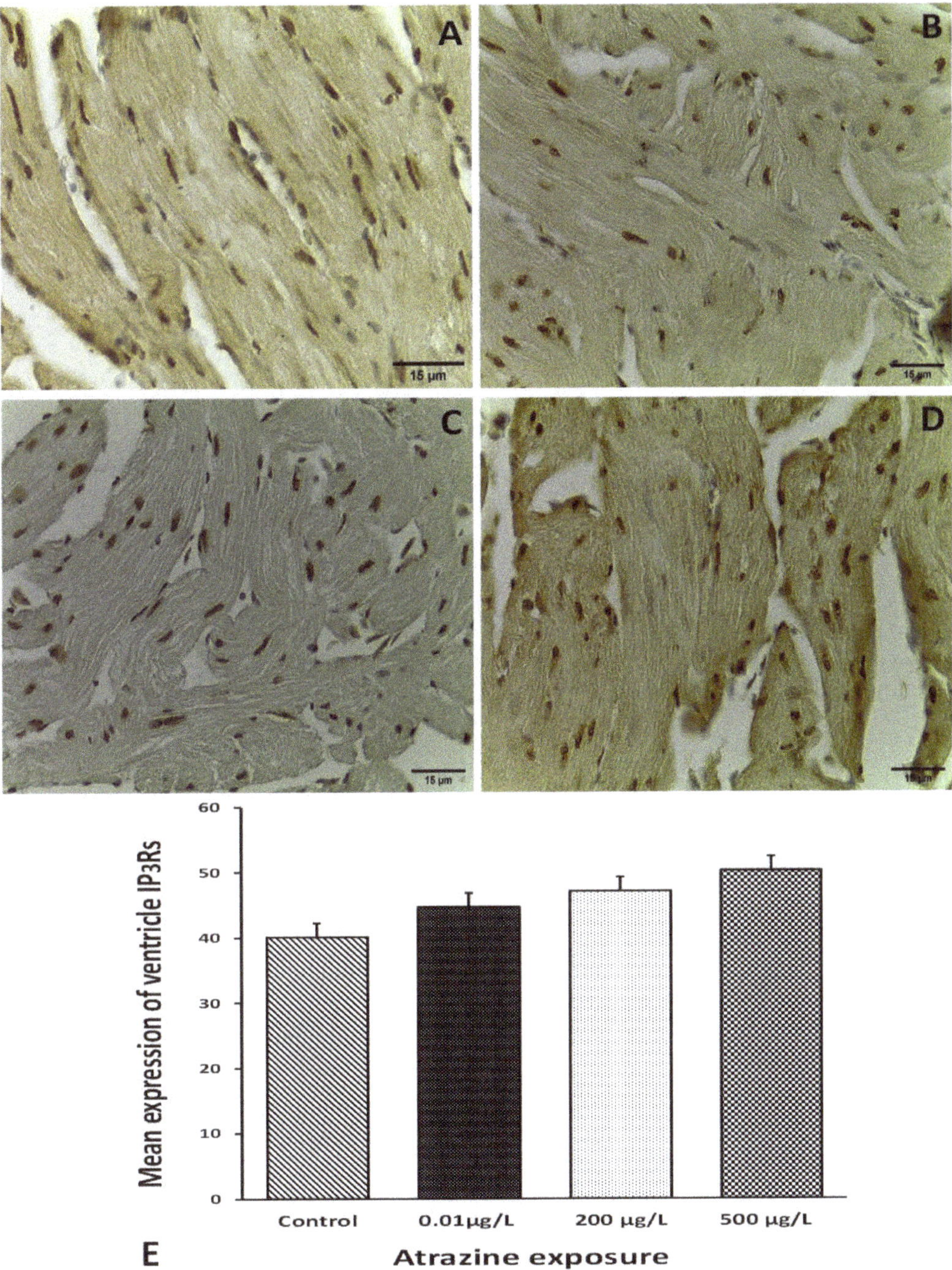

Figure 5. Photomicrograph of IP$_3$R expression in the cardiac ventricle adult frogs, showing the effects of atrazine in the different groups. (**A**) Control, (**B**) 0.01 µg L^{-1} group, (**C**) 200 µg L^{-1} group, and (**D**) 500 µg L^{-1} group, ×1000. Scale bar = 15 µm. (**E**) No significant differences between the groups (one-way ANOVA, p = 0.16), ×1000. Scale bar = 15 µm.

The mean number of atrial myocytes expressing IP$_3$R was statistically different between the treated and control groups (Kruskal–Wallis test, $p < 0.0001$). Furthermore, a post-hoc Duncan pairwise test showed significant ($p = 0.03$, 0.005, and 0.0001, respectively) increases in IP$_3$R expression in each of the atrazine-exposed groups compared to the control (Figure 4A–E). However, atrial IP$_3$Rs expression were non-significantly ($p > 0.05$) reduced in the 500 µg/L group relative to the 200 µg/L group (Figure 4A–E). On the contrary, IP$_3$Rs expression in ventricular myocyte non-significantly increased with increase in atrazine concentration (Figure 5A–E).

3.4. Histology of the Cerebellar Cortex

In the control group, pale staining of pear-shaped Purkinje neurons with euchromatin nuclei was observed in a densely packed cluster in the Purkinje layer (Figure 6A). The Purkinje layer gradually became slightly enlarged and less clustered in the 0.01 µg/L and 200 µg/L groups (Figure 6B,C). The Purkinje cell layer was reduced to a scattered scanty row of small (and a few enlarged) cells in the 500 µg/L-treated group (Figure 6D).

3.5. Quantification of Granule and Purkinje Cell Density

The mean density of Purkinje cells decreased in all atrazine-treated groups, but this decrease was significant in the 0.01 µg/L and 500 µg/L groups (Bonferroni's post-hoc test, $p = 0.0001$ and 0.01, respectively) (Figure 6E). However, the mean density of the granule cells in the cerebellum did not differ significantly across the treatment groups (Figure 6F).

3.6. Expression of IP$_3$R in the Cerebellar Cortex

The expression of IP$_3$Rs in the cerebellum was observed in the Purkinje cell layer (Figure 7A–D,F). A decrease in the number of expressed IP$_3$Rs was observed with an increase in atrazine concentration (inverse relationship). A significant difference in the mean cell count of expressed IP$_3$Rs was observed between the control and treated groups (one-way ANOVA test, $p < 0.00$). Meanwhile, the Bonferroni post-hoc test showed significant decreases in the number of cells that expressed IP$_3$Rs in the treated groups (0.01 µg/L, $p = 0.008$; 200 µg/L, $p = 0.001$; 500 µg/L, $p = 0.0001$) compared to the control (Figure 7E).

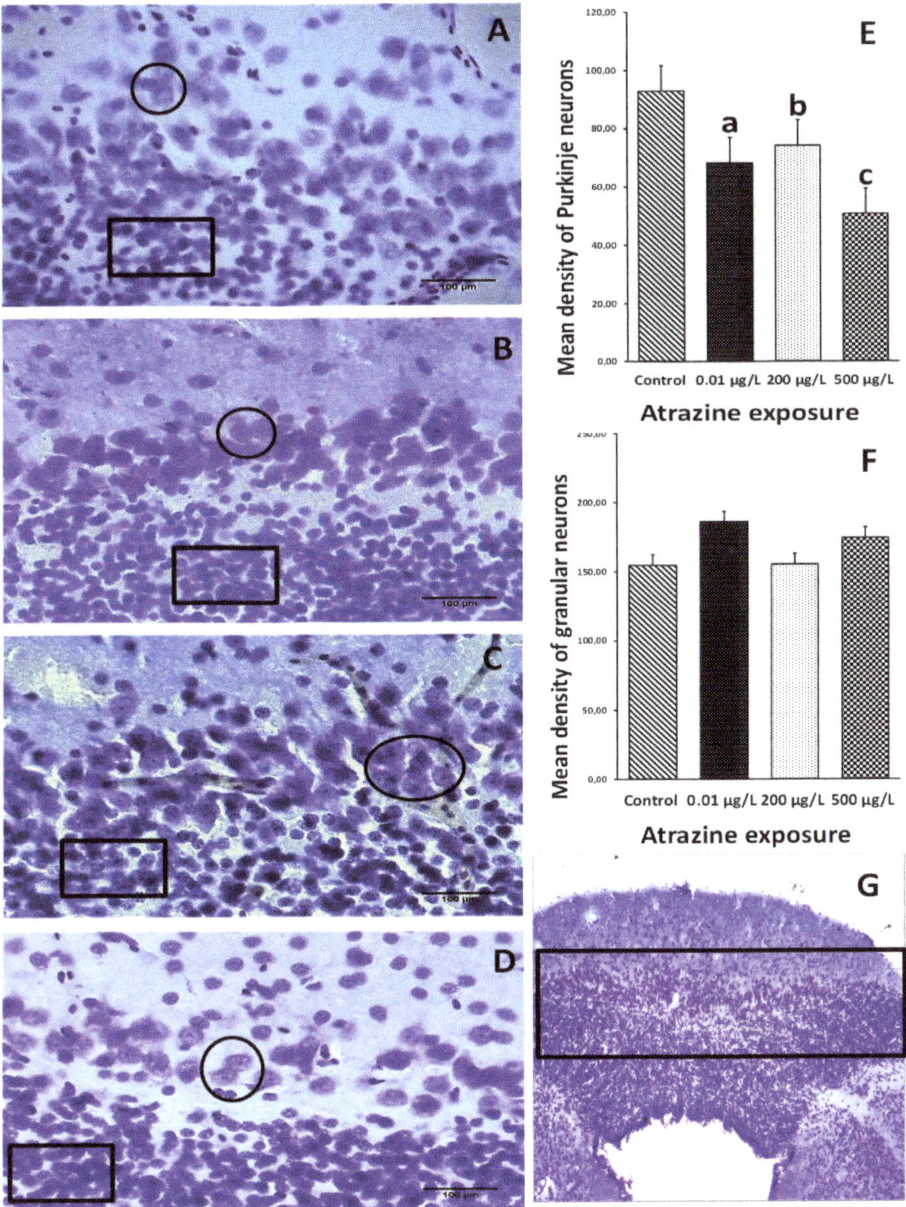

Figure 6. Photomicrographs of the cerebellar cortex of adult frogs, showing the effects of atrazine in the different groups. (**A**) Control group, with densely packed very wide layer of Purkinje cells (circles: Purkinje cells), reaching the molecular layer and small granule cells (square). (**B**) 0.01 µg/L group, with a narrow layer of enlarged Purkinje cells (circle). Granular cells appear slightly larger and clustered (square). (**C**) 200 µg/L group, with a slightly enlarged row of diffuse Purkinje cells (circles). (**D**) 500 µg/L, with diffuse and scanty Purkinje cells and larger clustered cells in the granular layer (square). Cresyl violet stain, ×63. Scale bar A–D = 100 µm. (**E**) "c" is significantly decreased compared to the control and "b," but not "a." Bonferroni post-hoc test, p = 0.0001 and 0.01, respectively. (**F**) No significant difference in density of granular neurons was observed between the groups. (**G**) Rectangle indicates the region of the cerebellar cortex from which photomicrographs were taken and quantification of densities were made, ×400. Scale bar = 100 µm.

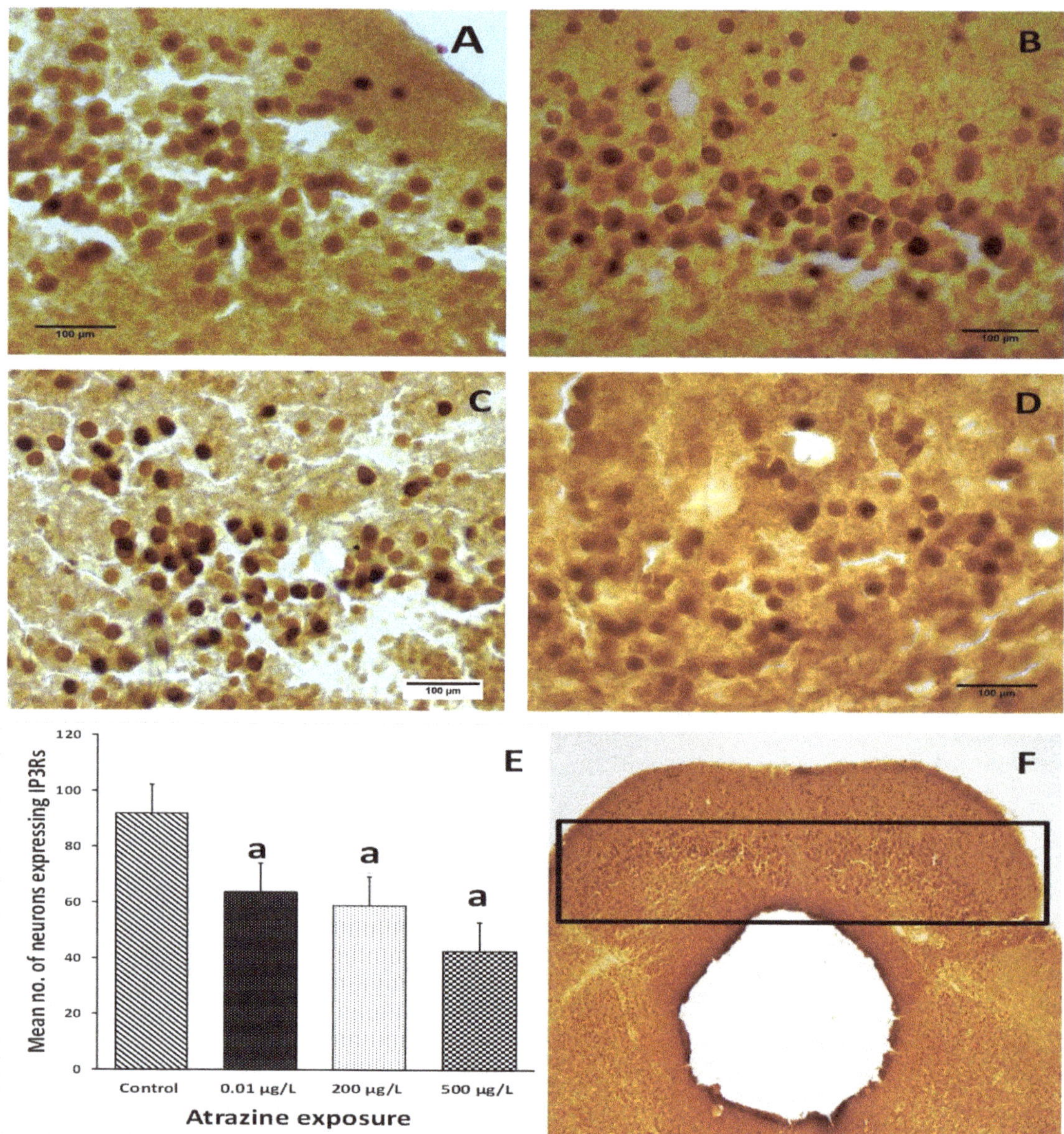

Figure 7. Photomicrographs showing atrazine-induced changes in the expression of IP$_3$R expression in the cerebellar cortex of adult frogs. (**A**) Control, with numerous Purkinje cells expressing IP$_3$Rs. (**B**) 0.01 µg/L group, showing several slightly enlarged cells expressing IP$_3$Rs. (**C**) 200 µg/L, showing fewer cells in clusters expressing IP$_3$Rs. (**D**) 500 µg/L, showing a diffused and diminished number of cells expressing IP$_3$Rs, ×63. Scale bar = 100 µm. (**E**) Significant reduction in the mean cell count of expressed IP$_3$Rs between the control and treated groups (one-way ANOVA test, $p < 0.0001$). Significant reduction in "a" relative to the control (Bonferroni post-hoc test, $p < 0.008$, 0.001, and 0.0001 respectively). (**F**) Rectangle indicates the region of the cerebella cortex with predominantly large Purkinje neurons that express IP$_3$Rs, and photomicrographs and quantifications were taken from this region, ×400. Scale bar = 100 µm.

4. Discussion

The gross morphological examination of frog hearts revealed perturbations in size and conformation, as well as structural configuration associated with chemical and pathological injury. The increased ventricular size in all atrazine-exposed groups, with the greatest increase in the 200 µg/L atrazine group, suggests cardiac ventricular hypertrophy and corelates with the previously reported significant increase in heart weight and size at this atrazine concentration [17]. The cerebellar cortex is morphologically integrated into the hind brain in frogs, and the gross examination showed normal brain morphology, but independent cerebellar examination was not possible.

The histomorphology observations of significant increases in the interstitial space area fraction complement the reduction in the myocytes' area fraction in the 0.01 µg/L (atrium only) and 200 µg/L (atrium and ventricle) groups compared to the control. Additionally, the significantly increased area fraction of myocytes in the 500 µg/L compared to the 0.01 µg/L (atrium) and 200 µg/L (ventricle) groups suggests dose-dependent effects. Meanwhile, the significant increase in thin and wavy myofibers in the atrazine-exposed groups compared to the control points to adverse effects of atrazine on the size of myocytes, which may interfere with their cellular integrity, especially with connective tissue infiltration, as previously reported [17]. These histomorphology results further agree with previously reported dilated cardiomyopathy [32] in the atrium and ventricle of the 200 µg/L group, hallmarked by reduced atrial and ventricular myocytes. However, hypertrophic cardiomyopathy is suspected in the slightly hypertrophied ventricular myocyte area fraction [33] of the 500 µg/L group.

While it was not possible to carry out quantitative analysis of cardiac Purkinje cells within the sub-endocardium, the results of the cerebellar Purkinje cell quantitative analysis indicate a significant reduction in the cerebellar Purkinje cell numbers in the 0.01 µg/L and 500 µg/L groups and an insignificant reduction in the 200 µg/L group compared to the control, which suggests differential effects of atrazine and possible disturbances of the electrophysiological function in the cerebellum. Furthermore, the reduction in Purkinje cell numbers may corroborate a previously reported reduced Purkinje cell inhibition on granule cells, with loss of Purkinje cells (reduction in number) in older mice [34,35] and a non-significant increase in the density of granular cells. These perturbations in neuronal density may affect cerebellar interconnections, consequently leading to disturbances of the neurotransmission function in the cerebellum, which may be associated with the locomotor mode of the life of adult *Xenopus laveis* frogs [36].

The regulation of neuronal synaptic plasticity and cardiac contraction and relaxation are achieved by an increase in cytosolic calcium [29,37]. In the brain, the influx of calcium triggers neurotransmission, signaling complexes in postsynaptic dendrites and their spines to induce long-term potentiation of cerebellar neurons [38]. As a pivotal coordinating center receiving input from several cortices and integrating them into a single response, the cerebellum is dependent on IP$_3$Rs for the release of calcium from intracellular storage [39]. Therefore, the significant decrease in cerebellar cortex IP$_3$Rs due to atrazine exposure may imply disturbances of the functional integrity of IP$_3$Rs and possibly consequences of calcium influx. This is consistent with previous reports on atrazine neural effects, such as atrazine disruption of cerebellar electrophysiology in a rodent in-vivo study [14], modulation of vestibular discharge in semi-circular canals [40], and alterations in Purkinje neuron IP$_3$Rs, which have been associated with spinocerebellar ataxia [28] and other brain pathologies [41] in human and rodent models.

Furthermore, cardiac ryanodine receptors (RYRs) and IP$_3$Rs are the two main families of calcium channels important for modulating calcium release for cardiac excitation contraction coupling (ECC) in atrial and ventricular cardiomyocytes in response to extracellular stimuli [29,42]. Though RYR expression is significantly higher than IP$_3$Rs in mammalian models, the presence of RYRs in the ventricle of frog hearts has not been established [42], suggesting that ECC in the ventricle of frog hearts is entirely dependent on the modulation of ventricular IP$_3$Rs, which has not been micromorphologically reported until now. In this study, IP$_3$Rs were successfully localized in the atrium and ventricles of frog hearts, and the results showed a significant increase in atrial IP$_3$Rs and a non-significant increase in ventricular IP$_3$Rs following atrazine treatment, which may be related to the differences in function of atrial and ventricular myocytes [43,44]. Additionally, elevated IP$_3$R expression is reputably observed in cardiac ventricular hypertrophy, heart failure, atrial fibrillation, and dilated cardiomyopathy [45,46], which mirrors our histomorphometry observations. These elevated cardiac IP$_3$R levels are also suggested to compensate for waning RYRs in human heart diseases [45,47].

The observed micromorphological changes in the heart and cerebellum following atrazine treatment and IP$_3$R perturbations in both organs (peripheral and central toxicity) indicates atrazine's potential to disrupt cardiac and cerebellar excitability with attendant dysfunctions. These histopathological observations together suggest impaired motor agility, mild ventricular cardiomyopathy, and progressed atrial arrythmias, all of which severely impair the ability of *Xenopus* to thrive in its environment. The results further emphasize that caution should be applied to atrazine use and application and regulatory guidelines should be enforced, as significant organ toxicities are evident at the environmentally relevant concentration (0.01 µg/L), which may have grave consequences in aquatic animal life conservation, the ecosystem, and general environmental health.

Author Contributions: Conceptualization, J.A.J. and E.F.M.; methodology, J.A.J. and E.F.M.; software, P.N.; validation, J.A.J. and E.F.M.; formal analysis, J.A.J., E.F.M. and P.N.; investigation, J.A.J. and E.F.M.; resources, J.A.J. and E.F.M.; data curation, J.A.J.; writing—original draft preparation, J.A.J. and E.F.M.; writing—review and editing, J.A.J., E.F.M. and P.O.; visualization, J.A.J., E.F.M. and P.O.; supervision, E.F.M.; project administration, J.A.J. and E.F.M.; funding acquisition, J.A.J., E.F.M. and P.N. All authors have read and agreed to the published version of the manuscript.

Funding: This work was partially supported by a Faculty Research Committee Individual Grant awarded to Jaclyn Asouzu Johnson in 2015 by the University of Witwatersrand, Faculty of Health Sciences (grant number; 001254842110151211015121105004985).

Institutional Review Board Statement: This study was conducted according to the guidelines of the approved standard procedures of the University of the Witwatersrand Animals Ethics Research Committee (2014/32/D) and the Gauteng Province Nature Conservation (CPF6 0115 (2015) and CPF6 0120, 2016).

Data Availability Statement: The supporting data reported in this study will be made available publicly during the course of publication, if this manuscript is accepted.

Acknowledgments: The authors especially acknowledge Hasiena Ali for her hands-on assistance at various stages of this research, and Ihunwo, Cornelius Rimayi, and Lynette Sena for their amazing collaboration and inspiration.

Conflicts of Interest: The authors declare no conflict of interest.

References

1. Bai, X.; Sun, C.; Xie, J.; Song, H.; Zhu, Q.; Su, Y.; Qian, H.; Fu, Z. Effects of atrazine on photosynthesis and defense response and the underlying mechanisms in Phaeodactylum tricornutum. *Environ. Sci. Pollut. Res.* **2015**, *22*, 17499–17507. [CrossRef]
2. Ribaudo, M.; Bouzaher, A. Atrazine: Environmental Characteristics and Economics of Management. 1994. Available online: http://www.ers.usda.gov/publications/pub-details/?pubid=40594 (accessed on 21 September 2021).
3. Barchanska, H.; Sajdak, M.; Szczypka, K.; Swientek, A.; Tworek, M.; Kurek, M. Atrazine, triketone herbicides, and their degradation products in sediment, soil and surface water samples in Poland. *Environ. Sci. Pollut. Res.* **2016**, *24*, 644–658. [CrossRef] [PubMed]
4. Dabrowski, J.M.; Shadung, J.M.; Wepener, V. Prioritizing agricultural pesticides used in South Africa based on their environmental mobility and potential human health effects. *Environ. Int.* **2014**, *62*, 31–40. [CrossRef]
5. Tiryaki, O.; Temur, C. The fate of pesticide in the environment. *J. Biol. Env. Sci.* **2010**, *4*, 29–38.
6. Scahill, J.L. Effects of atrazine on embryonic development of fathead minnows (Pimephales promelas) and Xenopus laevis. *BIOS* **2008**, *79*, 139–149. [CrossRef]
7. Langerveld, A.J.; Celestine, R.; Zaya, R.; Mihalko, D.; Ide, C.F. Chronic exposure to high levels of atrazine alters expression of genes that regulate immune and growth-related functions in developing Xenopus laevis tadpoles. *Environ. Res.* **2009**, *109*, 379–389. [CrossRef]

8. Sifkarovski, J.; Grayfer, L.; Andino, F.D.J.; Lawrence, B.P.; Robert, J. Negative effects of low dose atrazine exposure on the development of effective immunity to FV3 in Xenopus laevis. *Dev. Comp. Immunol.* **2014**, *47*, 52–58. [CrossRef]

9. Hayes, T.B.; Case, P.; Chui, S.; Chung, D.; Haeffele, C.; Haston, K.; Lee, M.; Mai, V.P.; Marjuoa, Y.; Parker, J.; et al. Pesticide Mixtures, Endocrine Disruption, and Amphibian Declines: Are We Underestimating the Impact? *Environ. Health Perspect.* **2006**, *114* (Suppl. 1), 40–50. [CrossRef] [PubMed]

10. Jooste, A.M.; Du Preez, L.H.; Carr, J.A.; Giesy, J.P.; Gross, T.S.; Kendall, R.J. Gonadal development of larval male Xenopus laevis exposed to atrazine in outdoor microcosms. *Environ. Sci. Technol.* **2005**, *39*, 5255–5261. [CrossRef] [PubMed]

11. Rimayi, C.; Odusanya, D.; Weiss, J.; de Boer, J.; Chimuka, L.; Mbajiorgu, F. Effects of environmentally relevant sub-chronic atrazine concentrations on African clawed frog (Xenopus laevis) survival, growth and male gonad development. *Aquat. Toxicol.* **2018**, *199*, 1–11. [CrossRef] [PubMed]

12. Kabra, A.N.; Ji, M.-K.; Choi, J.; Kim, J.R.; Govindwar, S.; Jeon, B.-H. Toxicity of atrazine and its bioaccumulation and biodegradation in a green microalga, Chlamydomonas mexicana. *Environ. Sci. Pollut. Res.* **2014**, *21*, 12270–12278. [CrossRef] [PubMed]

13. Lin, J.; Zhao, H.-S.; Qin, L.; Li, X.-N.; Zhang, C.; Xia, J.; Li, J.-L. Atrazine Triggers Mitochondrial Dysfunction and Oxidative Stress in Quail (Coturnix C. coturnix) Cerebrum via Activating Xenobiotic-Sensing Nuclear Receptors and Modulating Cytochrome P450 Systems. *J. Agric. Food Chem.* **2018**, *66*, 6402–6413. [CrossRef]

14. Podda, M.V.; Deriu, F.; Solinas, A.; Demontis, M.P.; Varoni, M.V.; Spissu, A.; Anania, V.; Tolu, E. Effect of atrazine Administration on Spontaneous and Evoked Cerebellar Activity in the RAT. *Pharmacol. Res.* **1997**, *36*, 199–202. [CrossRef]

15. Xia, J.; Qin, L.; Du, Z.-H.; Lin, J.; Li, X.-N.; Li, J.-L. Performance of a novel atrazine-induced cerebellar toxicity in quail (Coturnix C. coturnix): Activating PXR/CAR pathway responses and disrupting cytochrome P450 homeostasis. *Chemosphere* **2017**, *171*, 259–264. [CrossRef]

16. Lin, Z.; Dodd, C.A.; Xiao, S.; Krishna, S.; Ye, X.; Filipov, N.M. Gestational and Lactational Exposure to Atrazine via the Drinking Water Causes Specific Behavioral Deficits and Selectively Alters Monoaminergic Systems in C57BL/6 Mouse Dams, Juvenile and Adult Offspring. *Toxicol. Sci.* **2014**, *141*, 90–102. [CrossRef] [PubMed]

17. Asouzu Johnson, J.; Ihunwo, A.; Chimuka, L.; Mbajiorgu, E.F. Cardiotoxicity in African clawed frog (Xenopus laevis) sub-chronically exposed to environmentally relevant atrazine concentrations: Implications for species survival. *Aquat. Toxicol.* **2019**, *213*, 105218. Available online: https://www.sciencedirect.com/science/article/pii/S0166445X18308099 (accessed on 16 September 2021). [CrossRef] [PubMed]

18. Amir, S.; Shah, S.; Mamoulakis, C.; Docea, A.; Kalantzi, O.-I.; Zachariou, A.; Calina, D.; Carvalho, F.; Sofikitis, N.; Makrigiannakis, A.; et al. Endocrine Disruptors Acting on Estrogen and Androgen Pathways Cause Reproductive Disorders through Multiple Mechanisms: A Review. *Int. J. Environ. Res. Public Health* **2021**, *18*, 1464. [CrossRef]

19. Spealman, C.R. The action of ions on the frog heart. *Am. J. Physiol. Content* **1940**, *130*, 729–738. [CrossRef]

20. Fijorek, K.; Püsküllüoğlu, M.; Tomaszewska, D.; Tomaszewski, R.; Glinka, A.; Polak, S. Serum potassium, sodium and calcium levels in healthy individuals—literature review and data analysis. *Folia Med. Crac.* **2014**, *54*, 53–70.

21. Philipson, K.D.; Nicoll, D.A. Sodium-Calcium Exchange: A Molecular Perspective. *Annu. Rev. Physiol.* **2000**, *62*, 111–133. [CrossRef]

22. Bers, D.M.; Barry, W.H.; Despa, S. Intracellular Na+ regulation in cardiac myocytes. *Cardiovasc. Res.* **2003**, *57*, 897–912. [CrossRef]

23. Eisner, D.A.; Caldwell, J.L.; Kistamas, K.; Trafford, A.W. Calcium and Excitation-Contraction Coupling in the Heart. *Circ. Res.* **2017**, *121*, 181–195. [CrossRef] [PubMed]

24. Ghosh, A.; Greenberg, M.E. Calcium signaling in neurons: Molecular mechanisms and cellular consequences. *Science* **1995**, *268*, 239–247. [CrossRef] [PubMed]

25. Dellis, O.; Dedos, S.G.; Tovey, S.C.; Rahman, T.U.; Dubel, S.J.; Taylor, C.W. Ca 2+ Entry Through Plasma Membrane IP 3 Receptors. *Science* **2006**, *313*, 229–233. [CrossRef] [PubMed]

26. Dellis, O.; Rossi, A.M.; Dedos, S.G.; Taylor, C. Counting Functional Inositol 1,4,5-Trisphosphate Receptors into the Plasma Membrane. *J. Biol. Chem.* **2008**, *283*, 751–755. [CrossRef]

27. Koulen, P.; Janowitz, T.; Johnston, L.D.; Ehrlich, B.E. Conservation of localization patterns of IP3 receptor type 1 in cerebellar Purkinje cells across vertebrate species. *J. Neurosci. Res.* **2000**, *61*, 493–499. Available online: https://onlinelibrary.wiley.com/doi/abs/10.1002/1097-4547%2820000901%2961%3A5%3C493%3A%3AAID-JNR3%3E3.0.CO%3B2-9 (accessed on 21 September 2021). [CrossRef]

28. Ando, H.; Hirose, M.; Mikoshiba, K. Aberrant IP3 receptor activities revealed by comprehensive analysis of pathological mutations causing spinocerebellar ataxia 29. *Proc. Natl. Acad. Sci. USA* **2018**, *115*, 12259–12264. [CrossRef]

29. Garcia, M.I.; Boehning, D. Cardiac inositol 1,4,5-trisphosphate receptors. *Biochim. Biophys. Acta (BBA)—Bioenerg.* **2017**, *1864*, 907–914. [CrossRef]

30. Rohr, J.R.; McCoy, K.A. A qualitative meta-analysis reveals consistent effects of atrazine on freshwater fish and amphibians. *Environ. Health Perspect.* **2010**, *118*, 20. [CrossRef] [PubMed]

31. Farruggia, F.T.; Rossmeisl, C.M.; Hetrick, J.A.; Biscoe, M.; Branch, M.E.R., III. *Refined Ecological Risk Assessment for Atrazine*; US Environmental Protection Agency, Office of Pesticide Programs: Washington, DC, USA, 2016.

32. Mitrut, R.; Stepan, A.; Pirici, D. Histopathological Aspects of the Myocardium in Dilated Cardiomyopathy. *Curr. Health Sci. J.* **2018**, *44*, 243–249. [CrossRef]

33. Marshall, L.; Vivien, C.; Girardot, F.; Péricard, L.; Demeneix, B.A.; Coen, L.; Chai, N. Persistent fibrosis, hypertrophy and sarcomere disorganisation after endoscopy-guided heart resection in adult Xenopus. *PLoS ONE* **2017**, *12*, e0173418. [CrossRef] [PubMed]

34. Guo, C.; Witter, L.; Rudolph, S.; Elliott, H.L.; Ennis, K.A.; Regehr, W.G. Purkinje Cells Directly Inhibit Granule Cells in Specialized Regions of the Cerebellar Cortex. *Neuron* **2016**, *91*, 1330–1341. [CrossRef]

35. Ralcewicz, T.A.; Persaud, T. Purkinje and granule cells distribution in the cerebellum of the rat following prenatal exposure to low dose ionizing radiation. *Exp. Toxicol. Pathol.* **1994**, *46*, 443–452. [CrossRef]

36. Manzano, A.S.; Herrel, A.; Fabre, A.-C.; Abdala, V. Variation in brain anatomy in frogs and its possible bearing on their locomotor ecology. *J. Anat.* **2017**, *231*, 38–58. [CrossRef]

37. Redmond, L.; Ghosh, A. Regulation of dendritic development by calcium signaling. *Cell Calcium* **2005**, *37*, 411–416. [CrossRef]

38. Nanou, E.; Catterall, W.A. Calcium Channels, Synaptic Plasticity, and Neuropsychiatric Disease. *Neuron* **2018**, *98*, 466–481. [CrossRef]

39. Berridge, M.J.; Bootman, M.; Roderick, H. Calcium signalling: Dynamics, homeostasis and remodelling. *Nat. Rev. Mol. Cell Biol.* **2003**, *4*, 517–529. [CrossRef]

40. Rossi, M.L.; Prigioni, I.; Gioglio, L.; Rubbini, G.; Russo, G.; Martini, M.; Farinelli, F.; Rispoli, G.; Fesce, R. IP3 receptor in the hair cells of frog semicircular canal and its possible functional role. *Eur. J. Neurosci.* **2006**, *23*, 1775–1783. [CrossRef]

41. Hisatsune, C.; Mikoshiba, K. IP3receptor mutations and brain diseases in human and rodents. *J. Neurochem.* **2017**, *141*, 790–807. [CrossRef]

42. Fill, M.; Copello, J.A. Ryanodine Receptor Calcium Release Channels. *Physiol. Rev.* **2002**, *82*, 893–922. [CrossRef] [PubMed]

43. Tijskens, P.; Meissner, G.; Franzini-Armstrong, C. Location of Ryanodine and Dihydropyridine Receptors in Frog Myocardium. *Biophys. J.* **2003**, *84*, 1079–1092. [CrossRef]

44. Lipp, P.; Laine, M.; Tovey, S.C.; Burrell, K.M.; Berridge, M.J.; Li, W.; Bootman, M.D. Functional InsP3 receptors that may modulate excitation–contraction coupling in the heart. *Curr. Biol.* **2000**, *10*, 939–942. [CrossRef]

45. Signore, S.; Sorrentino, A.; Ferreira-Martins, J.; Kannappan, R.; Shafaie, M.; Del Ben, F.; Isobe, K.; Arranto, C.; Wybieralska, E.; Webster, A.; et al. Inositol 1,4,5-Trisphosphate Receptors and Human Left Ventricular Myocytes. *Circulation* **2013**, *128*, 1286–1297. [CrossRef] [PubMed]

46. Go, L.O.; Moschella, M.C.; Watras, J.; Handa, K.K.; Fyfe, B.S.; Marks, A.R. Differential regulation of two types of intracellular calcium release channels during end-stage heart failure. *J. Clin. Investig.* **1995**, *95*, 888–894. [CrossRef] [PubMed]

47. Harzheim, D.; Movassagh, M.; Foo, R.S.-Y.; Ritter, O.; Tashfeen, A.; Conway, S.J.; Bootman, M.D.; Roderick, H.L. Increased InsP3Rs in the junctional sarcoplasmic reticulum augment Ca2+ transients and arrhythmias associated with cardiac hypertrophy. *Proc. Natl. Acad. Sci. USA* **2009**, *106*, 11406–11411. [CrossRef] [PubMed]

Article

Cellular Stress Responses of the Endemic Freshwater Fish Species *Alburnus vistonicus* Freyhof & Kottelat, 2007 in a Constantly Changing Environment

Emmanouil Tsakoumis [1,2,†,‡], Thomais Tsoulia [1,2,†], Konstantinos Feidantsis [1], Foivos Alexandros Mouchlianitis [2], Panagiotis Berillis [3,*], Dimitra Bobori [2] and Efthimia Antonopoulou [1,*]

[1] Laboratory of Animal Physiology, Department of Zoology, School of Biology, Aristotle University of Thessaloniki, 541 24 Thessaloniki, Greece; manolis.tsakoumis@ebc.uu.se (E.T.); thomi.tsoulia@gmail.com (T.T.); kfeidant@bio.auth.gr (K.F.)

[2] Laboratory of Ichthyology, Department of Zoology, School of Biology, Aristotle University of Thessaloniki, 541 24 Thessaloniki, Greece; amouchl@bio.auth.gr (F.A.M.); bobori@bio.auth.gr (D.B.)

[3] Department of Ichthyology and Aquatic Environment, School of Agricultural Sciences, University of Thessaly, 384 46 Fytoko Volos, Greece

* Correspondence: pveril@uth.gr (P.B.); eantono@bio.auth.gr (E.A.)

† Equally contributed.

‡ Present address: Environmental Toxicology, Department of Organismal Biology, Evolutionary Biology Centre, Uppsala University, 75236 Uppsala, Sweden.

Citation: Tsakoumis, E.; Tsoulia, T.; Feidantsis, K.; Mouchlianitis, F.A.; Berillis, P.; Bobori, D.; Antonopoulou, E. Cellular Stress Responses of the Endemic Freshwater Fish Species *Alburnus vistonicus* Freyhof & Kottelat, 2007 in a Constantly Changing Environment. *Appl. Sci.* **2021**, *11*, 11021. https://doi.org/10.3390/app112211021

Academic Editor: Simona Picchietti

Received: 8 September 2021
Accepted: 18 November 2021
Published: 21 November 2021

Publisher's Note: MDPI stays neutral with regard to jurisdictional claims in published maps and institutional affiliations.

Abstract: Herein we investigated the cellular responses of the endemic fish species *Alburnus vistonicus* Freyhof & Kottelat, 2007, under the variation of several physico-chemical parameters including temperature (°C), salinity (psu), dissolved oxygen (mg/L), pH and conductivity (µS/cm), which were measured in situ. Monthly fish samplings (October 2014–September 2015) were conducted in Vistonis Lake in northern Greece, a peculiar ecosystem with brackish waters in its southern part and high salinity fluctuations in its northern part. Fish gills and liver responses to the changes of the physico-chemical parameters were tested biochemically and histologically. Heat shock protein levels appeared to be correlated with salinity fluctuations, indicating the adaptation of *A. vistonicus* to the particular environment. The latter is also enhanced by increased Na^+-K^+ ATPase levels, in response to salinity increase during summer. The highest mitogen activated protein kinases phosphorylation levels were observed along with the maximum mean salinity values. A variety of histological lesions were also detected in the majority of the gill samples, without however securing salinity as the sole stress factor. *A. vistonicus* cellular stress responses are versatile and shifting according to the examined tissue, biomarker and season, in order for this species to adapt to its shifting habitat.

Keywords: Vistonis Lake; physico-chemical parameters; gills; liver; HSPs; MARKs; Na^+-K^+ ATPase; histology

1. Introduction

The ecological stability of inland and transitional waters is more vulnerable to exogenous pressures (e.g., climate change) compared to marine ecosystems, as the physico-chemical characteristics of freshwater environments can be highly variable by season and by water body [1]. Moreover, these aquatic ecosystems, and especially lakes, are subject to constant anthropogenic pressures, with the impact of degradation of their ecological value [2]. The majority of published reports focus on individual parameters as a single stress factor [1,3,4]. However, in order to assess the risk of extinction and to secure populations of endemic inland fish species, it has been proposed to study the combined effect of stressors (e.g., increase in temperature and pH change) on the regulation of gene expression and enzymatic activity [1,5].

123

Fluctuations of the physico-chemical parameters may affect freshwater organisms on several levels of biological organization [6]. The fish responses towards environmental stressors are widely assessed in gills and liver due to the great sensitivity of these tissues in external stimuli [7]. Cellular Stress Response (CSR) represents a highly evolutionarily conserved physiological mechanism for handling cellular stress, due to changes in the environment. The response to the stress factor will determine the adaptation or not of the organism to the change [8]. CSR fundamental components might include the induction of macromolecules for protection and repair, as well as apoptosis, especially if cellular tolerance limits are exceeded. For instance, Heat shock proteins (HSPs) are widely used as a measure of cellular stress in fish [9,10], for a number abiotic factors [11], such as changes in salinity [12], pH and CO_2 [13], as well as environmental pollutants such as heavy metals [14], industrial waste [15] and pesticides [16]. Moreover, members of the Mitogen Activated Protein Kinases (MAPKs) superfamily are also related to various stressors, such as thermal stress, salinity fluctuations, as well as chemical agents [17–22]. It should be noted that the study of Na^+-K^+ ATPase is of utmost importance in experiments with salinity as the main stress factor [23], since this protein is mostly located in osmoregulation tissues, such as fish gills [24–26].

The freshwater fish species *Alburnus vistonicus* Freyhof & Kottelat, 2007 commonly known as Vistonis shemaja, is found in the basins of Vistonis Lake, and in the basins of Ismaris Lake and Filiouris River [27,28]. The species is classified as "critically endangered" (CR) (www.iucnredlist.org, accessed on 1 January 2008) and it is protected by national and international law. Specifically, it is included in the Annex II of the Directive 92/43/EEC on the conservation of natural habitats and of wild fauna and flora and it is also included in the Red Book of Threatened Species of Greece [29] and in the Berne Convention. Therefore, its conservation is of significant importance. Aspects of species biology and ecology, such as growth and reproductive biology, were studied recently [28,30,31]. Although the lacustrine environment is known to constantly change and affect inhabiting fish species ecology [32–35], the effect of Vistonis Lake's continuously changing physico-chemical parameters on the biochemical and physiological responses of *A. vistonicus* remains unknown.

2. Materials and Methods

2.1. Study Area

Vistonis Lake (Figure 1) is a natural, hypereutrophic, shallow lake in northern Greece and is considered as a peculiar ecosystem. The lake's northern part contains low salinity water, due to the freshwater inflows from the Kosynthos, Kompsatos and Travos Rivers. Meanwhile, its southern part receives seawater inflows from the North Aegean Sea, via a narrow artificial channel, the Porto Lagos Lagoon [36]. The distinctive hydrology of the system results in brackish waters in its southern part, while in the northern part the salinity levels are fluctuating, following the seasonal variations of the freshwater inflows from the rivers. This phenomenon is even more pronounced during the dry period, when the freshwater inflows from the rivers are minimized, resulting in increased salinity levels, even in the northern part of the lake [28,30,36].

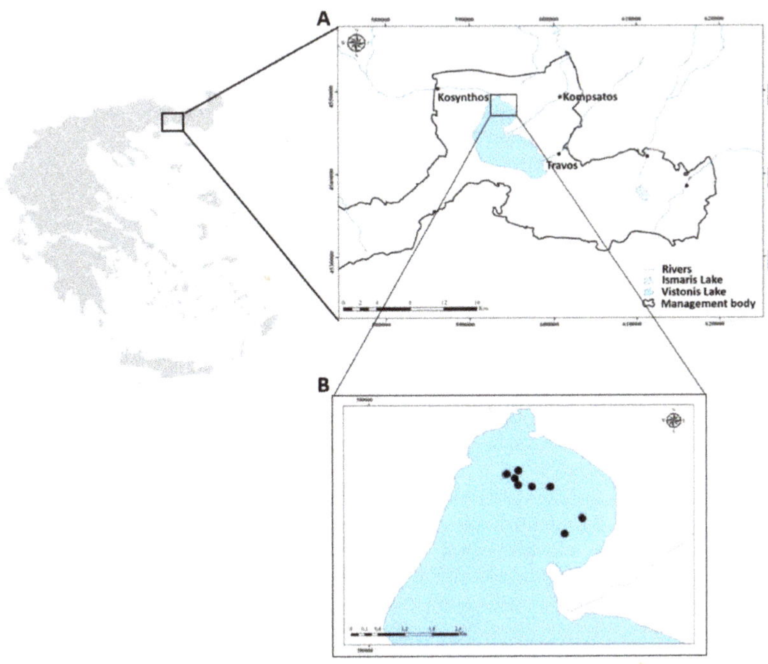

Figure 1. Map of the study area (Vistonis Lake, Northern Greece) (**A**) and sampling stations indicated with black dots (**B**) (Reprinted with permission from ref. [30]. Copyright 2108, Copyright Owners Assoc. Professors Dimitra Bobori and Efthimia Antonopoulou and Dr. Emmanouil Tsakoumis).

2.2. Sample Collection—Ethical Approval

Fish samplings were performed on a monthly basis throughout a full calendar year (October 2014–September 2015). In January 2015, due to flooding events in the study area, the access to Vistonis Lake was not feasible and therefore no samples are available from this month. Fish samplings were performed in the northern part of the lake, where water exhibits lower salinity levels, using Nordic-type benthic multi-mesh gill-nets following the requirements set by EU [37,38] in accordance with the EU (Directive 63/2010) legislation for the protection of animals used for scientific purposes. All necessary permissions were provided by the Management Body of the Delta Nestos and Lakes Vistonida—Ismarida. During each sampling, five individuals were collected and were all used for both histopathological and biochemical analysis (45 individuals in total). On the sampling days, fish were removed from the water and were immediately anesthetized by immersion in water with an overdose of concentrated solution of buffered ethyl 3-aminobenzoate methanesulfonate (MS-222). Five minutes after cessation of respiration, fish were removed from water and were euthanized, according to the protocols of the Canadian Council of Animal Care (Euthanasia of finfish) and the American Veterinary Medical Association, aiming to minimize fish psychophysical stress. Dead fish were placed on ice and from each sampled individual, one pair of gills was immediately fixed in 10% buffered formalin and stored for histopathological analysis. For the biochemical analysis, and immediately after dissection, the other pair of gills and liver tissue were dissected and retained in dry ice, until their transfer to the laboratory, where the samples were stored at $-80\,^{\circ}\mathrm{C}$ until further analysis. Fish gills are, besides their respiratory function, one of the most important osmoregulatory organs, playing a critical role in ionic regulation [39]. The responses of fish gills under the variation of the physico-chemical parameters can be tested either histologically or on biochemical level [40–42]. Moreover, the liver is also considered as an early stress indicator compared to other fish tissues, as shown by several researchers [21,43–45].

Parallel to fish sampling, water physico-chemical parameters such as temperature (°C), salinity (psu), dissolved oxygen (DO) (mg/L), pH and conductivity (µS/cm) were also recorded at eight stations in the lake using a portable multi sensor (AquaRead, AP-2000) (Figure 1). Recording of the physico-chemical parameters was performed at each sampling station, both at the surface and above the lake's bottom (n = 16 records per sampling).

2.3. Analytical Procedures

2.3.1. SDS-PAGE and Immunoblot Analysis

Frozen gill and liver samples were homogenized in 3 mL/g of cold lysis buffer [20 mM β-glycerophosphate, 50 mM NaF, 2 mM EDTA, 20 mM Hepes, 0.2 mM Na_3VO_4, 10 mM benzamidine, pH 7, 200 mM leupeptin, 10 mM transepoxy succinyl-L-leucylamido-(4 guanidino) butane, 5 mM dithiothreitol, 300 mM phenyl methyl sulfonyl fluoride (PMSF), 50 µg/mL pepstatin, 1% v/v Triton X-100], and extracted on ice for 30 min. The samples were then centrifuged (10,000× g, 10 min, 4 °C) and the supernatants were collected and boiled with 0.33 volumes of SDS/PAGE sample buffer (330 mM Tris-HCl, 13% v/v glycerol, 133 mM DTT, 10% w/v SDS, 0.2% w/v bromophenol blue). Protein concentrations were determined using the BioRad protein assay (BioRad, Hercules, CA, USA). Thereafter, equivalent amounts of proteins (50 µg) were separated either on 10% and 0.275% (w/v) acrylamide and bisacrylamide slab gels and transferred electrophoretically onto nitrocellulose membranes (0.45 µm, Schleicher and Schuell, Keene, NH, USA). All nitrocellulose membranes were dyed with Ponceau stain in order to assure a good quality of transfer and equal protein loading. Non-specific binding sites on the membranes were blocked by 5% (w/v) non-fat milk in TBST [20 mM Tris-HCl, pH 7.5, 137 mM NaCl, 0.1% (v/v) Tween 20], during a 30–45 min incubation at room temperature. Subsequently, the membranes were further incubated overnight with primary antibodies, including polyclonal rabbit anti-HSP70 (Cat. No. 4872, Cell Signaling, Beverly, MA, USA), polyclonal rabbit anti-HSP90 (Cat. No. 4874, Cell Signaling, Beverly, MA, USA), polyclonal rabbit anti-Na, K-ATPase (Cat. No. 3010, Cell Signaling, Beverly, MA, USA), polyclonal rabbit anti-p38MAP kinase (Cat. No. 9212, Cell Signaling, Beverly, MA, USA), monoclonal rabbit anti-HSP60 (Cat. No. 12165, Cell Signaling, Beverly, MA, USA), monoclonal rabbit anti-phospho-p38 MAPK (Cat. No. 4511, Cell Signaling, Beverly, MA, USA) and monoclonal rabbit anti-phospho-p44/42 MAPK (Cat. No. 4370, Cell Signaling, Beverly, MA, USA) and monoclonal rabbit anti-p44/42 MAPK (Cat. No. 4695, Cell Signaling, Beverly, MA, USA). The next day and after washing in TBST (3 time periods for 5 min each), the blots were incubated with horseradish peroxidase-linked secondary antibodies and washed again in TBST (3 time periods for 5 min each). The bands were then detected, using enhanced chemiluminescence (Chemicon, MA, USA) and were exposed to Fuji Medical X-ray films. Films were quantified by laser-scanning densitometry (GelPro Analyzer Software, Media Cybernetics/Image Studio Lite Software Ver 5.2).

2.3.2. Histopathological Analysis

Samples were fixed in 10% buffered formalin and dehydrated in graded series of ethanol and then immersion in xylol, and embedding in paraffin wax followed. Thin sections of 5–7 mm were mounted, deparaffinized, rehydrated, stained with Hematoxylin-Eosin, mounted with Cristal/Mount and examined for alterations with a microscope (Bresser Science TRM 301) under total magnification of 100× and 400×. A digital camera adjusted to the microscope (Bresser MikroCam 5.0 MP) was used for acquiring histopathological photomicrographs.

2.4. Statistical Analysis

Data are expressed as mean ± standard deviation (SD) of n = 5 biological samples. Significant differences ($p < 0.05$) between the monthly expression levels of the proteins analyzed were conducted by one-way analysis of variance (ANOVA), using SPSS version 27 (SPSS, Inc., Chicago, IL, USA). Further elaboration of the potential physiological and environmental codependency was assessed by Principal Component Analysis (PCA), using the FactoMineR package in R [46]. Graphs were prepared using GraphPad Prism 9 (San Diego, CA, USA).

3. Results

3.1. Water Parameters

The variation of the levels of the water parameters recorded in Vistonis Lake throughout the sampling period are shown in Figure 2. Temperature levels exhibited seasonal fluctuation, with the lowest value (4.7 °C) recorded in February and the highest (29.5 °C) in July (Figure 2A). The lowest salinity values (0.044 psu) were recorded in March and the highest (9.77 psu) in September (Figure 2B). Dissolved oxygen (DO) values also showed seasonal variation, with the lowest concentration (7.4 mg/L) observed in August and the highest (13.96 mg/L) in February (Figure 2C). The pH levels were quite stable throughout the sampling period (Figure 2D). The variation of conductivity followed a similar pattern to that of salinity, since during the colder months low levels were recorded, in contrast to the warmer months, when the levels were higher (Figure 2E).

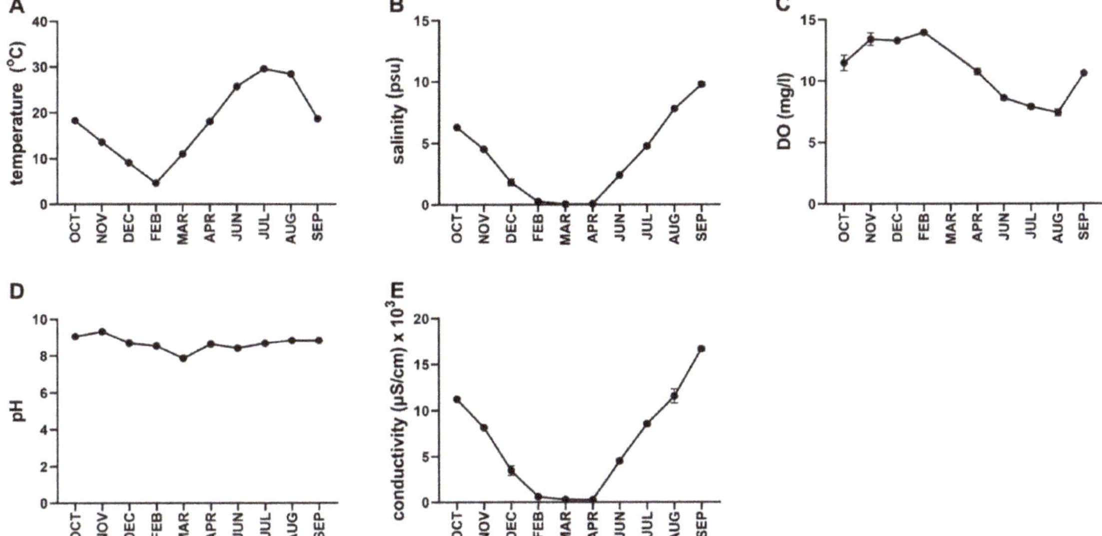

Figure 2. Mean values (±SD) of (**A**) temperature (°C), (**B**) salinity (psu), (**C**) dissolved oxygen (DO, mg/L), (**D**) pH and (**E**) conductivity (µS/cm) in Vistonis Lake (Northern Greece), during the sampling period (October 2014–September 2015). No sampling was conducted in January 2015.

3.2. Gills Histopathology

Several histological lesions were found in the gill samples by the histopathological examination (Figure 3). Specifically, epithelium detachment at the secondary lamella (edema) (Figure 3B), hyperplasia of the primary lamella (Figure 3B,C), aneurysms in the secondary lamella (Figure 3D), hyperplasia of the edge of the secondary lamella, hypoplasia of the secondary lamella and hyperplasia of the secondary lamella (Figure 3C) were the histological lesions detected in the gills of the species (Table 1).

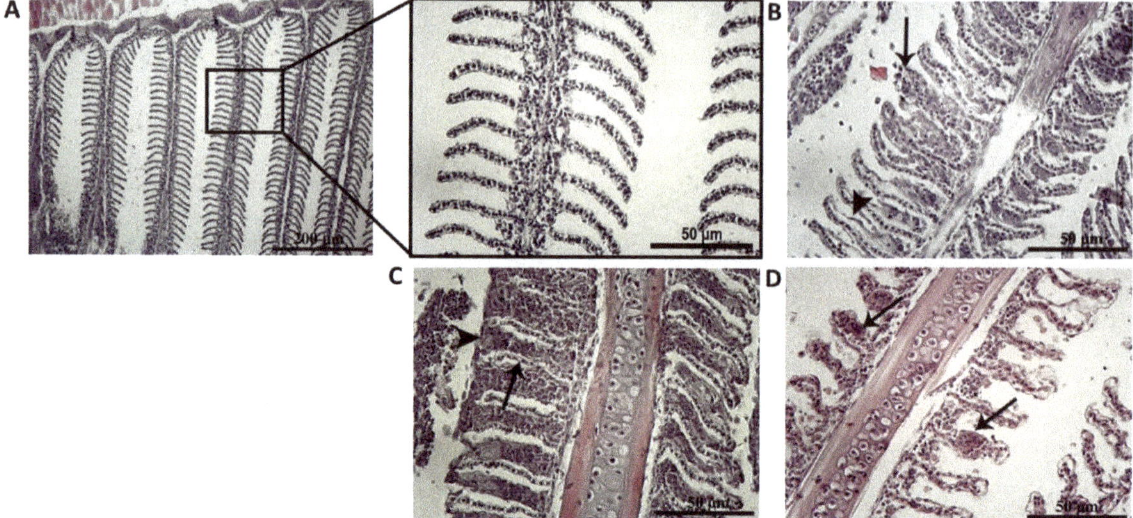

Figure 3. Histological lesions detected in the gills of the freshwater fish species *Alburnus vistonicus* in Vistonis Lake (Northern Greece), during the period October 2014–September 2015. (**A**) November sampling. Normal gills architecture. (**B**) April sampling. Hyperplasia of the primary lamella (arrow) and epithelium detachment (edema) of the secondary lamella (arrowhead). (**C**) April sampling. Hyperplasia of the primary lamella (arrowhead). Arrow shows a secondary lamella. (**D**) June sampling. Aneurysm of the secondary lamella (arrows).

Table 1. Presence (+) and absence (−) of histological lesions detected in gill samples of the freshwater fish species *Alburnus vistonicus* in Vistonis Lake (Northern Greece), during the period October 2014–September 2015 (no samples were obtained in January 2015 due to unfavorable weather conditions).

Histological Lesion	OCT	NOV	DEC	FEB	MAR	APR	JUN	JUL	AUG	SEP
Epithelium detachment (edema) at the secondary lamella	+(3)	-	+(4)	+(2)	+(3)	+(3)	+(2)	-	+(2)	+(4)
Hyperplasia of the primary lamella	+(3)	-	+(3)	+(2)	+(4)	+(2)	+(2)	+(1)	+(1)	+(3)
Hyperemia (aneurysm) of the secondary lamella	+(2)	-	-	-	-	+(2)	+(1)	+(1)	-	-
Hyperplasia of the edge of the secondary lamella	-	-	-	-	-	-	-	-	-	-
Hypoplasia (small size) of the secondary lamella	-	-	-	-	-	-	-	+(1)	-	-
Hyperplasia of the secondary lamella	-	-	-	+(2)	+(3)	-	-	+(1)	-	-

The number of fish in which lesions are observed is mentioned in parentheses.

3.3. HSPs

HSPs expression levels in the gills and liver of *A. vistonicus* are shown in Figure 4. In the gills, HSP70 levels exhibited their lowest in July and their highest in August, when they were also significantly higher compared to all the rest samplings (Figure 4A). In the liver, HSP70 lowest expression levels were observed in June and the highest in December, when they were also significantly higher compared to the rest of the samplings, with the exception of those in February (Figure 4A).

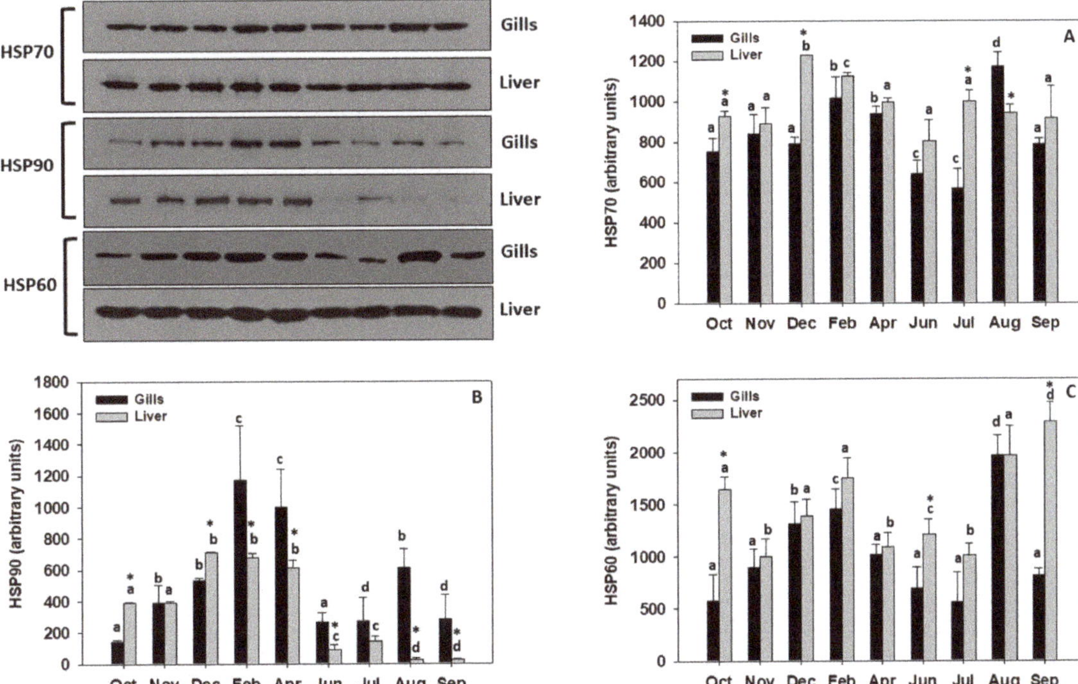

Figure 4. (**A**) HSP70, (**B**) HSP90 and (**C**) HSP60 levels (mean $\pm$ SD) in the gills and liver ($n = 5$) of *Alburnus vistonicus* in Vistonis lake (Northern Greece), during the sampling period (October 2014–September 2015). Samplings were performed on a monthly basis. Representative blots are presented. Significant differences ($p < 0.05$) are presented as: lower case letters—between sampling months and *—between gills and liver. No sampling was conducted in January 2015.

HSP90 levels in the gills exhibited seasonality, with their minimum in October and maximum in February. The expression levels in February were significantly higher compared to those in the rest of the months (Figure 4B). In the liver, HSP90 minimum expression levels were observed in August, while the maximum in December. HSP90 levels exhibited seasonality, with higher values during the winter and autumn months and lower during the summer ones (Figure 4B).

HSP60 minimum expression levels in the gills were found in July and their maximum in August, when they were also significantly higher compared to the rest months (Figure 4C). In the liver, HSP60 levels exhibited their lowest in July. Protein expression levels were significantly lower in April, June, July and November compared to those in October, December, February, August and September. Moreover, HSP60 levels in September were significantly higher compared to those from the rest months, with the exception of August (Figure 4C).

3.4. Na⁺-K⁺ ATPase

Na^+-K^+ ATPase was examined only in gills as shown in Figure 5. The lowest levels were observed in September and the maximum in August, which was also significantly higher compared to those from the rest months (Figure 5).

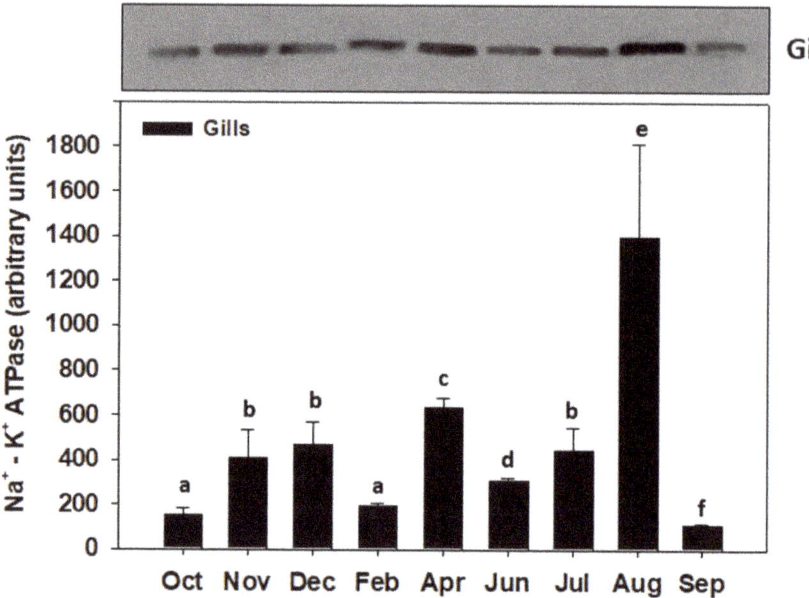

Figure 5. Na^+-K^+ ATPase levels (mean $\pm$ SD) in the gills ($n = 5$) of *Alburnus vistonicus* in Vistonis lake (Northern Greece), during the sampling period (October 2014–September 2015). Samplings were performed on a monthly basis. Representative blot is presented. Significant differences ($p < 0.05$) between sampling months are presented as lower case letters. No sampling was conducted in January 2015.

3.5. MAPKs

Phospho p38 MAPK/p38 MAPK ratio exhibited its highest levels, for both tissues (gills and liver) in September, which were significantly higher compared to those of the rest months (Figure 6A). In the gills, significantly lower ratio levels were observed in August compared to the other sampling months, while the ratio levels were also significantly different in the samples from April, June and August to those from all the rest sampling months (Figure 6A). In the liver, significantly lower ratio levels were observed in November compared to the rest sampling months, while those in February were also significantly higher than those from October, November, December, April and August (Figure 6A).

Phospho p44/42 MAPK/p44/42 MAPK ratio in the gills exhibited its minimum levels in August and maximum in December. Phosphorylation ratio levels in December were also significantly higher compared to those from the rest months. Phosphorylation ratio levels showed seasonality, with those in the dry season months (April–August) being significantly lower than those from the wet season months (October–February) (Figure 6B). In the liver, phosphorylation ratio levels exhibited mild fluctuations throughout the sampling period, with the exception of August and September, when the minimum and maximum ratio levels were found, respectively (Figure 6B).

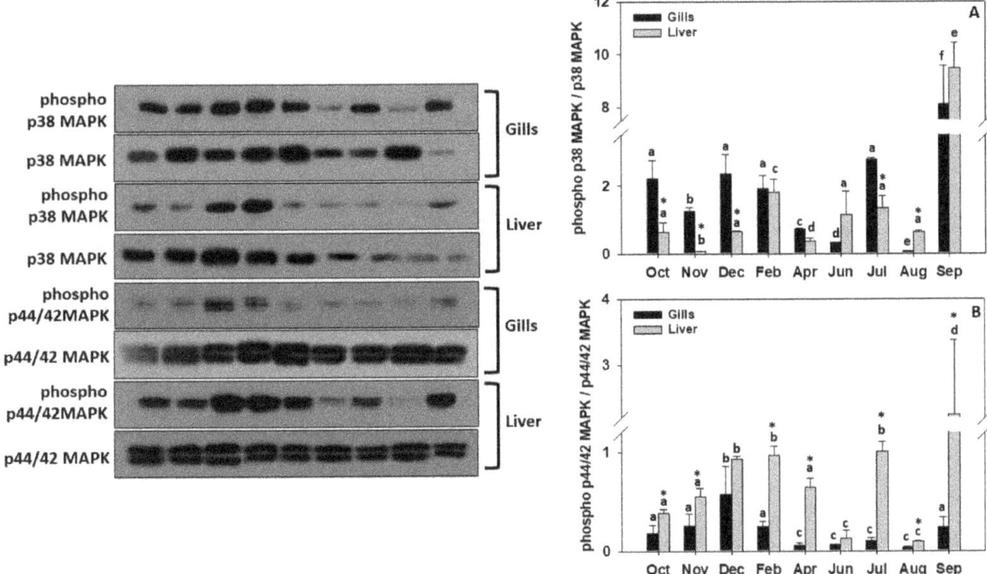

Figure 6. Phosphorylation ratio levels (mean ± SD) of (**A**) p38 MAPK (phospho p38 MAPK/p38 MAPK) and (**B**) p44/42 MAPK (phospho p44/42 MAPK/p44/42 MAPK) in the gills and liver (*n* = 5) of *Alburnus vistonicus* in Vistonis lake (Northern Greece), during the sampling period (October 2014–September 2015). Samplings were performed on a monthly basis. Representative blots are presented. Significant differences (*p* < 0.05) are presented as: lower case letters—between sampling months and *—between gills and liver. No sampling was conducted in January 2015.

3.6. Multivariate Analysis

PCA applied on the bioindicators of cellular stress responses and the average values of the water physico-chemical parameters measured in Vistonis Lake extract two significant components (PC1 and PC2) that explained 61.32% of the total variance (35.21% for PC1 and 26.11% for PC2). HSP70 in the gills and HSP90 in the liver and gills were positively correlated to PC1 axis while salinity and conductivity were negatively related (Figure 7 and inserted table). Accordingly, HSP60 in the liver, phospho p38 MAPK/p38 MAPK and phospho p44/42 MAPK/p44/42 MAPK in both tissues and dissolved oxygen (DO) were negatively related to PC2 axis, while Na$^+$-K$^+$ ATPase and temperature had positioned on the positive part of the axis (Figure 7 and inserted table).

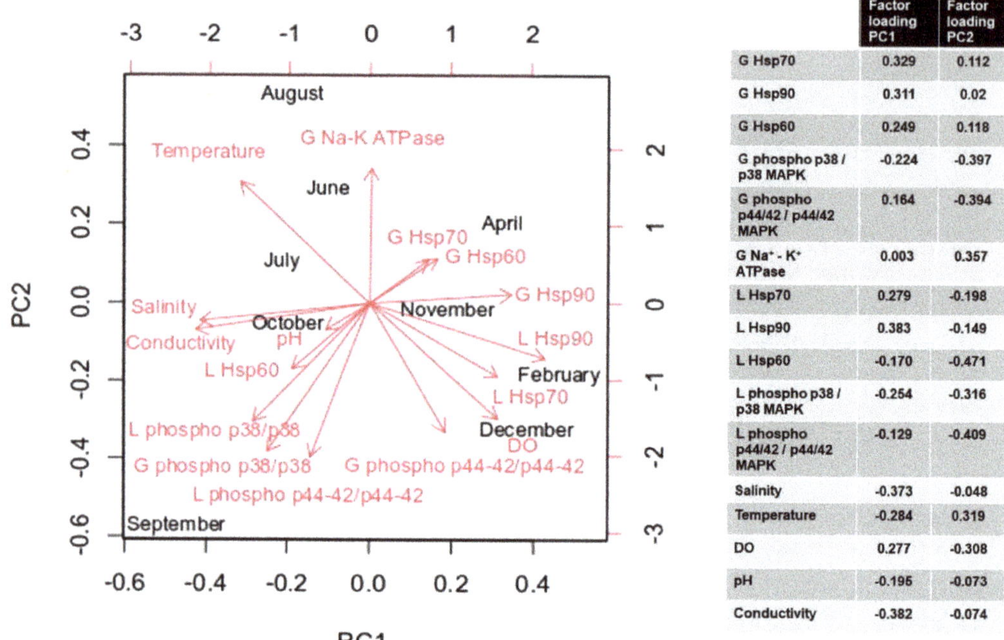

	Factor loading PC1	Factor loading PC2
G Hsp70	0.329	0.112
G Hsp90	0.311	0.02
G Hsp60	0.249	0.118
G phospho p38 / p38 MAPK	-0.224	-0.397
G phospho p44/42 / p44/42 MAPK	0.164	-0.394
G Na$^+$ - K$^+$ ATPase	0.003	0.357
L Hsp70	0.279	-0.198
L Hsp90	0.383	-0.149
L Hsp60	-0.170	-0.471
L phospho p38 / p38 MAPK	-0.254	-0.316
L phospho p44/42 / p44/42 MAPK	-0.129	-0.409
Salinity	-0.373	-0.048
Temperature	-0.284	0.319
DO	0.277	-0.308
pH	-0.195	-0.073
Conductivity	-0.382	-0.074

Figure 7. Principal Component Analysis biplot of the CSR biomarkers (G = gills, L = liver) measured in *Alburnus vistonicus* (*n* = 45, 5 fish per month) and mean values of the physico-chemical parameters measured in Vistonis Lake (northern Greece) during the sampling period (October 2014–September 2015). In the inserted table the factor loadings of each parameter on the two principal components (PC1 and PC2) are presented.

4. Discussion

The present study aimed to provide some insights into the physiological responses of the endemic freshwater fish species *A. vistonicus* in Vistonis Lake, in relation to the continuous fluctuation of the abiotic parameters in the lake.

Temperature values exhibited seasonal fluctuations, a phenomenon that is generally observed in aquatic freshwater ecosystems [47–49]. As expected from previous studies and due to the hydrology of the lake [50], salinity values followed a seasonal pattern, with low values recorded during the cold months December–February and higher during the hot months June–August. The periodic inflow of seawater in coastal lakes, such as Vistonis Lake, is a common, but alarming phenomenon, since even small increases in salinity might cause serious defects in the biodiversity of the habitats and threaten the survival of the species [51]. DO and pH values did not show any strong fluctuations and were ranging among the permitted limits for cyprinids [52]. pH values were constantly higher than 7, reflecting the alkaline character of the natural freshwater systems in Greece, and conductivity values were high, especially during the dry period, due to the seawater inflows in the lake [53].

The peculiar hydrology of Vistonis Lake system, leading often to salinization, especially during the dry season, has resulted in species population decline, threatening thus its survival [28]. Species with restricted distribution, such as *A. vistonicus*, may be especially prone to extinction than others with wider distribution range [54]. Therefore, the importance for acute implementation of management plans for the protection and preservation of the species has already been recommended [28,30]. Our results show that biomarkers of cellular stress responses could also be included in future management plans of the species, for achieving a better and more complete assessment of the biology of the species.

4.1. HSPs' Induction

HSP60 and HSP70 expression levels, both in the gills and in the liver, were maintained at high levels throughout the year. During colder and warmer months, when salinity and temperature showed minimum and maximum values, respectively, both HSP60 and HSP70 in the gills exhibited their highest levels. This was also depicted in the two dimensional plot of the results of PC analysis (Figure 7). Generally, HSP members in both gill and liver samples were positioned on the positive part of PC1, where the samples of the cold months are placed, having a negative correlation to salinity and conductivity. It has been demonstrated that several molecular chaperones are upregulated in fish gills when exposed to stress provoked by salinity variations [18,55,56]. Specifically, in black sea bream *Mylio macrocephalus* (Basilewsky, 1855), HSPs (HSP70, HSP90 and HSP60) were induced after acclimation to both plasma-hyperosmotic and plasma-hyposmotic salinity [57]. According to Protas et al. [58], low conductivity appears to elicit responses in fish similar to heat stress response [59–61]. In fact, the above claim has been confirmed by Rohner et al. [62], who showed that low water conductivity values were accompanied by increased HSP90 expression in the fish species *Astyanax mexicanus* (De Filippi, 1853). These indications are very similar to the ones obtained in the present work, since increased levels of HSP members in the gills were mostly negatively correlated to conductivity, as shown by the PCA. We cannot support with certainty whether salinity is the single stress factor, as PCA has shown that, in gill samples, HSP levels are positively related to temperature, since they are positioned on the positive part of the PC2 axis. However, the opposite stands for HSP measured in liver samples that were placed on the negative part of PC2. (Figure 7). Because cold and heat stress destabilizes the hydrophobic interactions of polypeptide chains, leading to non-functional proteins, HSPs provide protection against protein structural changes [63–66]. Similar results of Heat Shock Response (HSR) as a means of fish acclimatization to cold [67–70] as well as to increasing ambient temperatures were reported by Feidantsis et al. [71], who investigated seasonality effects in the gilthead sea bream *Sparus aurata* (Linnaeus, 1758). Regarding HSP90, its expression levels exhibited seasonality in both gills and liver. Specifically, in the liver, its levels were consistently high during the colder and low in the warmer months. It has been argued that HSP induction in fish has a protective role against a number of environmental stressors [72,73]. However, according to Iwama et al. [8], HSPs increased levels in fish may be more of a stress indication rather than a measure of its degree. Differences observed between gills and liver HSR can be attributed to the well-known tissue specificity observed in fish species (e.g., [13,71]).

4.2. MAPKs Phosphorylation

MAPKs involvement in HSP induction has been shown in various tissues of *S. aurata* [22,71]. In some published reports, MAPKs activation by various stressors has been suggested [17–19,74,75]. However, data on the seasonal pattern of their activation in fish are very limited (e.g., [71]). According to Kültz and Avila [18], MAPKs phosphorylation is related to the process of osmoregulation of large fish, especially in gill cells. The above observation seems to depict part of the results of the present study, since in September, along with the maximum mean salinity values, the highest MAPKs phosphorylation levels were observed. Specifically, p38 MAPK higher activation levels were observed, both for gills and liver in the same period. The latter coincides well with increased water conductivity levels and increased p38 MAPK phosphorylation in the gills in December and February. However, literature concerning the relation between MAPK pathway and water conductivity is extremely limited. In contrast, p44/42 MAPK, both in the liver and the gills, exhibit their higher phosphorylation levels in September, and are probably more a result of high summer temperatures and less of high salinity levels. Seasonal activation of the MAPK members was also observed in several examined tissues of *S. aurata* after exposure to increased environmental and laboratory temperature, as a response to heat stress [13,22,71]. However, on the PCA biplot a different pattern is evident. MAPKs measured in both liver and gill samples were placed close to September samples when salinity values were the highest, having negative correlation with temperature.

4.3. Gills' Physiology

A. vistonicus belongs to the Cyprinidae family [76]. Cyprinids show lack of secondary lamellae in the gills [77,78] under normoxic water (DO > 3 mg/L) and temperatures below 20 °C. This strange structure makes them strikingly different from most other fish species. The specific gill structure (known as "hyperplasia of the primary lamella") was more or less observed at all sampling months, except November (Table 1). The DO values during the sampling period were above 8 ml/L. So, the observed "hyperplasia of the primary lamella" may be a normal *A. vistonicus* gill structure and not a gill lesion. During the sample period the salinity of the lake was very high (>3.5 psu) for 5 months (October, November, July, August and September) and the pH was over 8.5 (except March). These conditions could be described as "stress conditions" for the fish and could explain the observed histological lesions to the gills.

According to Tine et al. [12], HSP70 high expression levels are accompanied by increased levels of Na^+-K^+ ATPase in fish, as a response to hyper-osmotic stress. Similarly, Herrera et al. [79] have also observed that Na^+-K^+ ATPase in the gills of Senegalese sole *Solea senegalensis* (Kaup, 1858) has been only increased when fish were transferred from seawater to 55 g/kg salinity, and not from seawater to low 5 g/kg salinity. These data indicate that active ion transport is only significantly increased when the proper mechanisms are already in place i.e., when ionic and osmotic gradients are not reversed. In fact, this observation could be a possible explanation for the survival of fish in habitats with extreme salinity values. In the case of Vistonis Lake, the results presented herein, exhibited a similar pattern regarding HSP70 expression levels and Na^+-K^+ ATPase in the gills, since both exhibited their maximum levels during increasing ambient water temperature, when mean salinity values increased (also confirmed by PCA analysis). This observation could be an indication of the adaptability of the fish to the particular lake ecosystem, with Na^+-K^+ ATPase responding to the increase in salinity, balancing the ion gradient between plasma and the extracellular environment with active Cl^- excretion [23,80,81].

5. Conclusions

In conclusion, it seems that *A. vistonicus* can still cope with the constantly changing environment of Vistonis Lake. The latter is concluded by the fact that *A. vistonicus* CSR is versatile and varies according to the examined tissue, biomarker and season, in order for this species to adapt to its shifting habitat. Therefore, HSPs' induction, Na^+-K^+ ATPase activity and MAPKs activation seem to consist of a successful adaptive mechanism (Figure 8). However, additional studies on the species' physiological responses, including additional biomarkers, would contribute both to safer conclusions and to further understanding of the species' physiology. At the same time, the estimation of the range, as well as the lower and upper limits, of each environmental parameter, where critically endangered *A. vistonicus* exhibits its optimal physiological capacity (e.g., Oxygen- and capacity-limited thermal tolerance hypothesis according to Pörtner et al. [82]) is of great importance. However, further research of the potential correlation between the changing physico-chemical parameters and the synergistic response of different proteins, along with expansion of histological studies, will contribute to understanding the physiology of the endangered species *A. vistonicus*.

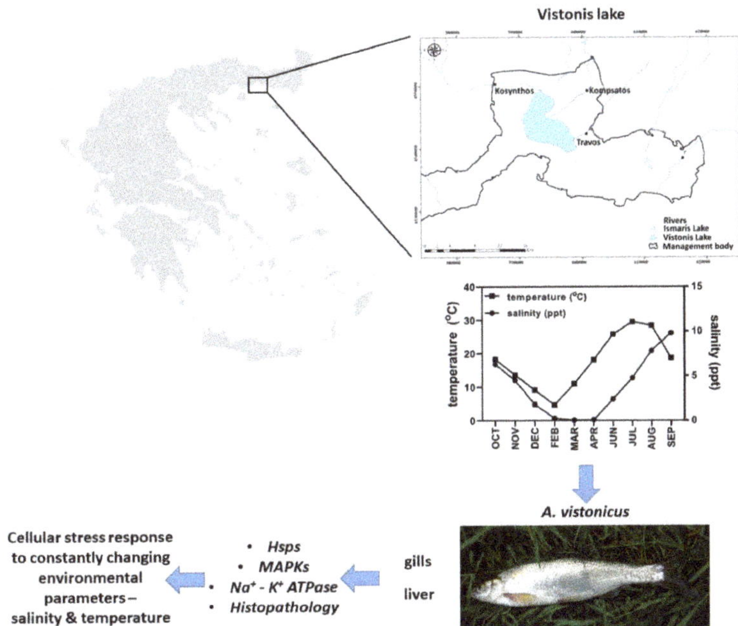

Figure 8. Summarized model of cellular stress responses of the freshwater fish species *Alburnus vistonicus in Vistonis Lake (Northern Greece) under the seasonal variation of the physico-chemical parameters.*

Author Contributions: D.B. and E.A. conceived and designed the study. E.T. and F.A.M. conducted fieldwork and collected the samples. T.T., E.T. and K.F. performed the biochemical analyses, and P.B. the histological analysis. K.F., E.T., T.T. and D.B. performed statistical analysis. E.T., T.T. and K.F. visualized results. E.T., T.T. and K.F. wrote the first draft of the manuscript. E.A. and D.B. corrected and revised the manuscript. All authors have read and agreed to the published version of the manuscript.

Funding: The study was funded by the Management Body of Delta Nestos and Lakes Vistonis-Ismarida, as part of the research project "Study and management proposals for the two endemic and under extinction fish species of lakes Vistonida and Mitrikou (*Alosa vistonica* & *Alburnus vistonicus*)".

Institutional Review Board Statement: All necessary permissions were provided by the Management Body of the Delta Nestos and Lakes Vistonida-Ismarida in the frame of the implementation of the project titled "Study and management proposals for the two endemic and under extinction fish species of lakes Vistonida and Mitrikou (*Alosa Vistonica* & *Alburnus vistonicus*)" (code 90981) financed by the operational program "Environmental and Sustainable Development" in Greece for the period 2007–2013. Fish handling and treatment were in accordance with the local guidelines for treating animals which comply with the Official Journal of the Greek Government No. 106/30 April 2013 on the protection of animals used for scientific purposes. Animals received proper care in compliance with the "Guidelines for the Care and Use of Laboratory Animals" published by US National Institutes of Health (NIH publication No 85–23, revised 1996) and the "Principles of laboratory animal care" published by the Greek Government (160/1991) based on EU regulations (86/609).

Informed Consent Statement: Not applicable.

Data Availability Statement: The data presented in this study are available on request from the corresponding author. The data are not publicly available due to local authorities' privacy restrictions.

Acknowledgments: The authors would like to thank the Management Body of the Delta Nestos and Lakes Vistonida-Ismarida for providing all the necessary permissions for fish sampling and for their technical and financial support. Moreover, the authors would like to specially thank Nesdet, the fisherman who with safety helped us throughout the fieldwork.

Conflicts of Interest: The authors declare no conflict of interest.

References

1. Jesus, T.F.; Moreno, J.M.; Repolho, T.; Athanasiadis, A.; Rosa, R.; Almeida-Val, V.M.; Coelho, M.M. Protein analysis and gene expression indicate differential vulnerability of Iberian fish species under a climate change scenario. *PLoS ONE* **2017**, *12*, e0181325. [CrossRef]
2. Latinopoulos, D.; Ntislidou, C.; Kagalou, I. Multipurpose plans for the sustainability of the Greek lakes: Emphasis on multiple stressors. *Environ. Process.* **2016**, *3*, 589–602. [CrossRef]
3. Eliason, E.J.; Clark, T.D.; Hague, M.J.; Hanson, L.M.; Gallagher, Z.S.; Jeffries, K.M.; Gale, M.K.; Patterson, D.A.; Hinch, S.G.; Farrell, A.P. Differences in thermal tolerance among sockeye salmon populations. *Science* **2011**, *332*, 109–112. [CrossRef]
4. Veilleux, H.D.; Ryu, T.; Donelson, J.M.; Van Herwerden, L.; Seridi, L.; Ghosheh, Y.; Beruman, M.L.; Leggat, W.; Ravasi, T.; Munday, P.L. Molecular processes of transgenerational acclimation to a warming ocean. *Nat. Clim. Chang.* **2015**, *5*, 1074–1078. [CrossRef]
5. Pimentel, M.S.; Faleiro, F.; Diniz, M.; Machado, J.; Pousão-Ferreira, P.; Peck, M.A.; Pörtner, H.O.; Rosa, R. Oxidative stress and digestive enzyme activity of flatfish larvae in a changing ocean. *PLoS ONE* **2015**, *10*, e0134082. [CrossRef]
6. Boeuf, G.; Payan, P. How should salinity influence fish growth? *Comp. Biochem. Physiol. C* **2001**, *130*, 411–423. [CrossRef]
7. Tabassum, H.; Ashafaq, M.; Khan, J.; Shah, M.Z.; Raisuddin, S.; Parvez, S. Short term exposure of pendimethalin induces biochemical and histological perturbations in liver, kidney and gill of freshwater fish. *Ecol. Indic.* **2016**, *63*, 29–36. [CrossRef]
8. Iwama, G.K.; Thomas, P.T.; Forsyth, R.B.; Vijayan, M.M. Heat shock protein expression in fish. *Rev. Fish Biol. Fish.* **1998**, *8*, 35–56. [CrossRef]
9. Hofmann, G.E. Patterns of Hsp gene expression in ectothermic marine organisms on small to large biogeographic scales. *Integr. Comp. Biol.* **2005**, *45*, 247–255. [CrossRef]
10. Tomanek, L. Variation in the heat shock response and its implication for predicting the effect of global climate change on species' biogeographical distribution ranges and metabolic costs. *J. Exp. Biol.* **2010**, *213*, 971–979. [CrossRef]
11. Sørensen, J.G.; Kristensen, T.N.; Loeschcke, V. The evolutionary and ecological role of heat shock proteins. *Ecol. Lett.* **2003**, *6*, 1025–1037. [CrossRef]
12. Tine, M.; Bonhomme, F.; McKenzie, D.J.; Durand, J.D. Differential expression of the heat shock protein Hsp70 in natural populations of the tilapia, *Sarotherodon melanotheron*, acclimatised to a range of environmental salinities. *BMC Ecol.* **2010**, *10*, 11. [CrossRef]
13. Feidantsis, K.; Pörtner, H.O.; Antonopoulou, E.; Michaelidis, B. Synergistic effects of acute warming and low pH on cellular stress responses of the gilthead seabream *Sparus aurata*. *J. Comp. Physiol. B* **2015**, *185*, 185–205. [CrossRef]
14. Duffy, L.K.; Scofield, E.; Rodgers, T.; Patton, M.; Bowyer, R.T. Comparative baseline levels of mercury, Hsp 70 and Hsp 60 in subsistence fish from the Yukon-Kuskokwim delta region of Alaska. *Comp. Biochem. Physiol. C* **1999**, *124*, 181–186. [CrossRef]
15. Vijayan, M.M.; Pereira, C.; Kruzynski, G.; Iwama, G.K. Sublethal concentrations of contaminant induce the expression of hepatic heat shock protein 70 in two salmonids. *Aquat. Toxicol.* **1998**, *40*, 101–108. [CrossRef]
16. Sanders, B.M.; Jenkins, K.D.; Nichols, J.L.; Imber, B.E. Accumulation of Heat Shock Proteins for Evaluating Biological Damage due to Chronic Exposure of an Organism to Sublethal Levels of Pollutants. U.S. Patent No. 5232833; A, 3 August 1993.
17. Hashimoto, H.; Fukuda, M.; Matsuo, Y.; Yokoyama, Y.; Nishida, E.; Toyohara, H.; Sakaguchi, M. Identification of a nuclear export signal of MKK6, an activator of the carp p38 mitogen activated protein kinases. *Eur. J. Biochem.* **2000**, *267*, 4362–4371. [CrossRef]
18. Kültz, D.; Avila, K. Mitogen-activated protein kinases are in vivo transducers of osmosensory signals in fish gill cells. *Comp. Biochem. Physiol. B* **2001**, *129*, 821–829. [CrossRef]
19. Marshall, W.S.; Ossum, C.G.; Hoffmann, E.K. Hypotonic shock mediation by p38 MAPK, JNK, PKC, FAK, OSR1 and SPAK in osmosensing chloride secreting cells of killifish opercular epithelium. *J. Exp. Biol.* **2005**, *208*, 1063–1077. [CrossRef]
20. Leal, R.B.; Ribeiro, S.J.; Posser, T.; Cordova, F.M.; Rigon, A.P.; Zaniboni Filho, E.; Bainy, A.C. Modulation of ERK1/2 and p38 MAPK by lead in the cerebellum of Brazilian catfish *Rhamdia quelen*. *Aquat. Toxicol.* **2006**, *77*, 98–104. [CrossRef]
21. Feidantsis, K.; Pörtner, H.O.; Lazou, A.; Kostoglou, B.; Michaelidis, B. Metabolic and molecular stress responses of the gilthead sea bream *Sparus aurata* during long term exposure to increasing temperatures. *Mar. Biol.* **2009**, *156*, 797–809. [CrossRef]
22. Feidantsis, K.; Pörtner, H.O.; Markou, T.; Lazou, A.; Michaelidis, B. Involvement of p38 MAPK in the induction of Hsp70 during acute thermal stress in red blood cells of the gilthead sea bream, *Sparus aurata*. *J. Exp. Zool. A* **2012**, *317*, 303–310. [CrossRef]
23. Karnaky, J.R.; Kinter, L.B.; Kinter, W.B.; Stirling, C.E. Na, K-ATPase in Killifish *Fundulus heteroclitus* Adapted to Low and High Salinity Environments. *J. Cell Biol.* **1976**, *70*, 157–177. [CrossRef] [PubMed]
24. Maetz, J. Fish gills: Mechanisms of salt transfer in fresh water and sea water. *Philos. Trans. R. Soc. Lond. B Biol. Sci.* **1971**, *262*, 209–249. [CrossRef]
25. Motais, R.; Garcia-Romeu, F. Transport mechanisms in the teleostean gill and amphibian skin. *Annu. Rev. Physiol.* **1972**, *34*, 141–176. [CrossRef] [PubMed]
26. Maetz, J.; Bornancin, M. Biochemical and biophysical aspects of salt excretion by chloride cells in teleosts. *Fortschr. Der Zool.* **1975**, *23*, 322–362.
27. Freyhof, J.; Kottelat, M. *Alburnus vistonicus*, a new species of shemaya from eastern Greece, with remarks on *Chalcalburnus chalcoides macedonicus* from Lake Volvi (Teleostei: Cyprinidae). *Ichthyol. Explor. Freshw.* **2007**, *18*, 205–212.

28. Bobori, D.C.; Leonardos, I.; Ganias, K.; Sapounidis, A.; Petriki, O.; Ntislidou, C.; Mouchlianitis, F.; Tsakoumis, E.; Polyzou, C. *Study and Management Proposals for the Two Endemic and under Extinction Fish Species of Lakes Vistonida and Mitrikou* (Alosa vistonica and Alburnus vistonicus); Final Technical Report; Aristotle University of Thessaloniki: Thessaloniki, Greece, 2015.

29. Legakis, A.; Maragou, P. *The Red Data Book of Threatened Animals of Greece*; Hellenic Zoological Society: Athens, Greece, 2009.

30. Bobori, D.C.; Tsakoumis, E.; Mouchlianitis, F.A.; Antonopoulou, E.; Ganias, K. Growth and Reproductive Ecology of the Endemic Freshwater Fish *Alburnus vistonicus* Freyhof & Kottelat, 2007 (Actinopterygii: Cyprinidae) in Lake Vistonis System, Northern Greece. *Acta Zool. Bulg.* **2018**, *70*, 569–574.

31. Mouchlianitis, F.A.; Bobori, D.; Tsakoumis, E.; Sapounidis, A.; Kritikaki, E.; Ganias, K. Does fragmented river connectivity alter the reproductive behavior of the potamodromous fish *Alburnus vistonicus*? *Hydrobiologia* **2021**, *848*, 4029–4044. [CrossRef]

32. Manyala, J.O.; Ojuok, J.E. Survival of the Lake Victoria *Rastrineobola argentea* in a rapidly changing environment: Biotic and abiotic interactions. *Aquat. Ecosyst. Health Manag.* **2007**, *10*, 407–415. [CrossRef]

33. Miller, M.J.; Capriles, J.M.; Hastorf, C.A. The fish of Lake Titicaca: Implications for archaeology and changing ecology through stable isotope analysis. *J. Archaeol. Sci.* **2010**, *37*, 317–327. [CrossRef]

34. Njiru, M.; Mkumbo, O.C.; van der Knaap, M. Some possible factors leading to decline in fish species in Lake Victoria. *Aquat. Ecosyst. Health Manag.* **2010**, *13*, 3–10. [CrossRef]

35. Yue, H.; Guo-Xiang, L.; Guo-Feng, P. Effect of artificial macrocosms on water characteristics and benthic diatom communities in Donghu Lake, China. *J. Freshw. Ecol.* **2016**, *31*, 533–542. [CrossRef]

36. Koutrakis, E.T.; Kamidis, N.I.; Leonardos, I.D. Age, growth and mortality of a semi-isolated lagoon population of sand smelt, *Atherina boyeri* (Risso, 1810) (Pisces: Atherinidae) in an estuarine system of northern Greece. *J. Appl. Ichthyol.* **2004**, *20*, 382–388. [CrossRef]

37. Appelberg, M.; Berger, H.M.; Hesthagen, T.; Kleiven, E.; Kurkilahti, M.; Raitaniemi, J.; Rask, M. Development and intercalibration of methods in Nordic freshwater fish monitoring. *Water Air Soil Pollut.* **1995**, *85*, 401–406. [CrossRef]

38. CEN; European Committee for Standardization. *Water Quality-Sampling of Fish with Multi-Mesh Gillnets*; 2005; pp. 3–26. Available online: https://infostore.saiglobal.com/preview/is/en/2005/i.s.en14757-2005.pdf?sku=675315Bates (accessed on 17 November 2021).

39. Hirai, N.; Tagawa, M.; Kaneko, T.; Seikai, T.; Tanaka, M. Distributional changes in branchial chloride cells during freshwater adaptation in Japanese sea bass *Lateolabrax japonicus*. *Zool. Sci.* **1999**, *16*, 43–49. [CrossRef]

40. Pan, F.; Zarate, J.M.; Tremblay, G.C.; Bradley, T.M. Cloning and characterization of salmon hsp90 cDNA: Upregulation by thermal and hyperosmotic stress. *J. Exp. Zool.* **2000**, *287*, 199–212. [CrossRef]

41. Azizi, S.; Kochanian, P.; Peyghan, R.; Khansari, A.; Bastami, K.D. Chloride cell morphometrics of common carp, *Cyprinus carpio*, in response to different salinities. *Comp. Clin. Path.* **2011**, *20*, 363–367. [CrossRef]

42. Berillis, P.; Mente, E.; Nikouli, E.; Makridis, P.; Grundvig, H.; Bergheim, A.; Gausen, M. Improving aeration for efficient oxygenation in sea bass sea cages. Blood, brain and gill histology. *Open Life Sci.* **2016**, *11*, 270–279. [CrossRef]

43. Dias, P.; Gupta, A.; Manna, S.K. Heat shock protein 70 expression in different tissues of *Cirrhinus mrigala* (Ham) following heat stress. *Aquac. Res.* **2005**, *36*, 525–529. [CrossRef]

44. Deng, D.F.; Wang, C.; Lee, S.; Bai, S.; Hung, S.S.O. Feeding rates affects heat shock protein levels in liver of larval white sturgeon (*Acipenser transmontanus*). *Aquaculture* **2009**, *287*, 223–226. [CrossRef]

45. Antonopoulou, E.; Kentepozidou, E.; Roufidou, C.; Despoti, S.; Feidantsis, K.; Chatzifotis, S. Starvation and re-feeding affect the expression of Hsp, MAPK and antioxidative enzymes of European sea bass (*Dicentrarchus labrax*). *Comp. Biochem. Physiol. A* **2013**, *165*, 79–88. [CrossRef]

46. Lê, S.; Josse, J.; Husson, F. FactoMineR: A Package for Multivariate Analysis. *J. Stat. Softw.* **2008**, *25*, 1–18. [CrossRef]

47. Livingstone, D.M.; Lotter, A.F. The relationship between air and water temperatures in lakes of the Swiss Plateau: A case study with pal\sgmaelig; olimnological implications. *J. Paleolimnol.* **1998**, *19*, 181–198. [CrossRef]

48. Mohseni, O.; Stefan, H.G. Stream temperature/air temperature relationship: A physical interpretation. *J. Hydrol.* **1999**, *218*, 128–141. [CrossRef]

49. Jackson, M.C.; Loewen, C.J.; Vinebrooke, R.D.; Chimimba, C.T. Net effects of multiple stressors in freshwater ecosystems: A meta-analysis. *Glob. Chang. Biol.* **2016**, *22*, 180–189. [CrossRef]

50. Koutrakis, E.T. Biology and Population Dynamics of Grey Mullets (Pisces: Mugilidae) in the Lake Vistonis and the Lagoon of Porto–Lagos. Ph.D. Dissertation, Aristotle University of Thessaloniki, Thessaloniki, Greece, 1994.

51. Schallenberg, M.; Hall, C.J.; Burns, C.W. *Climate Change Alters Zooplankton Community Structure and Biodiversity in Coastal Wetlands*; Report of Freshwater Ecology Group; University of Otago: Hamilton, New Zealand, 2001.

52. Directive 2006/44/EC of 6 September 2006, on the quality of fresh waters needing protection or improvement in order to support fish life. *Off. J. Eur. Union L* **2006**, *264*, 20–31.

53. Skoulikidis, N.T.; Bertahas, I.; Koussouris, T. The environmental state of freshwater resources in Greece (rivers and lakes). *Environ. Geol.* **1998**, *36*, 1–17. [CrossRef]

54. Sodhi, N.S.; Brook, B.W.; Bradshaw, C.J. Causes and consequences of species extinctions. *Princet. Guide Ecol.* **2009**, *1*, 514–520.

55. Kültz, D. Osmotic regulation of DNA activity and the cell cycle. In *Environmental Stressors and Gene Responses*; Storey, K.B., Storey, J.M., Eds.; Elsevier: Amsterdam, The Netherlands, 2000; pp. 157–180.

56. Kültz, D.; Somero, G.N. Differences in protein patterns of gill epithelial cells of the fish *Gillichthys mirabilis* after osmotic and thermal acclimation. *J. Comp. Physiol. B* **1996**, *166*, 88–100. [CrossRef] [PubMed]
57. Deane, E.E.; Kelly, S.P.; Luk, J.C.Y.; Woo, N.Y.S. Chronic salinity adaptation modulates hepatic heat shock protein and insulin-like growth factor I expression in black sea bream. *Mar. Biotechnol.* **2002**, *4*, 193–205. [CrossRef]
58. Protas, M.; Conrad, M.; Gross, J.B.; Tabin, C.; Borowsky, R. Regressive evolution in the Mexican cave tetra, *Astyanax mexicanus*. *Curr. Biol.* **2007**, *17*, 452–454. [CrossRef]
59. Grosell, M.; Nielsen, C.; Bianchini, A. Sodium turnover rate determines sensitivity to acute copper and silver exposure in freshwater animals. *Comp. Biochem. Physiol. C* **2002**, *133*, 287–303. [CrossRef]
60. Monserrat, J.M.; Martínez, P.E.; Geracitano, L.A.; Amado, L.L.; Martins, C.M.; Pinho, G.L.L.; Chaves, I.S.; Ferreira-Cravo, M.; Ventura-Lima, J.; Bianchini, A. Pollution biomarkers in estuarine animals: Critical review and new perspectives. *Comp. Biochem. Physiol. C* **2007**, *146*, 221–234. [CrossRef]
61. Saglam, D.; Atli, G.; Dogan, Z.; Baysoy, E.; Gurler, C.; Eroglu, A.; Canli, M. Response of the antioxidant system of freshwater fish (*Oreochromis niloticus*) exposed to metals (Cd, Cu) in differing hardness. *Turkish J. Fish. Aquat. Sci.* **2014**, *14*. [CrossRef]
62. RohnerNJarosz, D.F.; Kowalko, J.E.; Yoshizawa, M.; Jeffery, W.R.; Borowsky, R.L.; Lindquist, S.; Tabin, C.J. Cryptic variation in morphological evolution: HSP90 as a capacitor for loss of eyes in cavefish. *Science* **2013**, *342*, 1372–1375. [CrossRef]
63. Somero, G.N. Temperature relationships from molecules to biogeography. Hopkins Marine Station, Stanford University, Pacific Grove, California. In *Handbook of Physiology, Vol. II, Section 13: Comparative Physiology*; Oxford University Press: Oxford, UK, 1997; Chapter 19; pp. 1392–1444.
64. Pörtner, H.O.; Knust, R. Climate change affects marine fishes through the Oxygen Limitation of Thermal Tolerance. *Science* **2007**, *315*, 95–97. [CrossRef] [PubMed]
65. Kassahn, K.S.; Crozier, R.S.; Pörtner, H.O.; Caley, M.J. Animal performance and stress responses and tolerance limits at different levels of biological organisation. *Biol. Rev.* **2009**, *84*, 277–292. [CrossRef]
66. Pörtner, H.O.; Peck, M.A. Climate change effects on fishes and fisheries: Towards a cause-and-effect understanding. *J. Fish Biol.* **2010**, *77*, 1745–1779. [CrossRef] [PubMed]
67. Ju, Z.; Dunham, R.A.; Liu, Z. Differential gene expression in the brain of channel catfish (*Ictalurus punctatus*) in response to cold acclimation. *Mol. Genet. Genom.* **2002**, *268*, 87–95. [CrossRef]
68. Ali, K.S.; Dorgai, L.; Abraham, M.; Hermesz, E. Tissue- and stressor-specific differential expression of two hsc70 genes in carp. *Biochem. Biophys. Res. Commun.* **2003**, *307*, 503–509. [CrossRef]
69. Place, S.P.; Zippay, M.L.; Hofmann, G.E. Constitutive roles for inducible genes: Evidence for the alteration in expression of the inducible hsp70 gene in Antarctic notothenioid fishes. *Am. J. Physiol.* **2004**, *287*, R429–R436. [CrossRef]
70. Place, S.P.; Hofmann, G.E. Comparison of Hsc70 orthologs from polar and temperate notothenioid fishes: Differences in prevention of aggregation and refolding of denatured proteins. *Am. J. Physiol.* **2005**, *288*, R1195–R1202. [CrossRef] [PubMed]
71. Feidantsis, K.; Antonopoulou, E.; Lazou, A.; Pörtner, H.O.; Michaelidis, B. Seasonal variations of cellular stress response of the gilthead sea bream (*Sparus aurata*). *J. Comp. Physiol. B* **2013**, *183*, 625–639. [CrossRef] [PubMed]
72. Currie, S.; Moyes, C.D.; Tufts, B. The effects of heat shock and acclimation temperature on hsp 70 and hsp30 mRNA expression in rainbow trout: In vivo and in vitro comparisons. *J. Fish Biol.* **2000**, *56*, 398–408. [CrossRef]
73. Sherry, J.P. The role of biomarkers in the health assessment of aquatic ecosystems. *Aquat. Ecosyst. Health Manag.* **2003**, *6*, 423–440. [CrossRef]
74. Lewis, T.S.; Shapiro, P.S.; Ahn, N.G. Signal Transduction through MAP kinase cascades. *Adv. Cancer Res.* **1998**, *74*, 49–139. [CrossRef]
75. Kyriakis, J.M.; Avruch, J. Mammalian mitogen activated protein kinase signal transduction pathways activated by stress and inflammation. *Physiol. Rev.* **2001**, *81*, 807–869. [CrossRef]
76. Froese, R.; Pauly, D. (Eds.) FishBase. World Wide Web Electronic Publication 2016. Version (01/2016). Available online: www.fishbase.org (accessed on 22 May 2021).
77. Sollid, J.; De Angelis, P.; Gundersen, K.; Nilsson, G.E. Hypoxia induces adaptive and reversible gross morphological changes in crucian carp gills. *J. Exp. Biol.* **2003**, *206*, 3667–3673. [CrossRef]
78. Sollid, J.; Weber, R.E.; Nilsson, G.E. Temperature alters the respiratory surface area of crucian carp (*Carassius carassius*) and goldfish (*Carassius auratus*). *J. Exp. Biol.* **2005**, *208*, 1109–1116. [CrossRef]
79. Herrera, M.; Aragao, C.; Hachero, I.; Ruiz-Jarabo, I.; Vargas-Chacoff, L.; Miguel Mancera, J.; Conceicao, L.E.C. Physiological shortterm response to sudden salinity change in the Senegalese sole (*Solea senegalensis*). *Fish Physiol. Biochem.* **2012**, *38*, 1741–1751. [CrossRef]
80. McCormick, S.D. Methods for nonlethal gill biopsy and measurement of Na^+, K^+-ATPase activity. *Can. J. Fish. Aquat. Sci.* **1993**, *50*, 656–658. [CrossRef]
81. Weng, C.F.; Chiang, C.C.; Gong, H.Y.; Chen, M.H.C.; Lin, C.J.F.; Huang, W.T.; Cheng, C.Y.; Hwang, P.P.; Wu, J.L. Acute changes in gill Na+-K+-ATPase and creatine kinase in response to salinity changes in the euryhaline teleost, tilapia (*Oreochromis mossambicus*). *Physiol. Biochem. Zool.* **2002**, *75*, 29–36. [CrossRef] [PubMed]
82. Pörtner, H.O.; Lucassen, M.; Storch, D. Metabolic biochemistry: Its role in thermal tolerance and in the capacities and in the capacities of physiological and ecological function. *Fish Physiol.* **2005**, *22*, 79–118. [CrossRef]

MDPI

Article

Atrazine-Induced Hepato-Renal Toxicity in Adult Male *Xenopus laevis* Frogs

Lynette Sena [1], Jaclyn Asouzu Johnson [1,*], Pilani Nkomozepi [2] and Ejikeme Felix Mbajiorgu [1]

[1] School of Anatomical Sciences, University of the Witwatersrand, Johannesburg 2193, South Africa; rufaro.sena@gmail.com (L.S.); ejikeme.mbajiorgu@wits.ac.za (E.F.M.)

[2] Department of Human Anatomy and Physiology, University of Johannesburg, Johannesburg 2006, South Africa; pilanin@uj.ac.za

[*] Correspondence: jaclyn.johnson@wits.ac.za

Abstract: Atrazine (ATZ) is an herbicide commonly detected in groundwater. Several studies have focused on its immunological and endocrine effects on adult *Xenopus laevis* species. However, we investigated the impact of atrazine on the renal and hepatic biochemistry and histomorphology in adult male frogs. Forty adult male frogs were allocated to four treatment groups (control, one ATZ (0.01 µg/L), two ATZ (200 µg/L) and three ATZ (500 µg/L), 10 animals per group, for 90 days. Alanine aminotransferase (ALT) and creatinine levels increased significantly ($p < 0.05$) in the 200 and 500 µg/L groups but malondialdehyde only in the 500 µg/L group ($p < 0.05$). Histopathological observations of derangement, hypertrophy, vascular congestion and dilation, infiltration of inflammatory cells incursion, apoptosis and hepatocytes cell death were observed with atrazine exposure, mostly in the 500 µg/L group. Additionally, histochemical labelling of caspase-3 in the sinusoidal endothelium was observed in all the treated groups, indicating vascular compromise. Evaluation of renal histopathology revealed degradation and atrophy of the glomerulus, vacuolization, thick loop of Henle tubule epithelial cells devolution and dilation of the tubular lumen. Furthermore, expression of caspase-3 indicates glomerular and tubular apoptosis in atrazine-exposed animals. These findings infer that environmentally relevant atrazine doses (low or high) could induce hepatotoxicity and nephrotoxicity in adult male *Xenopus laevis* frogs and potentially related aquatic organisms.

Keywords: atrazine; hepatorenal; pathology; toxicosis; biomarkers; adult *Xenopus laevis*

Citation: Sena, L.; Asouzu Johnson, J.; Nkomozepi, P.; Mbajiorgu, E.F. Atrazine-Induced Hepato-Renal Toxicity in Adult Male *Xenopus laevis* Frogs. *Appl. Sci.* **2021**, *11*, 11776. https://doi.org/10.3390/app112411776

Academic Editors: Panagiotis Berillis and Božidar Rašković

Received: 30 October 2021
Accepted: 6 December 2021
Published: 11 December 2021

1. Introduction

Atrazine (ATZ) as (2-chloro-4-ethylamino-6-isopropylamino-s-triazine) is a broad-spectrum herbicide used extensively for agricultural production and for farming activities around the world for the control of selective pre- and post-emergence weeds [1,2]. Atrazine studies have been focused on evaluating environmental and ecological toxicity because of the detection of environmentally relevant (0.1 µg/L), lethal and/or sub-lethal concentrations detected as chemical contaminants of rivers, streams and wells, which are common habitats for aquatic organisms and water sources for humans [3,4], and might constitute a public health problem.

Frogs, toads and the anuran group of amphibians are hugely responsive to ecological pollutants because of their semi-permeable skin and are thus suitable as bio-indicators for detecting the toxic effects of herbicides [5,6]. Over the last few decades, a global increase in endangered amphibian species has been reported [7], such as the *Xenopus laevis*, due to exposure to agrochemicals. The African clawed frog comes from the genus Xenopus, and they have a significant role in maintaining the water and terrestrial ecosystem where they can be both prey and predators, metamorphosing from an herbivorous tadpole to a carnivorous adult [8,9]. Although the *Xenopus laevis* has a lifespan of 15 years, there are some concerns as their population is threatened by water pollution.

The reported adverse effects of ATZ in anuran populations inhabiting agricultural sites and non-agricultural sites (with water bodies linked to herbicide use areas) include, amongst others, disruption of reproductive potential [10,11], histopathology changes [12–14], changes in oxidative stress parameters [15–17] and disruption of endocrine physiology [18,19]. However, the mechanisms of toxicity of atrazine and its attendant histopathology in metabolic organs such as the kidneys and liver are not yet fully understood.

Atrazine is widely believed to alter the production of adenosine triphosphate (ATP), thereby adjusting mitochondrial structure and impairing antioxidant defensive systems by causing oxidative damage in aquatic organisms through the formation of reactive oxidative species (ROS) [17,20]. This ROS-mediated damage is often indicated by biomarkers such as the levels of malondialdehyde (MDA) and lipid peroxidation levels (LPO) [21,22]. Additionally, LPO is known as a key step that modulates toxicities of a wide range of herbicides [15]. Furthermore, mitochondrial and oxidative damage culminates in necrosis and apoptosis by the release of pro-apoptotic proteins such as caspase-3 [23,24].

Atrazine-induced liver and kidney oxidative damage [15,17,25] and caspase-3 immunolocalization have been reported in other vertebrates and *Xenopus laevis* tadpoles, respectively, but reports on atrazine toxicities relative to hepato-renal surrogate markers such as AST, ALKp, ALT, blood urea nitrogen (BUN), creatinine (CREA) and the associated caspase-3 immuno-expression in the liver and kidney of post-metamorphic African clawed frogs (*Xenopus laevis*) are lacking. It is anticipated that atrazine-induced oxidative stress and caspase-3 toxicity observed in tadpoles and adult rats may be present in adult frogs. However, reports of atrazine-induced effects on hepatorenal biomarkers in juvenile and adult rats are inconclusive [25–27]. These conflicting reports between effects observed in tadpoles and rats indicate the differential effects of atrazine effects in both species (frogs and rats). Therefore, this study was designed against this backdrop to clearly delineate the hepatorenal histopathological impairment incurred by these environmentally sensitive adult frogs during atrazine use.

2. Materials and Methods

2.1. Chemical and Reagents

Atrazine obtained from Accu Standard Inc. (New Haven, CT, USA), 1, 1, 3, 3-tetraethoxypropane (TEP) and HPLC solvents (Sigma-Aldrich, Johannesburg, South Africa), KOH, KH2 PO4, perchloric acid (HClO4) and HPLC-grade water (Darmstadt, Germany) were used in this study. Kits for hepatic and renal function tests, alanine amino transferase (ALT), alkaline phosphate (ALKp) and aspartate amino transferase (AST) and renal creatinine (CREA) and blood urea nitrogen (BUN), were purchased from IDEXX Laboratories (Parktown, South Africa).

2.2. Preparation of Treatment Solutions for Selected Atrazine Concentrations

Twenty milligrams of 98.9% pure atrazine was used in preparing a standard stock solution of 4000 μg/L in dechlorinated tap water (pH 7.3). This stock was further diluted accordingly to obtain the concentrations: 0.01 μg/L, 200 μg/L and 500 μg/L of atrazine. Atrazine doses were chosen on earlier anuran reproductive toxicity study [28]. Thus, while 0.01 μg/L ATZ concentration resulted in disruptions of the endocrine system in frogs [29,30], 200 μg/L seemingly mimicked the effects reported in 0.01 μg/L ATZ concentration exposure [31], and 450 μg/L of ATZ and higher concentrations have also been used [28] as well. Though the highest dose of 500 μg/L or ppb of ATZ concentration used in the present study represents a midpoint between high and low dose points reported in literature for one to two weeks of exposure, this dose (500 μg/L or ppb) seems high enough to induce toxicity, if indeed ATZ is toxic, especially for sub-chronic exposure.

2.3. Animals and Housing

Forty (40) healthy, post-anuran metamorphosis male *Xenopus laevis* frogs used for this study were procured from African Xenopus Facility farm, Knysner, (Western Cape,

South Africa). The animals were randomly divided into four groups in four stainless steel tanks (225 cm × 24 cm × 12.5 cm) with water at the University of the Witwatersrand Animal Research Facility (UARF). They were allowed to acclimatize in tanks at UARF for 30 days prior to treatment, maintained a 12:12 h light/dark cycle, a room temperature of 22 ± 2 °C, oxygen saturation exceeding 70% and pH 6.5, according to methods of [32,33]. All experimental animal treatments were performed according to the ethical principles for animal research approved by the University Animal Research Ethics Committee (certificate number 2014/14/D) and Gauteng Nature Conservation (certificate 0115 and 0120).

2.4. Experimental Design and Procedure

The stainless-steel test aquaria were each filled with 60 L of water, according to the dose concentrations of ATZ and control group without ATZ. The frogs (40) were distributed at a density of 10 frogs per tank. The first tank contained zero ATZ concentration and the animals served as control, while the second, third and fourth tanks contained 0.01 µg/L, 200 µg/L and 500 µg/L ATZ concentrations, respectively, and housed the treated animals during the 3 months of sub-chronic exposure. Food, Kori TM Frog Brittle pellets (from Daro Pet Products, Johannesburg, South Africa) was provided ad libitum. Tank water was consistently aerated while recycling of the tank water occurred two times a week, to ensure a sterile environment throughout the length of the exposure/treatment. To regularly determine the atrazine concentration in each tank, once a week, 50 mL of water obtained from each tank was injected into a 200 mg Bond Elut Plexa solid phase extraction cartridge linked to gas chromatography–mass spectrometry (GC-MS, GC 6890, MS 5975, Agilent Technologies, CA, USA). Control tanks recorded zero atrazine, while treated groups maintained atrazine concentrations within the required levels in each tank as previously reported [32–34]. Thereafter, 0.02% benzocaine inhalation in a glass jar was used to anaesthetize the animals and blood was collected (through cardiac puncture) into specimen plain tubes and stored at room temperature and subsequently in the freezer (−80 °C) after clotting, pending laboratory analysis.

2.5. Lipid Peroxidation (LPO) Evaluation

The levels of oxidized lipids in vitro were marked by malondialdehyde (MDA) and were assessed with sensitive high-performance liquid chromatography (HPLC) in line with the method reported by Karatas et al. [35]. Samples were briefly analyzed with Bischoff HPLC device attached to a UV detector, collimated at 254 nm, and the analytical Prontosil column, Bischoff Chromatography (Leonberg, Germany) (12.5 cm × 4.0 mm, 5 µm particle size), was utilized for detection and measurement of MDA level. Sample injection volume was 50 µL, and flow rate was kept constant at 1 mL/min, while acetonitrile distilled water (50:50, *v/v*) was used for the mobile phase composition. MDA peaks were recorded relative to its retention time and established by spiking with added exogenous standard. MDA serum concentrations were calculated with the standard curve, prepared from 1, 1, 3, 3-tetraethoxypropane (TEP) and expressed as µg/mL.

2.6. Measurement of Serum Biochemical Markers

Appropriate kits were used to determine the levels of liver ALT, AST and ALKp, as well as urea and creatinine biomarkers of renal function, as set out by the manufacturers' procedures. Blood samples were centrifuged at 4000× *g* at 4 °C for 10 min, and obtained serum was subjected to spectrophotometric measurement of liver enzymes, ALT and AST, by the method of Reitman et al. [36]. Alkaline phosphatase (ALKp) was determined using an enzymatic colorimetric method as previously reported [37], while creatinine (CREA) and blood urea nitrogen (BUN) biomarkers were, respectively, determined according to methods previously described [38,39].

2.7. Samples Preparation for Histological Analysis

Liver and kidney fixed-tissue samples routinely processed overnight using automated tissue processor (Shandon Citadel 1000) were embedded in paraffin wax. Tissue blocks sectioned at 5 μm thickness using 2035 Biocut microtome (Leica, Germany) and sections stained with Mayer's haematoxylin and eosin and Van Gieson's stain were cover-slipped and analyzed for histo-morphological and connective tissue changes analysis, respectively.

2.8. Morphometry

Histological images were captured with a Leica ICC50 HD camera mounted to a light microscope (Leica DM500) linked to the Leica application suite (LAS EZ, version 3.0.0, Heerbrugg, Switzerland) imaging software through HP (intel) Xeon computer. Hepatocyte diameters were measured using the Fiji image analysis software [40] line width tool. Measurements of 20 hepatocytes per animal at $\times 100$ magnification were restricted to only those with a visible nucleus and continuous outer lining. Only hepatocytes with visible nucleus and complete peripheral outline were measured at $100\times$ magnification. Furthermore, the Fiji's image threshold plugin was used in measuring the percentage area of melanomacrophage at $10\times$ magnification.

In the kidney, twenty photomicrographs per animal were used to measure corpuscle perimeter and epithelial cell height of both proximal convoluted and thick loop of Henle tubules, at $40\times$ magnification. Tubules with a distinct lumen only were measured using Fiji's imaging software freehand tool. Connective tissue area fraction was measured by superimposing a grid (area per point = 500 pixels^2) in a single camera field (2048×1536 pixels^2) photomicrograph at $40\times$ magnification and using the multi-point tool of Fiji image analysis software to count the number points that were stained for collagen. Eight randomly selected sections per group were analyzed and average counts were used for comparative analysis between groups.

2.9. Immunohistochemical Labelling of Caspase-3

Liver and kidney sections picked up on silane-coated slides were dewaxed, rehydrated and washed in 1 M phosphate buffer solution (PBS; pH 7.4) for 5 min. Thereafter, sections were incubated in 3% hydrogen peroxide in methanol for 30 min, washed three times in PBS and then incubated for one hour in 5% normal goat serum. After which, the sections were incubated overnight at 4 °C with anti-caspase-3 primary antibody (1:100 dilution, Ab 4059, Abcam). Tissue sections were washed three times in PBS and then (30 min room temperature) incubated with biotinylated goat anti-rabbit secondary antibody (1:1000 dilution, Vectastain, Vector Laboratories, Burlingame, CA, USA). Sections were washed three times in PBS followed by incubation in avidin–biotin complex (Vectastain ABC kit, Vector Laboratories, Burlingame, CA, USA) for 30 min. Sections were washed three times in PBS and then incubated with diaminobenzidine 3, 3′ tetrachloride (DAB) working solution for 5 min. Tissue sections were rinsed in running tap water, counterstained in haematoxylin, hydrated in alcohol, cleared in xylene and finally cover-slipped with Entellan new (Merck). The specificity of the anti-caspase-3 antibody was determined using tonsil tissue as the positive control. The negative control included omission of the primary antibody and substituting with PBS.

2.10. Statistical Analyses

Statistical analysis was carried out using Statistica TM (StatSoft) and Graph Pad prism software for Windows (version 7.0). All data were expressed as mean $\pm$ standard deviation (SD). One-way ANOVA was carried out, followed by post hoc Tukey's multiple comparison tests for statistical comparisons among the groups. The data obtained were presented in graphs with Microsoft Excel and Graph Pad prism software for Windows, respectively. Statistical significance was set at $p < 0.05$.

3. Results

3.1. Serum Levels of Liver, Kidneys Biomarkers and Lipid Peroxidation

In the 200 µg/L and 500 µg/L ATZ-treated groups, serum levels of ALT ($p < 0.0$ and $p < 0.01$, respectively) and creatinine ($p < 0.03$ and $p < 0.002$, respectively) were significantly increased in comparison to the control group (Figure 1A,D), with no significant difference ($p > 0.05$) in creatinine levels between the 200 µg/L and 500 µg/L treated groups. However, the levels of AST (Figure 1B), ALKp (Figure 1C) and BUN (Figure 1E) increased non-significantly ($p > 0.05$) in all the treated groups relative to the control group. The serum levels of MBA in the 500 µg/L ATZ-treated group significantly increased relative to the control group (Figure 1F).

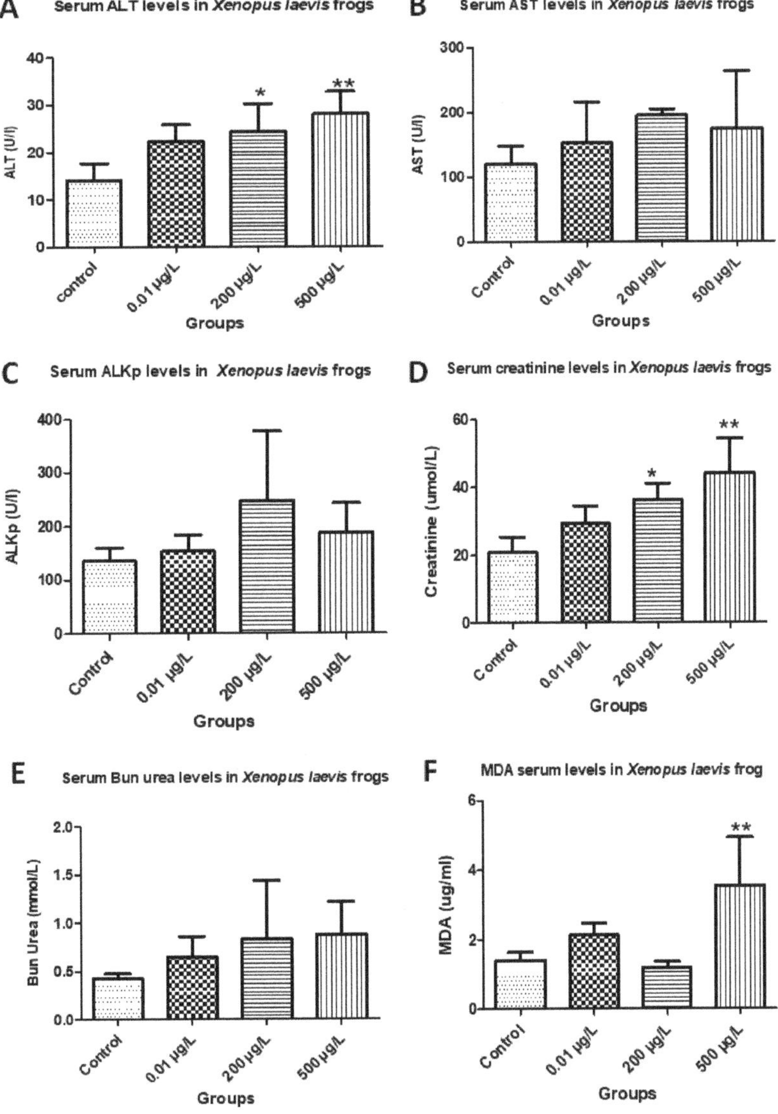

Figure 1. Effects of atrazine on serum levels of MDA, liver and kidney biomarkers in adult male *Xenopus laevis* frogs. (**A**) Serum ALT levels: * increased significantly in comparison to control $p < 0.05$;

** increased significantly compared to control $p < 0.01$. (**B**) Serum AST levels had no significant change between the control and treated groups. (**C**) Serum ALK$_P$ levels had no significant change between the control and treated groups. (**D**) Serum creatinine levels: * increased significantly ($p < 0.05$) compared to control; ** significantly increased ($p < 0.01$) compared to control. (**E**) Serum BUN levels had no significant change between control and treated groups. (**F**) Serum MDA levels: ** significantly increased ($p < 0.01$) relative to control. Data represented as mean $\pm$ SD ($n = 10$). ALT: alanine aminotransferase; AST: aspartate aminotransferase; ALKp: alkaline phosphatase.

3.2. Histopathology

3.2.1. Liver Histopathology

The liver parenchyma revealed a normal microanatomy in the control group with radiating anastomosing cords of hepatocyte (arising from the central veins (Cv) (Figure 2A)) having one or two pale basophilic rounded nuclei and scanty cytoplasm (Figure 2I, red arrow). Between hepatic cords, narrow irregular hepatic sinusoids containing nucleated red blood cells were observed (Figure 2I, black arrow). At the portal triad, branches of the hepatic artery, hepatic portal vein and bile duct lined by simple cuboidal epithelium (Figure 2A, circled) and some melanomacrophages were observed (Figure 2A, red arrow).

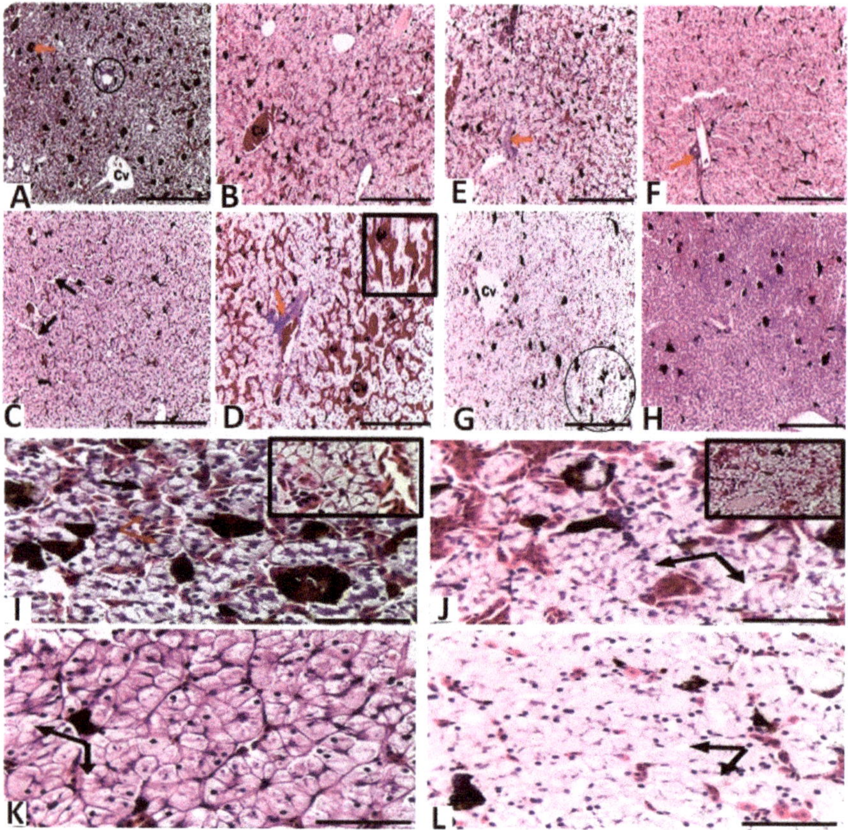

Figure 2. *Cont.*

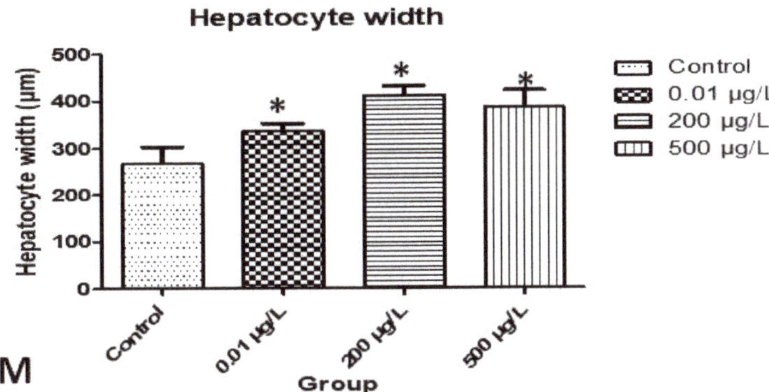

Figure 2. Photomicrographs of liver histopathology of frogs exposed to atrazine. (**A**) Control group with normal histology, central vein (Cv), portal triad (circle) and melanomacrophages (red arrow). (**B**) 0.01 µg/L group with clogged central vein (Cv). (**C**) 200 µg/L group with dilated sinusoids (black arrow). (**D**) 500 µg/L group with clogged central vein (Cv), sinusoidal congestion (insert) and portal inflammation (red arrow). (**E**) 0.01 µg/L group with portal inflammation (red arrow). (**F**) 200 µg/L group with portal inflammation (red arrow). (**G**) 500 µg/L group with parenchymal (circle) and central vein (Cv) necrosis. (**H**) 500 µg/L group with abundance of inflammatory cells. (**I**) Control group with normal hepatocyte histology (red arrow) alternating cords and sinusoid arrangement (insert) with nucleated red blood cells (black arrow). (**J**) 0.01 µg/L group with absence of normal hepatic cord arrangement (insert) and vacuolated hepatocytes (black arrow). (**K**) 200 µg/L group with highest observed hypertrophy with some vacuolation (black arrow). (**L**) 500 µg/L group with absence of hepatic cord arrangement and highly vacuolated hepatocytes (black arrows). (**M**) hepatocyte width with significantly increased * in comparison to control ($p = 0.003$). Values expressed as mean $\pm$ SD ($n = 10$). H&E stain. Scale bar in A to H = 240 µm ($10\times$ magnification). Scale bar in I to L = 38 µm ($40\times$ magnification).

In the atrazine-exposed groups, the 200 µg/L atrazine-exposed group revealed a dilated sinusoid (Figure 2C, black arrows), while in the 500 µg/L sinusoidal congestion with blood numerous haematopoietic cells were present (Figure 2D, insert), and infiltrating inflammatory cells were observed obscuring the liver tissues (Figure 2H). Parenchyma necrotic areas were observed in the 500 µg/L group (Figure 2C, circled), and portal inflammation was observed in all atrazine-exposed groups (Figure 2D–F). Hepatocyte cord disarrangement was observed in the 0.01 µg/L (Figure 2J) and 500 µg/L (Figure 2L) groups. Hepatocyte vacuolation was also observed in the 0.01 µg/L, 200 µg/L and 500 µg/L atrazine (Figure 2J–L, black arrows). Additionally, ATZ exposure of 200 µg/L and 500 µg/L significantly increased hepatocyte width ($p < 0.0003$) compared to the control (Figure 2).

Collagen tissue was increased in all the atrazine-treated groups periportally (Figure 3B–D: red arrows) and sinusoids (Figure 3F–H: red arrows) but not significantly increased relative to the control group (Figure 3I,J; $p > 0.050$). There was no collagen deposition observed within the hepatic lobules.

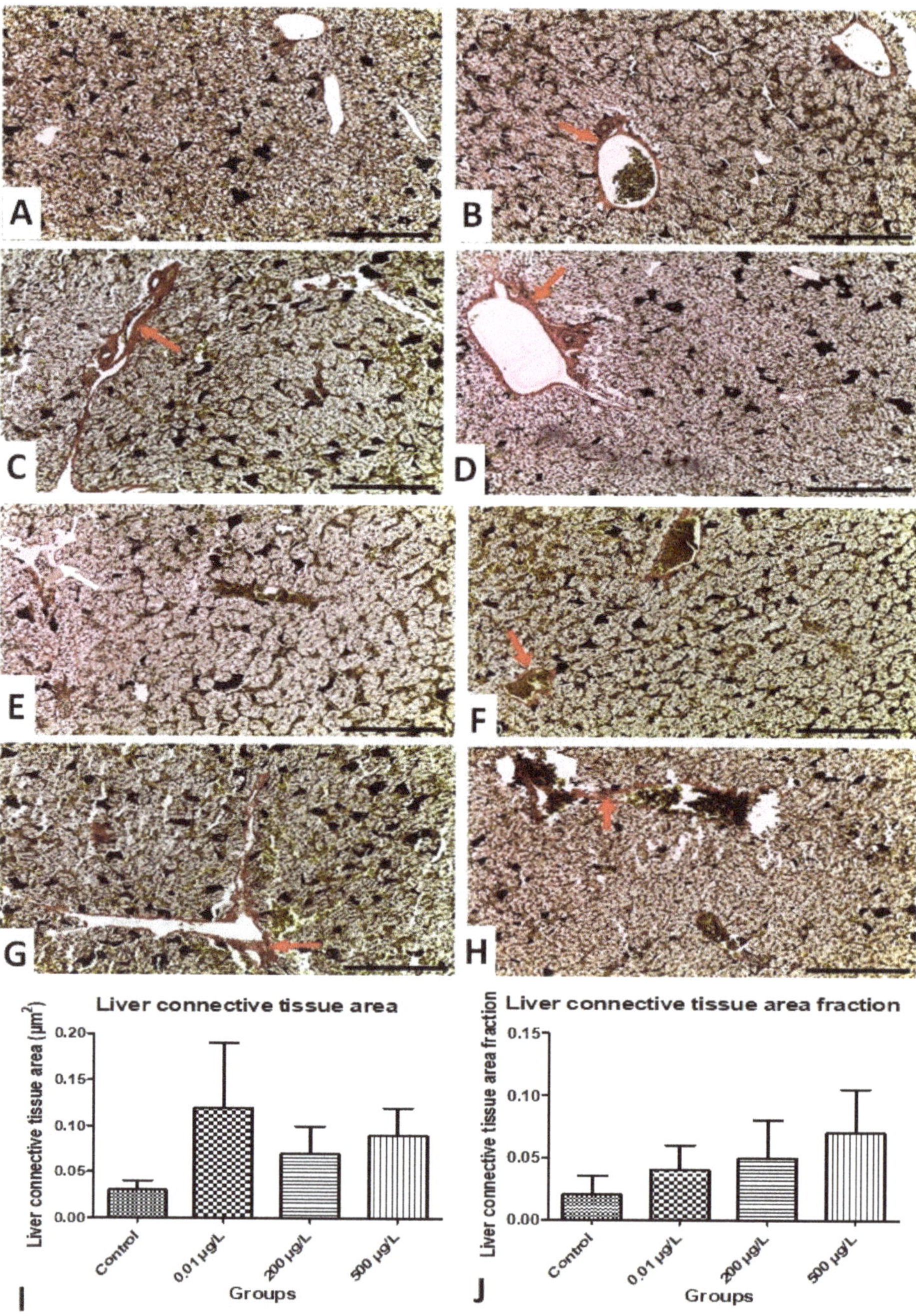

Figure 3. Photomicrographs of connective tissue profile in atrazine-exposed groups. (**A**) Normal connective tissue profile in control group. In (**B–D**), 0.01 µg/L, 200 µg/L and 500 µg/L groups, respectively, peri-portal fibrosis (red arrow). (**E**) Control group with normal connective tissue content and absence of peri-sinusoidal fibrosis. In (**F–H**), 0.01 µg/L, 200 µg/L and 500 µg/L groups, correspondingly, peri-sinusoidal fibrosis (red arrow). Van Gieson's stain, scale bar in A to D = 38 µm (40× magnification), scale bar in E to H = 38 µm (40× magnification). (**I,J**) No significant changes in liver connective tissue area and connective tissue area fraction.

3.2.2. Kidney Histopathology

The control group showed normal kidney histology, revealing normal proximal tubules (Figure 4A, black arrow), thick loop of Henle (Th) and distal tubules lined by simple short columnar and simple cuboidal epithelial cells, respectively (Figure 4A, red arrowhead). Collecting tubules with stratified cuboidal epithelial cells (white arrow) were observed. Hematopoietic cells were observed in the inter-tubular space with several red nucleated blood corpuscles (Figure 4A, red arrow). Vacuolized epithelial cells were noted within the tubules of the 0.01 µg/L group (Figure 4B, black arrow). In the 0.01 µg/L and 200 µg/L atrazine-exposed animals, tubular degeneration and loss of the epithelial cell–cell border of the thick loop of Henle were observed (Figure 4B,C, arrowhead). In the 500 µg/L group, the integrity of the thick loop of Henle tubule brush border was disrupted (Figure 4D, arrowhead).

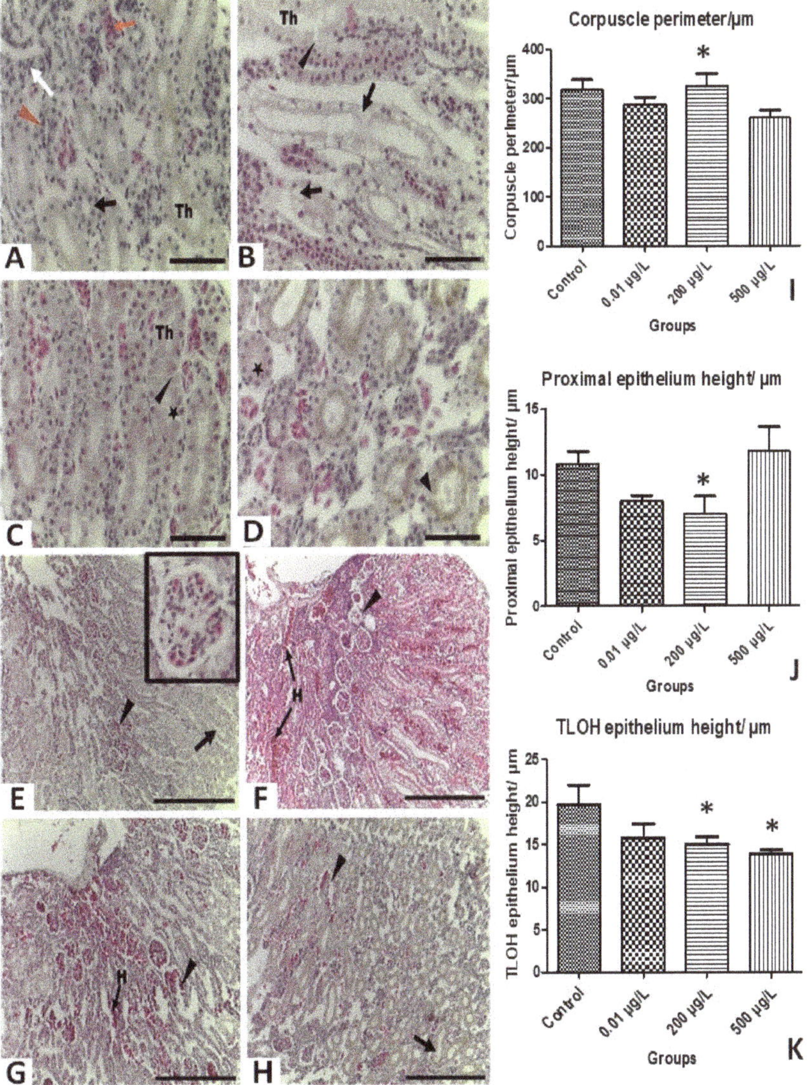

Figure 4. Representative photomicrographs of histopathologic findings in frog kidneys following

sub-chronic atrazine exposure. (**A**) Control with normal histology of proximal (black arrow), distal (red arrowhead), thick loop of Henle (Th) and collecting tubules (white arrow) and inter-tubular space with nucleated red blood cell (red arrow). (**B**) 0.01 μg/L group with degeneration of thick loop of Henle tubule (Th) and vacuolated tubules (black arrowhead). (**C**) 200 μg/L group with epithelial cell–cell boarder loss (black arrowhead) in thick loop of Henle, eosinophilic deposits (asterisk) inside the lumen of thick loop of Henle tubules (Th). (**D**) 500 μg/L group with loss of epithelial cell apical brush boarder (black arrowhead) and eosinophilic deposits (asterisk) in the proximal tubules. (**E**) Control with normal histology glomerulus (insert) and tubules (arrowhead). (**F,G**) 0.01 μg/L and 200 μg/L groups with glomeruli atrophy and necrosis (arrowhead) and peritubular hemorrhaging (**H**). (**H**) In 500 μg/L, mild glomerular atrophy (arrowhead) and dilated tubules (black arrow). (**I–K**) Corpuscle perimeter with proximal tubule epithelium height and thick loop of Henle epithelial height, respectively. * significantly decreased compared to control ($p < 0.05$; $p < 0.03$; $p < 0.01$, correspondingly). TLOH: thick loop of Henle. H&E stain, scale bar in A to D = 38 μm (40× magnification). Scale bar in E to G = 240 μm (10× magnification). Values are mean ± SD ($n = 10$).

The cortex of the control group showed normal renal corpuscles (Figure 4E, insert), but degeneration of the renal corpuscles, glomerular atrophy (Figure 4F–H, black arrowhead), dilation of tubular lumen in the 500 μg/L group (Figure 4H, black arrow) and vacuolized epithelial cells were noted within the tubule of the 200 μg/L group (Figure 4B, black arrow). Furthermore, there was shrinking of some tubules with eosinophilic material infiltration (Figure 4C,D, asterisk) and mild peritubular hemorrhaging (H) in the 0.01 μg/L and 200 μg/L groups (Figure 4F,G). The corpuscle perimeter and TLOH epithelial height were significantly reduced in the 500 μg/L group, while the proximal epithelium height and TLOH were significantly reduced in the 200 μg/L groups (Figure 4I–K). The 500 μg/L group significantly decreased in corpuscle perimeter compared to the control (Figure 4I, $p < 0.03$). The epithelium height of proximal tubules in the 200 μg/L group and thick loop of the Henle tubules in the 200 μg/L and 500 μg/L groups significantly decreased (Figure 4J,K, $p < 0.03$ and $p < 0.01$, correspondingly) relative to the control group.

3.2.3. Liver and Kidney Caspase-3 Immunohistochemistry

The expression of caspase-3 was predominantly in the epithelial and immune cells of the liver and not in the cytoplasm of hepatocytes. Controls showed no positive immuno-reaction (Figures 5A and 6A). In the liver, caspase-3 expression was observed in Kupffer cells within the peri-sinusoidal space surrounding the hepatocytes (Figure 5B–D; red arrow) and in monocytes within the sinusoids (Figure 5B,D; black arrow) and sinusoidal endothelial cells (Figure 5B–D; blue arrowhead). Caspase expression was also observed in the bile duct epithelial cells of the 0.01 μg/L and 500 μg/L groups (Figure 5E,F, blue arrowhead). The number of melanomacrophages (MMCs) significantly decreased in the 0.01 μg/L and 500 μg/L groups ($p < 0.0001$) and in the 200 μg/L group ($p < 0.03$) compared with the control group (Figure 5G). In the kidney, caspase-3 expression was not observed in the control group (Figure 6A) but in the epithelial cells of cortical proximal tubules (Figure 6B,C,E; black arrow), distal tubule (Figure 6B,C,E; red arrow) and collecting tubules (Figure 6B,C,E white arrow) in all the atrazine-exposed groups instead. However, weak expression in the glomeruli was observed in the 0.01 μg/L group (Figure 6B,G). In the 200 μg/L (Figure 6D) and 500 μg/L (Figure 6E) groups, intensely expressed caspase-3 was observed in the podocytes (black arrowhead), simple squamous epithelial cells (white arrowhead) of the parietal layer and within the macula densa cells (circle) of the glomeruli.

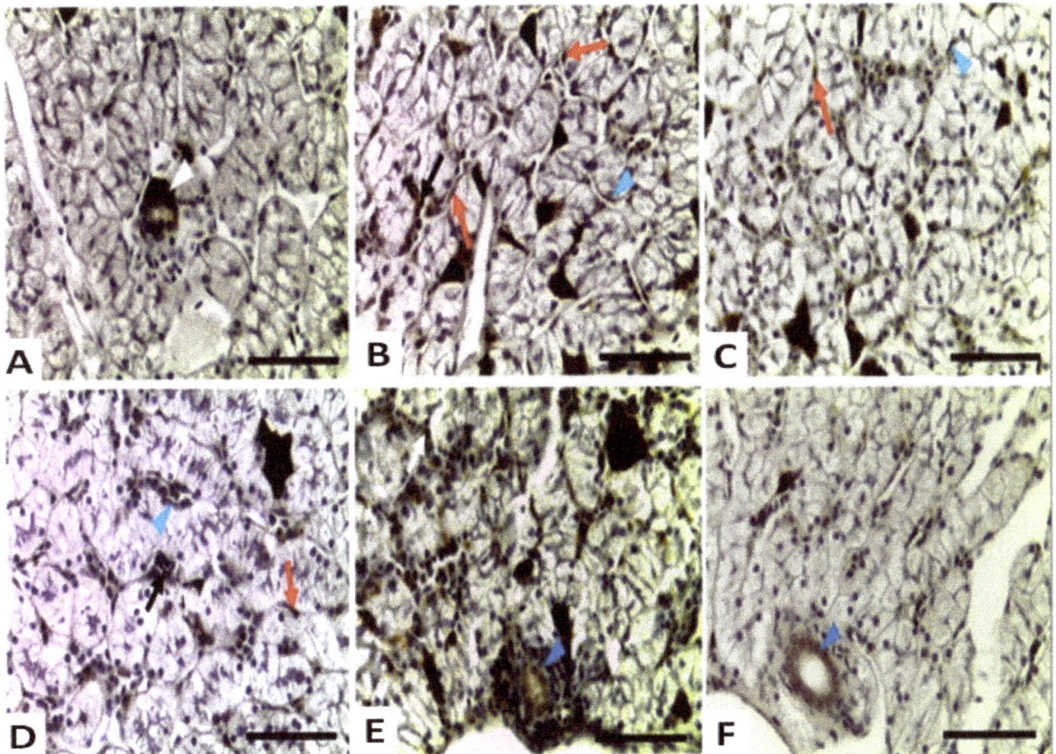

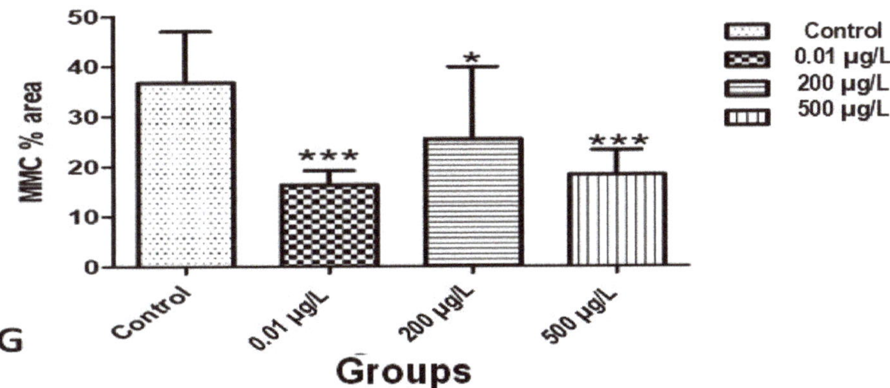

Figure 5. Expression of caspase-3 photomicrographs in *Xenopus laevis* liver cells with sub-chronic exposure to atrazine. (**A**) Liver tissue section showing melanomacrophages (white arrowhead) and lack of positive caspase-3 expression. (**B**) 0.01 µg/L, (**C**) 200 µg/L and (**D**) 500 µg/L) caspase-3 expression in monocytes in sinusoids (black arrow), Kupffer cells in the peri-sinusoidal space (red arrow) and flattened squamous endothelial cells lining sinusoids (blue arrowhead). (**E,F**) Caspase-3 expression in epithelial cells of bile ducts of the 0.01 µg/L and 500 µg/L groups, respectively. (**G**) The population percentage of melanomacrophages (MMC) in liver sections, * significantly decreased ($p < 0.05$) in comparison to control group; *** significantly decreased ($p < 0.001$) in comparison to the control group. Scale bar in A to F = 38 µm ($40\times$ magnification). Data presented as mean ± SD ($n = 10$).

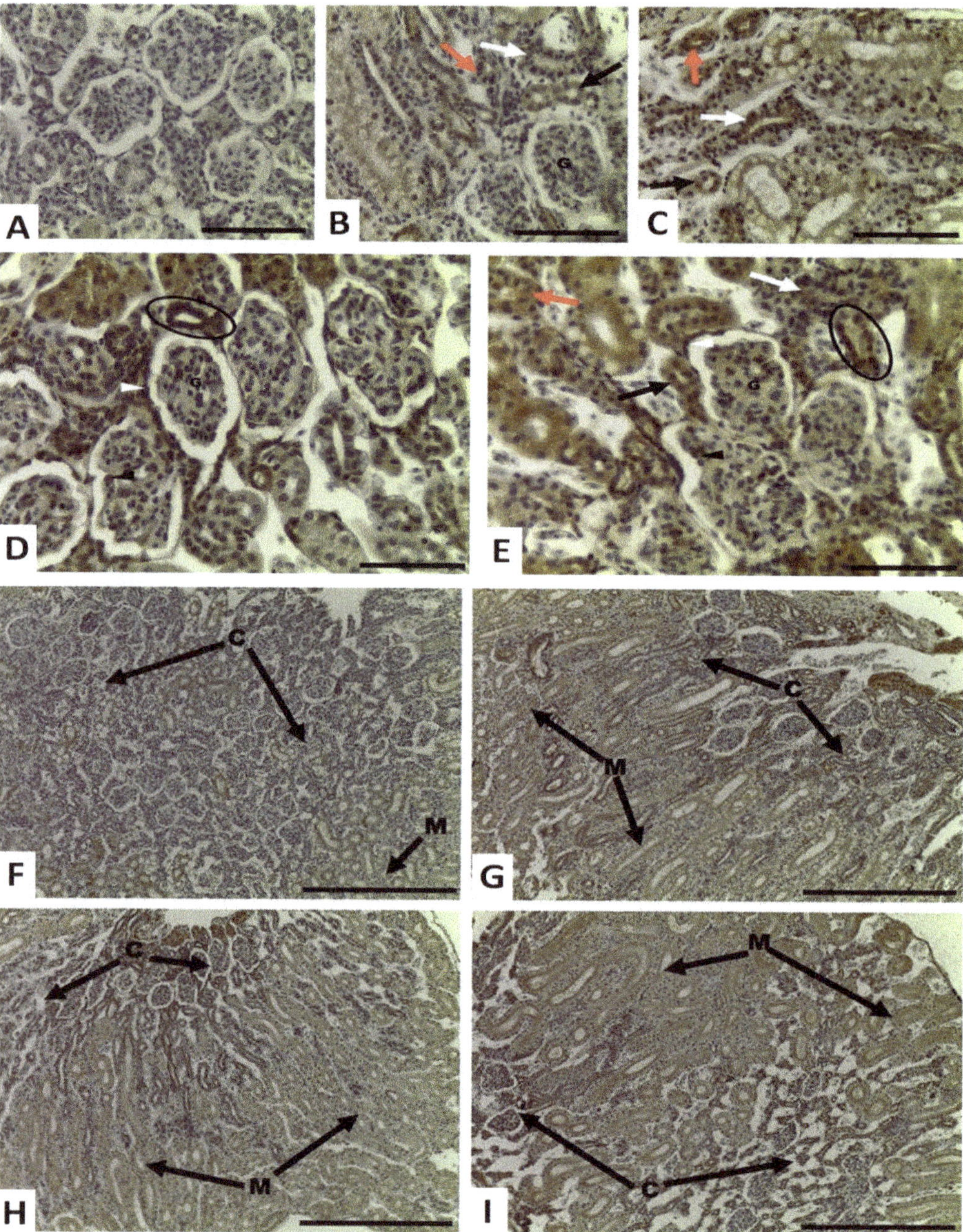

Figure 6. Photomicrographs of the expression of caspase-3 in frog kidneys following chronic exposure to atrazine. (**A**) Control with absence of caspase-3 expression. (**B**) 0.01 μg/L, (**C**) 200 μg/L and (**E**) 500 μg/L caspase-3 expression in epithelial cells of proximal tubules (black arrow), distal tubules (red arrow) and collecting tubules (white arrow). (**D**) 200 μg/L and (**E**) 500 μg/L caspase-3 expression in the glomeruli, macula densa cells (circle), podocytes (black arrowhead) and epithelial cells within the parietal layer (white arrowhead). (**F**) Control with low power absence of immuno-positive caspase-3 expression. (**G**) 0.01 μg/L, (**H**) 200 μg/L and (**I**) 500 μg/L caspase-3 expression in the cortex and medulla. Scale bar in A to E = 38 μm (40× magnification). Scale bar in F to I = 240 μm (10× magnification). C: cortex; M: medulla; G: glomerulus.

4. Discussion

The results have revealed the hepato-renal cytotoxicity of sub-chronic exposure to the pesticide atrazine in adult *Xenopus* frogs. Interestingly, the various doses of atrazine used showed variable significant levels of toxicity in the liver and kidneys, suggesting that, though this pesticide may be useful to increased agricultural productivity, caution must be applied in its application.

The significantly elevated serum MDA (500 μg/L exposed group) corroborates previous reports of elevated MDA in the liver, gills and muscles of bullfrog tadpoles [41] and in rats [25,42] exposed to atrazine. Correspondingly, the presence of hepatic cell injury with consequent functional implications is predicated on the significant increases in serum levels of ALT in the 200 μg/L and 500 μg/L exposure groups, as ALT levels function in the maintenance of cell membrane integrity [25,43]. However, the mildly reduced MDA level following 200 μg/L sub-chronic exposure suggests homeostatic adjustment adaptation in this group, as well as minimal adverse effects resulting from the non-significantly increased serum levels of AST and ALKp in all the treated groups, signifying the variable atrazine effects. This study utilized the sensitive high-performance liquid chromatography method, ensuring the accuracy of the measurements. The serum levels of ALT, AST and ALKp are markers of cell membrane integrity [26,43]. These results suggest hepatotoxicity and therefore disturbances of liver function following ATZ exposure.

Additionally, the assessment of renal function is based on the serum levels of urea and creatinine as biomarkers above or below the normal physiological level. Therefore, the non-significantly increased urea levels in all exposure groups and the significantly elevated creatinine levels in the 200 μg/L and 500 μ/L treated groups relative to the control group suggest renal glomeruli filtration dysfunction and/or impairment in all the treated groups but more severely in the 200 μg/L and 500 μ/L atrazine-exposed groups. The observed general trend in serum creatinine levels with a concentration-dependent effect in atrazine exposure suggests a greater adverse effect relative to urea as a renal biomarker.

Histopathology observations of hepatocyte perturbations are similar to reports observed in other atrazine-exposed species [44,45]. Atrazine-induced histological alteration is initiated by hypertrophy, as observed in the 200 μg/L and 500 μg/L exposed groups, as was previously reported [46] and which has been suggested to be an adaptive response to atrazine exposure [47]. Hepatocyte vacuolization observed in all our atrazine-exposed groups, particularly the 200 μg/L and 500 μg/L cohorts, has been reported in fish [45] and mice with attendant effects in lipid metabolism. Additionally, the observed central vein and sinusoidal congestion in the 0.01 μg/L and 500 μg/L groups suggest obstruction by immune cell migration and cellular breakdown debris. This corroborates with observations of vascular hemorrhaging indicated by erythrocyte infiltration into the hepatocytes reported in frogs exposed to 500 μg/L of ATZ. The red blood cell (erythrocytes) extravasation suggests compromised sinusoidal endothelium and may be a consequence of increased intra-sinusoidal pressure arising from vascular congestion [48]. Additionally, atrazine-induced perturbations in the immunology of amphibians are widely reported [49,50]. Similarly, the observed apoptosis of neutrophils and Kupffer cells confirmed by the expression of caspase-3 in the liver of atrazine-treated groups suggests the potential for atrazine to suppress phagocytic process or action and the gathering of monocytes directly or indirectly to sites of inflammation through induction of apoptosis in the immune cells. Contrastingly, a significant decrease in the population of MMCs observed in all the ATZ-exposed groups correlates with a field study report of aggregated melanomacrophages in the livers of *L. pipiens* exposed to atrazine [51]. Considering the role of MMCs in immunity (humoral and inflammatory responses) [52], the adult *Xenopus laevis* frogs' immunity may be adversely affected by different concentrations of atrazine.

The aforementioned histopathological changes seen in the kidneys of all the treated groups are similar to the effects reported following exposure to pollutants and pesticides [53,54]. Evidence of macula densa, podocyte and parietal cell (PECs) apoptosis is observed in the 200 μg/L and 500 μg/L atrazine-exposed groups. Apoptosis in podocytes is an indicator of several kidney diseases [55], while parietal epithelial cell (PECs) apoptosis is speculated to serve as a means of regularizing glomerular cell number as reported by [56]. Therefore, the observed glomerular atrophy may be due to apoptosis of PECs and podocytes, and these cellular alterations may result in dysfunctions of glomerular filtration. Furthermore, apoptosis in the macula densa may lead to inhibition of auto-regulatory osmolarity reactions or processes. Together, glomeruli atrophy and macula densa apoptosis may explain the significantly increased levels of creatinine in frogs treated with high atrazine concentrations.

The significant reduction in tubular epithelium height and consequent dilation in the thick loop of Henle of the 500 µg/L atrazine-exposed group suggest a compromised regulation of extracellular fluid volume, urine concentration, calcium, magnesium, bicarbonate and ammonium homeostasis and urine protein composition [57]. Although caspase-3 was not expressed in the thick loop of Henle in the atrazine-exposed groups, caspase-3 was expressed in the proximal, distal and collecting tubule epithelial cells and increased with increasing atrazine concentrations, suggesting atrazine-induced apoptosis in the proximal and distal tubules but degeneration of the thick loop of Henle tubules, which may be induced by other mechanism such as activation of caspase-6 and 7 or extrinsic apoptotic pathways initiated by death receptors as reported by [58].

5. Conclusions

Atrazine-induced perturbations in the liver and kidney histomorphology and serum biomarker levels (ALT and creatinine) suggest severe hepatorenal toxicosis due to environmentally relevant atrazine concentrations. The observed hepatorenal histopathology cumulates in compromised metabolic, immunological and vascular function. Further associated pathophysiological effects associated with atrazine exposure at electron microscopic and molecular levels are required to elaborate on mechanisms through which these effects may impact the survival of frogs with continuous atrazine exposure. Thus, the constant presence of atrazine in the environment needs to be cautiously managed, as it remains a risk to the existence of *Xenopus laevis* frog populations and other animal species. Therefore, atrazine use, handling and disposal should be monitored and assessed regularly to prevent environmental health consequences on non-targeted aquatic species, especially amphibian biodiversity.

Author Contributions: Conceptualization, E.F.M., L.S. and J.A.J.; methodology, L.S., J.A.J. and E.F.M.; software, P.N., L.S. and J.A.J.; validation, E.F.M., L.S. and J.A.J.; formal analysis, L.S., J.A.J., E.F.M. and P.N.; investigation, L.S., J.A.J. and E.F.M.; resources, E.F.M., L.S. and J.A.J.; data curation, L.S., J.A.J., E.F.M. and P.N.; writing—original draft preparation, L.S. and J.A.J.; writing—review and editing, E.F.M., L.S., J.A.J. and E.F.M.; visualization, L.S., J.A.J. and E.F.M.; supervision, E.F.M.; project administration, E.F.M., L.S. and J.A.J.; funding acquisition, E.F.M., P.N., L.S. and J.A.J. All authors have read and agreed to the published version of the manuscript.

Funding: This work was supported by the University of Witwatersrand, Faculty of Health Science faculty research grant awarded to Lynette Sena.

Institutional Review Board Statement: The guidelines of the University of the Witwatersrand Animals Ethics Research Committee (2014/32/D) and the Gauteng Province Nature Conservation (CPF6 0115 (2015) and CPF6 0120, 2016) were followed in this study.

Informed Consent Statement: Not applicable.

Data Availability Statement: Data reported in this study will be made available on request, on acceptance of manuscript.

Acknowledgments: The authors specially acknowledge Hasiena Ali for her hands-on assistance at various stages of this research and Amadi Ihunwo, Luke Chimuka and Cornelius Rimayi for their amazing collaboration and inspiration.

Conflicts of Interest: The authors declare no conflict of interest.

References

1. Singh, A.; Kaur, C. Weed control with pre and post emergence herbicides application in spring planted sugarcane. *Sugar Tech.* **2004**, *6*, 93–94. [CrossRef]
2. Singh, A.; Virk, A.S.; Singh, J.; Singh, J. Comparison of pre- and post-emergence application of herbicides for the control of weeds in sugarcane. *Sugar Tech.* **2001**, *3*, 109–112. [CrossRef]
3. Mast, M.A.; Foreman, W.T.; Skaates, S.V. Current-use pesticides and organochlorine compounds in precipitation and lake sediment from two high-elevation national parks in the Western United States. *Arch. Environ. Contam. Toxicol.* **2007**, *52*, 294–305. [CrossRef] [PubMed]
4. Sousa, A.S.; Duaví, W.C.; Cavalcante, R.M.; Milhome, M.A.L.; do Nascimento, R.F. Estimated levels of environmental contamination and health risk assessment for herbicides and insecticides in surface water of Ceará, Brazil. *Bull. Environ. Contam. Toxicol.* **2016**, *96*, 90–95. [CrossRef] [PubMed]
5. Parmar, T.K.; Rawtani, D.; Agrawal, Y.K. Bioindicators: The natural indicator of environmental pollution. *Front. Life Sci.* **2016**, *9*, 110–118. [CrossRef]
6. Zaghloul, A.; Saber, M.; Gadow, S.; Awad, F. Biological indicators for pollution detection in terrestrial and aquatic ecosystems. *Bull. Natl. Res. Cent.* **2020**, *44*, 127. [CrossRef]
7. IUCN SSC Amphibian Specialist Group (IUCN). IUCN Red List of Threatened Species: *Xenopus laevis*. 2016. Available online: https://www.iucnredlist.org/en (accessed on 22 November 2021).
8. African Clawed Frog. *AZ Animals*. Available online: https://a-z-animals.com/animals/african-clawed-frog/ (accessed on 22 November 2021).
9. Garvey, N. *Xenopus laevis* (African Clawed Frog). *Animal Diversity Web*. Available online: https://animaldiversity.org/accounts/Xenopus_laevis/ (accessed on 22 November 2021).
10. Hayes, T.B.; Khoury, V.; Narayan, A.; Nazir, M.; Park, A.; Brown, T.; Adame, L.; Chan, E.; Buchholz, D.; Stueve, T.; et al. Atrazine induces complete feminization and chemical castration in male African clawed frogs (*Xenopus laevis*). *Proc. Natl. Acad. Sci. USA* **2010**, *107*, 4612–4617. [CrossRef]
11. Jooste, A.M.; Du Preez, L.H.; Carr, J.A.; Giesy, J.P.; Gross, T.S.; Kendall, R.J.; Smith, E.E.; Van Der Kraak, G.L.; Solomon, K.R. Gonadal development of larval male *Xenopus laevis* exposed to atrazine in outdoor microcosms. *Environ. Sci. Technol.* **2005**, *39*, 5255–5261. [CrossRef]
12. Coady, K.; Murphy, M.; Villeneuve, D.; Hecker, M.; Jones, P.; Carr, J.; Solomon, K.; Smith, E.; Van Der Kraak, G.; Kendall, R.; et al. Effects of atrazine on metamorphosis, growth, and gonadal development in the green frog (*Rana clamitans*). *J. Toxicol. Environ. Health* **2004**, *67*, 941–957. [CrossRef]
13. Murphy, M.; Hecker, M.; Coady, K.; Tompsett, A.; Jones, P.; Du Preez, L.; Everson, G.; Solomon, K.R.; Carr, J.; Smith, E.; et al. Atrazine concentrations, gonadal gross morphology and histology in ranid frogs collected in Michigan agricultural areas. *Aquat. Toxicol.* **2006**, *76*, 230–245. [CrossRef]
14. Papoulias, D.M.; Tillitt, D.E.; Talykina, M.G.; Whyte, J.J.; Richter, C.A. Atrazine reduces reproduction in Japanese medaka (*Oryzias latipes*). *Aquat. Toxicol.* **2014**, *154*, 230–239. [CrossRef]
15. Dornelles, M.F.; Oliveira, G.T. Effect of atrazine, glyphosate and quinclorac on biochemical parameters, lipid peroxidation and survival in bullfrog tadpoles (*Lithobates catesbeianus*). *Arch. Environ. Contam. Toxicol.* **2014**, *66*, 415–429. [CrossRef]
16. Gao, S.; Wang, Z.; Zhang, C.; Jia, L.; Zhang, Y. Oral Exposure to Atrazine Induces Oxidative Stress and Calcium Homeostasis Disruption in Spleen of Mice. *Oxidative Med. Cell. Longev.* **2016**, *2016*, 7978219. [CrossRef] [PubMed]
17. Semren, T.Ž.; Žunec, S.; Pizent, A. Oxidative stress in triazine pesticide toxicity: A review of the main biomarker findings. *Arch. Ind. Hyg. Toxicol.* **2018**, *69*, 109–125. [CrossRef]
18. Rohr, J.R. Atrazine and amphibians: Data re-analysis and a summary of the controversy. *bioRxiv* **2017**, 164673. [CrossRef]
19. Wirbisky, S.E.; Freeman, J.L. Atrazine Exposure and Reproductive Dysfunction through the Hypothalamus-Pituitary-Gonadal (HPG) Axis. *Toxics* **2015**, *3*, 414–450. [CrossRef]
20. Slaninova, A.; Smutna, M.; Modra, H.; Svobodova, Z. A review: Oxidative stress in fish induced by pesticides. *Neuroendocrinol. Lett.* **2009**, *30* (Suppl. 1), 2–12. [PubMed]
21. Lykkesfeldt, J. Malondialdehyde as biomarker of oxidative damage to lipids caused by smoking. *Clin. Chim. Acta* **2007**, *380*, 50–58. [CrossRef] [PubMed]
22. Grotto, D.; Maria, L.S.; Valentini, J.; Paniz, C.; Schmitt, G.; Garcia, S.; Pomblum, V.J.; da Rocha, J.B.T.; Farina, M. Importance of the lipid peroxidation biomarkers and methodological aspects FOR malondialdehyde quantification. *Química Nova* **2009**, *32*, 169–174. [CrossRef]
23. Redza-Dutordoir, M.; Averill-Bates, D.A. Activation of apoptosis signalling pathways by reactive oxygen species. *Biochim. Biophys. Acta* **2016**, *1863*, 2977–2992. [CrossRef]
24. Bayir, H.; Kagan, V.E. Bench-to-bedside review: Mitochondrial injury, oxidative stress and apoptosis—There is nothing more practical than a good theory. *Crit. Care* **2008**, *12*, 206. [CrossRef] [PubMed]
25. Liu, W.; Du, Y.; Liu, J.; Wang, H.; Sun, D.; Liang, D.; Zhao, L.; Shang, J. Effects of atrazine on the oxidative damage of kidney in Wister rats. *Int. J. Clin. Exp. Med.* **2014**, *7*, 3235–3243.
26. Jestadi, D.B.; Phaniendra, A.; Babji, U.; Srinu, T.; Shanmuganathan, B.; Periyasamy, L. Effects of Short Term Exposure of Atrazine on the Liver and Kidney of Normal and Diabetic Rats. *J. Toxicol.* **2014**, *2014*, e536759. [CrossRef] [PubMed]

27. Campos-Pereira, F.D.; Oliveira, C.A.; Pigoso, A.A.; Silva-Zacarin, E.C.; Barbieri, R.; Spatti, E.F.; Marin-Morales, M.A.; Severi-Aguiar, G.D. Early cytotoxic and genotoxic effects of atrazine on Wistar rat liver: A morphological, immunohistochemical, biochemical, and molecular study. *Ecotoxicol. Environ. Saf.* **2012**, *78*, 170–177. [CrossRef]

28. Freeman, J.L.; Rayburn, A.L. Developmental impact of atrazine on metamorphing *Xenopus laevis* as revealed by nuclear analysis and morphology. *Environ. Toxicol. Chem.* **2005**, *24*, 1648–1653. [CrossRef] [PubMed]

29. Kloas, W.; Lutz, I.; Urbatzka, R.; Springer, T.; Krueger, H.; Wolf, J.; Holden, L.; Hosmer, A. Does atrazine affect larval development and sexual differentiation of South African clawed frogs? *Ann. N. Y. Acad. Sci.* **2009**, *1163*, 437–440. [CrossRef] [PubMed]

30. Cooper, R.L.; Stoker, T.E.; Tyrey, L.; Goldman, J.M.; McElroy, W.K. Atrazine disrupts the hypothalamic control of pituitary-ovarian function. *Toxicol. Sci.* **2000**, *53*, 297–307. [CrossRef]

31. Zaya, R.M.; Amini, Z.; Whitaker, A.S.; Kohler, S.L.; Ide, C.F. Atrazine exposure affects growth, body condition and liver health in *Xenopus laevis* tadpoles. *Aquat. Toxicol.* **2011**, *104*, 243–253. [CrossRef] [PubMed]

32. Rimayi, C.; Odusanya, D.; Weiss, J.M.; de Boer, J.; Chimuka, L.; Mbajiorgu, F. Effects of environmentally relevant sub-chronic atrazine concentrations on African clawed frog (*Xenopus laevis*) survival, growth and male gonad development. *Aquat. Toxicol.* **2018**, *199*, 1–11. [CrossRef] [PubMed]

33. Asouzu Johnson, J.; Ihunwo, A.; Chimuka, L.; Mbajiorgu, E.F. Cardiotoxicity in African clawed frog (*Xenopus laevis*) sub-chronically exposed to environmentally relevant atrazine concentrations: Implications for species survival. *Aquat. Toxicol.* **2019**, *213*, 105218. [CrossRef]

34. Asouzu Johnson, J.; Nkomozepi, P.; Opute, P.; Felix, M. Cardiac and Cerebellar Histomorphology and Inositol 1,4,5-Trisphosphate (IP3R) Perturbations in Adult *Xenopus laevis* Following Atrazine Exposure. *Appl. Sci.* **2021**, *11*, 10006. [CrossRef]

35. Karatas, F.; Karatepe, M.; Baysar, A. Determination of free malondialdehyde in human serum by high-performance liquid chromatography. *Anal. Biochem.* **2002**, *311*, 76–79. [CrossRef]

36. Reitman, S.; Frankel, S. A colorimetric method for the determination of serum glutamic oxalacetic and glutamic pyruvic transaminases. *Am. J. Clin. Pathol.* **1957**, *28*, 56–63. [CrossRef] [PubMed]

37. Tietz, N.W.; Burtis, C.A.; Duncan, P.; Ervin, K.; Petitclerc, C.J.; Rinker, A.D.; Shuey, D.; Zygowicz, E.R. A reference method for measurement of alkaline phosphatase activity in human serum. *Clin Chem.* **1983**, *29*, 751–761. [CrossRef] [PubMed]

38. Larsen, K. Creatinine assay in the presence of protein with LKB 8600 Reaction Rate Analyser. *Clin. Chim. Acta* **1972**, *38*, 475–476. [CrossRef]

39. Coulombe, J.J.; Favreau, L. A New Simple Semimicro Method for Colorimetric Determination of Urea. *Clin. Chem.* **1963**, *9*, 102–108. [CrossRef]

40. Schindelin, J.; Arganda-Carreras, I.; Frise, E.; Kaynig, V.; Longair, M.; Pietzsch, T.; Preibisch, S.; Rueden, C.; Saalfeld, S.; Schmid, B.; et al. Fiji: An open-source platform for biological-image analysis. *Nat. Methods* **2012**, *9*, 676–682. [CrossRef] [PubMed]

41. Dornelles, M.F.; Oliveira, G.T. Toxicity of atrazine, glyphosate, and quinclorac in bullfrog tadpoles exposed to concentrations below legal limits. *Environ. Sci. Pollut. Res. Int.* **2016**, *23*, 1610–1620. [CrossRef]

42. Abarikwu, S.O. Protective Effect of Quercetin on Atrazine-Induced Oxidative Stress in the Liver, Kidney, Brain, and Heart of Adult Wistar Rats. *Toxicol. Int.* **2014**, *21*, 148–155. [CrossRef]

43. Soror, A.; Hozyen, H.; Eldebaky, H.; Soror, A.H.; Shalaby, H.M. Protective Role of Selenium Against Adverse Effects of Atrazine Toxicity in Male Rats: Biochemical, Histopathological and Molecular Changes. *Glob. Vet.* **2015**, *15*, 357–365. [CrossRef]

44. Wani, G.P.; Vibhandik, A.M. Effect of Atrazine (herbicide) on Histology and Protein Content of the Freshwater Teleost *Barbus carnaticus*. *J. Ecobiol.* **2011**, *29*, 257–265.

45. Mela, M.; Guiloski, I.; Doria, H.B.; Randi, M.; Ribeiro, C.O.; Pereira, L.; Maraschi, A.; Prodocimo, V.; Freire, C.; de Assis, H.S. Effects of the herbicide atrazine in neotropical catfish (*Rhamdia quelen*). *Ecotoxicol. Environ. Saf.* **2013**, *93*, 13–21. [CrossRef] [PubMed]

46. Ziegler, U.; Groscurth, P. Morphological features of cell death. *Physiology* **2004**, *19*, 124–128. [CrossRef] [PubMed]

47. Kaware, M. Changes in Liver and Body Weight of Mice Exposed to Toxicant. *Int. J. Sci. Eng.* **2013**, *3*, 92–95.

48. Kakar, S.; Kamath, P.S.; Burgart, L.J. Sinusoidal dilatation and congestion in liver biopsy: Is it always due to venous outflow impairment? *Arch. Pathol. Lab. Med.* **2004**, *128*, 901–904. [CrossRef]

49. Houck, A.; Sessions, S.K. Could Atrazine Affect the Immune System of the Frog, *Rana pipiens*? *Bios* **2006**, *77*, 107–112. [CrossRef]

50. Brodkin, M.A.; Madhoun, H.; Rameswaran, M.; Vatnick, I. Atrazine is an immune disruptor in adult northern leopard frogs (*Rana pipiens*). *Environ. Toxicol. Chem.* **2007**, *26*, 80–84. [CrossRef]

51. Rohr, J.R.; Schotthoefer, A.M.; Raffel, T.R.; Carrick, H.J.; Halstead, N.; Hoverman, J.; Johnson, C.; Johnson, L.B.; Lieske, C.; Piwoni, M.D.; et al. Agrochemicals increase trematode infections in a declining amphibian species. *Nature* **2008**, *455*, 1235–1239. [CrossRef] [PubMed]

52. Agius, C.; Roberts, R.J. Melano-macrophage centres and their role in fish pathology. *J. Fish Dis.* **2003**, *26*, 499–509. [CrossRef]

53. Khan, M.Z.; Yasmeen, G.; Parveen, S.; Akbar, A.; Zehra, A.; Hussain, B. Induced Effect of Permakil (Pyrethroid) and Sandaphos (Organophosphate) on Liver and Kidney Cells of *Euphlyctis cyanophlyctis*. *Can. J. Pure Appl. Sci.* **2010**, *5*, 1615.

54. Medina, M.F.; González, M.E.; Klyver, S.M.R.; Odstrcil, I.M.A. Histopathological and biochemical changes in the liver, kidney, and blood of amphibians intoxicated with cadmium. *Turk. J. Biol.* **2016**, *40*, 229–238. [CrossRef]

55. Matovinovic, M.S. 3. Podocyte injury in glomerular diseases. *EJIFCC* **2009**, *20*, 21–27. [PubMed]

56. Ohse, T.; Pippin, J.W.; Chang, A.M.; Krofft, R.D.; Miner, J.H.; Vaughan, M.R.; Shankland, S.J. The enigmatic parietal epithelial cell is finally getting noticed: A review. *Kidney Int.* **2009**, *76*, 1225–1238. [CrossRef] [PubMed]
57. Mount, D.B. Thick Ascending Limb of the Loop of Henle. *Clin. J. Am. Soc. Nephrol.* **2014**, *9*, 1974–1986. [CrossRef] [PubMed]
58. Alkhouri, N.; Carter-Kent, C.; Feldstein, A.E. Apoptosis in nonalcoholic fatty liver disease: Diagnostic and therapeutic implications. *Expert Rev. Gastroenterol. Hepatol.* **2011**, *5*, 201–212. [CrossRef]

MDPI
St. Alban-Anlage 66
4052 Basel
Switzerland
Tel. +41 61 683 77 34
Fax +41 61 302 89 18
www.mdpi.com

Applied Sciences Editorial Office
E-mail: applsci@mdpi.com
www.mdpi.com/journal/applsci